Aquatic Ecosystems: Biodiversity and Conservation

Aquatic Ecosystems: Biodiversity and Conservation

Guest Editors

Marcos Gomes Nogueira
Douglas Donald Kane

MDPI

Basel • Beijing • Wuhan • Barcelona • Belgrade • Novi Sad • Cluj • Manchester

Guest Editors
Marcos Gomes Nogueira
Biodiversity and Biostatistics
State University of São Paulo
Botucatu
Brazil

Douglas Donald Kane
Department of Biology and
Environmental Science and
National Center for Water
Quality Research
Heidelberg University
Tiffin, OH
USA

Editorial Office
MDPI AG
Grosspeteranlage 5
4052 Basel, Switzerland

This is a reprint of the Special Issue, published open access by the journal *Water* (ISSN 2073-4441), freely accessible at: https://www.mdpi.com/journal/water/special_issues/5M5X88HFWY.

For citation purposes, cite each article independently as indicated on the article page online and as indicated below:

Lastname, A.A.; Lastname, B.B. Article Title. *Journal Name* **Year**, *Volume Number*, Page Range.

ISBN 978-3-7258-5261-1 (Hbk)
ISBN 978-3-7258-5262-8 (PDF)
https://doi.org/10.3390/books978-3-7258-5262-8

Cover image courtesy of Arthur Padial da Mota

Contents

water

MDPI

Editorial

Aquatic Ecosystems: Biodiversity and Conservation

Marcos Gomes Nogueira [1,*] and Douglas Donald Kane [2]

1 Instituto de Biociências (IBB), Campus de Botucatu, Universidade Estadual Paulista (UNESP), Botucatu CEP 18618-689, Brazil
2 Department of Biology and Environmental Science and National Center for Water Quality Research, Heidelberg University, Tiffin, OH 44883, USA; dkane@heidelberg.edu
* Correspondence: maarcos.nogueira@unesp.br

1. Introduction to the Special Issue

The structure and functioning of marine and inland water ecosystems are highly dependent on living organisms. The biodiversity of any particular habitat emerges from the complex relationships among the organisms themselves and their interaction with physical and chemical factors. Biodiversity reflects changes over long-term periods. Estimates on the actual number of living species on Earth are still controversial; however, it is agreed that much life and many ecological interactions are yet to be discovered. Furthermore, time is short and with the present worldwide political and economic volatility, conditions are very much unfavorable for preserving biodiversity. Currently, a quarter of the freshwater fauna is estimated to be at risk of extinction [1], while their aquatic habitats often experience a severe state of eutrophication, which is frequently dominated by pervasive cyanobacteria blooms [2,3]. Historically, freshwater ecosystems have been neglected by the official systems of protected areas [4,5], even in megadiverse continental countries [6]. In oceans and coastal waters, the widespread plastic contamination [7,8] and the extinction risk of certain taxonomic groups, such as sharks and rays [9–11], are also dramatic.

Biodiversity assessment is essential for different scientific approaches, as well as for conservation strategies and ecosystem management. This Special Issue collected papers that address the distinct spatial and temporal scales, as well as the past and present processes, that affect the conservation of the biodiversity of aquatic ecosystems.

2. Overview of the Articles Published in This Special Issue

Lim et al. tested conceptual models of the effect of salinity on macroinvertebrates in twelve estuaries in Southwestern Australia. They found that few species were associated with low salinities, which differed from conceptual models, while species richness was the greatest at mid-level salinities, declining at high salinities. Because hypersaline conditions are predicted to increase with global climate change, this research may help predict how benthic macroinvertebrate diversity in Mediterranean climates could decline in the future.

Spieles et al. examined the effect of a variety of land use-related parameters on macroinvertebrate diversity and biotic integrity in thirty low-order streams in Central Ohio. They found that the percentage of coverage of developed land at the watershed scale was the strongest predictor of both macroinvertebrate diversity and biotic integrity, while high-intensity development had a greater negative correlation with these measures than low-intensity development or agriculture. Taken together, these results indicate that impervious land cover at the scales of watershed and the Active River Area are important in affecting stream aquatic macroinvertebrate communities.

Demiray et al. used the transformer deep learning model (among other models) to determine its effectiveness in predicting the presence of harmful algal blooms (HABs) in Lake Erie. The target variable of chlorophyll *a*, as a surrogate for phytoplankton biomass, was used with a variety of physical, chemical, and biological water quality measurements taken from the western basin of Lake Erie as influencing parameters. Furthermore, using SHapley Additive exPlanations (SHAP) values as an explainable artificial intelligence (XAI) tool, it was determined that Particulate Organic Carbon (POC), Particulate Organic Nitrogen (PON), and Total Phosphorus (TP) were critical factors influencing HAB prediction in Lake Erie. This study demonstrates that predictive HAB models can be used successfully and may be applied to other systems in the future.

Simović et al. used a Convolutional Neural Network (CNN), which is a type of deep learning model, to identify aquatic insect species from three orders (Ephermeroptera (mayflies), Plecoptera (stone flies), and Trichoptera (caddisflies)) from an unbalanced data set. This data set contained between 10 and 80 individuals from each species. The authors found very little decrease in classification accuracy, as long as at least 30 individuals of the species were used. Strongly imbalanced data sets only decreased the classification accuracy by 2%, while moderately imbalanced data sets did not significantly affect classification accuracy. This is important to know because there are often wildly different numbers of individual taxa collected in aquatic macroinvertebrate biomonitoring.

Chung et al. examined riverine phytoplankton (also known as potamoplankton) in the Yeongsan River, following the construction of two weirs on this major river in South Korea. They examined Bacillariophyceae (diatoms), Chlorophyceae (green algae), Cyanophyceae (bluegreen algae), and other phytoplankton from 2019 to 2023, investigating how phytoplankton community structure and riverine physicochemical factors were related. They used several multivariate statistical techniques (e.g., canonical correspondence analysis (CCA) and principal components analysis (PCA)) to elucidate these relationships. PCA showed significant correlations with other phytoplankton and chlorophyll *a* (Chl-*a*) in spring, as well as with Cyanophyceae and water temperature in summer. CCA showed that the diatom *Stephanodiscus* sp. was associated with nitrogen-based nutrients, while the blue–green algae *Microcystis* sp. and *Dolichospermum* sp. were associated with water temperature and phosphate phosphorus (PO_4-P). This is important because *Stephanodiscus* can cause the clogging of waterworks and both *Microcystis* and *Dolichospermum* are toxin-producing genera.

Ergović et al. applied the Chironomid Pupal Exuvial Technique (CPET) to 28 man-made lakes in the Pannonian Lowlands and Dinaric Western Balkan Ecoregion of Croatia. This biomonitoring technique uses shed chironomid pupal skin (exuvia) to assess water quality. Of the 5698 chironomid pupal skins collected, the authors identified members of 141 taxa (including 97 species) belonging to five subfamilies. Physicochemical factors that influenced chironomid community assemblages differed from the Pannonian Lowlands (organic carbon (TOC) and orthophosphates (PO_4^{3-})) and the Dinaric Western Balkan (conductivity) ecoregions. The authors attribute these differing results to greater anthropogenic impacts in the Pannonian Lowlands. This study shows that in the future, the CPET can be applied to biomonitoring in lakes in Croatia and elsewhere.

Zhang et al. examined the fungal biodiversity and ecological function of fungal groups in the Tangchi Hot Springs in China. Illumina MiSeq high-throughput sequencing technology paired with bioinformatic techniques led the researchers to find that the dominant fungal phylum was Ascomycota, followed by Basidiomycota, Chytridiomycota, and Olpidiomycota, while the dominant fungal genera were *Rhizophydium*, *Aureobasidium*, *Rhodotorula*, and *Sclerotinia*. FUNGuild functional analysis showed that the dominant guilds in the hot springs were plant pathogens, followed by a variety of undefined sapotrophs. This research is important because most microbial investigations of hot springs have previously focused on bacteria, while this study adds to the knowledge regarding the fungal communities in these extreme ecosystems.

Li et al. examined the functional traits and niche characteristics of plant guilds in two ecosystems in the Xiangxi River basin in China. The reservoir water level fluctuation zones (RWLFZs) and the natural riparian zones (NRZs) were studied under different flooding regimes. The authors found 78 plant species in these zones; the dominant species were annuals and their percentage increased from 65.79% in the NRZs to 67.34% in the RWLFZs. Most of these annuals in the RWLFZs adopted the R adaptation strategy (sensu Grime). Furthermore, the Simpson dominance index of the RWLFZs was significantly higher than that of the NRZ. Finally, highly adaptable and widely distributed species with larger niche breadths and high-importance values usually had a greater niche overlap value in the RWLFZs than in the NRZs, which has implications both for competition after flooding events and the ecological restoration of these ecosystems.

Urbanski and Nogueira compared the effects of nutrient pollution on the fish fauna of Tietê River basin in São Paulo, Brazil, during dry and wet conditions. They measured physicochemical parameters and fish assemblages in three sites on the Tietê River and three sites on its tributaries. The Tietê River sites were classified from hypereutrophic to supereutrophic, while the tributaries were typically classified as mesotrophic during wet conditions and variable during dry conditions. Dissolved oxygen was less than 2 mg/L at all three sites in the Tietê River during wet conditions and >2 mg/L during dry conditions. With respect to fish, the richness per sample in the tributaries (11 to 14 spp.) was greater compared to that of the Tietê River proper (3 to 4 spp.). Siluriformes with accessory breathing dominated the Tietê proper, while the highly tolerant and detritivorous *Prochilodus lineatus* was the only species found at all sites. The authors conclude that eutrophication has created a chemical barrier to fish dispersal in this system, which needs to be addressed in order to restore the fish assemblage of the Tietê River.

Song et al. investigated the water quality and the macroinvertebrate community of the Ulungu River basin in northwest China between May and August 2022, as well as in October 2023. Along with sampling physicochemical parameters, the researchers collected 6101 macrobenthic organisms belonging to 3 phyla, 7 classes, 14 orders, 57 families, and 117 genera, with >85% of species being arthropods. Using these data, they calculated the biological monitoring working party (BMWP), family-level biotic index (FBI), and Shannon–Wiener Index. The mean values for the three indices (Shannon–Wiener, FBI, and BMWP) were greater for tributaries (light pollution, poor, and good, respectively) than the Ulungu River proper (moderate, poor, and general, respectively), varying according to sampling date.

Hu et al. examined the evolutionary mechanisms and morphological differences in geographically distinct populations of *Gymnodiptychus dybowskii*, which is a member of the carp family, in the Turks and Manas rivers in Xinjiang, China, between 2020 and 2021. The researchers performed morphometric analyses on 158 fish and found that the populations of the Turks and Manas rivers were different with one-way ANOVAs, showing 22 highly significant and 1 significant difference among the 35 morphological traits examined. The authors suggest that these differences are sufficient to make the fish in the two rivers different subspecies.

Hutorowicz used hydroacoustic sampling techniques in 2008 and 2021 to assess fish in Lake Dejguny (Poland). The target strength distribution was determined, along with temperature and oxygen concentrations, at 2 m depth intervals. A decline in the target strength in deep waters (>24 m) was associated with low oxygen levels (<2.5 mg/L). Finally, the researcher developed an index based on the number of large fish as a percentage of all acoustically recorded fish echoes; this was a modification of the Large Fish Index (LAFI), which compared the ratio of large fish to small fish.

Burandt et al. studied phytocenoses (plant communities) in riparian wetlands from 2017 to 2019 in the floodplains of northeastern Poland. They found that both the hydrological regime and geomorphological conditions controlled both the structure and pattern of the plant communities. Using linear discriminant analysis (LDA), four habitat types (wet, semi-wet, semi-dry, and dry zones) were identified. Indicator species analysis (ISA) showed that of the species with high water requirements, reed canary grass (*Phalaris arundinacea*) and common reed (*Phragmites australis*) dominated. Finally, flood pulses were found to be important in structuring these riparian wetland plant communities.

3. Conclusions

Researchers from eight different countries (Australia, Brazil, China, Croatia, Poland, Serbia, South Korea, and the United States of America) shared the results of their scientific investigations carried out on streams, rivers, lakes, artificial reservoirs, and estuary ecosystems. Ecological processes considered distinct spatial and temporal scales—local, local to regional, and regional geographic scales—as well as seasonal and long-term time series. The studied biota included phytoplankton/harmful algal blooms, plant guilds/riparian vegetation, and benthic macroinvertebrates, especially aquatic insects, fungal groups, and ichthyofauna. The main stressors affecting biotic integrity were global climate change, the intensification of land use, eutrophication, excessive pollution, and hydrological regime disturbance. Biotic communities' responses were evaluated through distinct methodological strategies, such as deep learning models, bioinformatics, DNA genomics, and morphological taxonomy. The authors also revisited integrated data sets produced by water quality and biomonitoring programs, suggesting innovative approaches. This Special Issue provides a comprehensive view of the predictors of the biodiversity changes in freshwater and estuarine environments from five continents. The effective inclusion of aquatic ecosystems in the conservation and restoration agendas is crucial, while knowledge about their extraordinary biodiversity helps the healthy reconnection of people with the natural world.

Author Contributions: Writing—original draft preparation: M.G.N. and D.D.K.; writing—review and editing: M.G.N. and D.D.K. All authors have read and agreed to the published version of the manuscript.

Funding: This research received no external funding.

Acknowledgments: Thanks to all the individual authors of the papers in this Special Issue.

Conflicts of Interest: The authors declare no conflicts of interest.

List of Contributions

1. Lim, R.; Fourie, S.A.; Stout, E.J.; Roots, B.J.; Cronin-O'Reilly, S.; Rodgers, E.M.; Tweedley, J.R. Testing the Remane Diagram: Occurrences of Benthic Macroinvertebrates in Oligohaline to Hyperhaline Salinities. *Water* **2025**, *17*, 1642. https://doi.org/10.3390/w17111642.
2. Spieles, D.; Krashes, Z.; Nguyen, K.; Rodgers, S.; Ruiz, L.; Vigilante, M. Relationships Between Land Use and Stream Macroinvertebrate Biotic Integrity in Central Ohio, USA. *Water* **2025**, *17*, 895. https://doi.org/10.3390/w17060895.
3. Demiray, B.Z.; Mermer, O.; Baydaroğlu, Ö.; Demir, I. Predicting Harmful Algal Blooms Using Explainable Deep Learning Models: A Comparative Study. *Water* **2025**, *17*, 676. https://doi.org/10.3390/w17050676.
4. Simović, P.; Milosavljević, A.; Stojanović, K.; Savić-Zdravković, D.; Petrović, A.; Predić, B.; Milošević, D. The Effects of Data Quality on Deep Learning Performance for Aquatic Insect Identification: Advances for Biomonitoring Studies. *Water* **2025**, *17*, 21. https://doi.org/10.3390/w17010021.

5. Chung, H.; Son, M.; Kim, T.; Park, J.; Lee, W.-S. Correlations Between Spatiotemporal Variations in Phytoplankton Community Structure and Physicochemical Parameters in the Seungchon and Juksan Weirs. *Water* **2024**, *16*, 2976. https://doi.org/10.3390/w16202976.
6. Ergović, V.; Čerba, D.; Vučković, N.; Mihaljević, Z. Chironomid Pupal Exuviae Technique in Ecological Research of Man-Made Water Bodies. *Water* **2024**, *16*, 2917. https://doi.org/10.3390/w16202917.
7. Zhang, F.-Q.; Liu, J.; Chen, X.-J. Diversity and Ecological Functions of Fungal Communities in Tangchi Hot Spring in Lujiang (China). *Water* **2024**, *16*, 2308. https://doi.org/10.3390/w16162308.
8. Li, X.; Yi, W.; Xu, S.; He, D.; Min, Q.; Chen, G.; Yang, J.; Deng, D.; Yang, Z.; Huang, G.; et al. Assessment of the Divergent Influence of Natural and Non-Seasonal Hydrological Fluctuations on Functional Traits and Niche Characteristics of Plant Guilds along the Xiangxi River, China. *Water* **2024**, *16*, 1808. https://doi.org/10.3390/w16131808.
9. Urbanski, B.; Nogueira, M. Excessive Eutrophication as a Chemical Barrier for Fish Fauna Dispersion: A Case Study in the Emblematic Tietê River (São Paulo, Brazil). Water 2024, 16, 1383. https://doi.org/10.3390/w16101383.
10. Song, Y.; Huo, Q.; Zi, F.; Ge, J.; Qiu, X.; Yun, L.; Serekbol, G.; Yang, L.; Wang, B.; Chen, S. Macrobenthic Community Structure and Water Quality Evaluation in Ulungu River Basin (Northwest China). *Water* **2024**, *16*, 918. https://doi.org/10.3390/w16070918.
11. Hu, L.; Yao, N.; Wang, C.; Yang, L.; Serekbol, G.; Huo, B.; Qiu, X.; Zi, F.; Song, Y.; Chen, S. Analyses of Morphological Differences between Geographically Distinct Populations of *Gymnodiptychus dybowskii*. *Water* **2024**, *16*, 755. https://doi.org/10.3390/w16050755.
12. Hutorowicz, A. Use of Hydroacoustic Methods to Assess Ecological Status Based on Fish: A Case Study of Lake Dejguny (Poland). *Water* **2024**, *16*, 282. https://doi.org/10.3390/w16020282.
13. Burandt, P.; Grzybowski, M.; Glińska-Lewczuk, K.; Gotkiewicz, W.; Szymańska-Walkiewicz, M.; Obolewski, K. Hydrology as a Determinant of Riparian Habitat Structure in Lowland River Floodplains. *Water* **2024**, *16*, 164. https://doi.org/10.3390/w16010164.

References

1. Sayer, C.A.; Fernando, E.; Jimenez, R.R.; Macfarlane, N.B.W.; Rapacciuolo, G.; Böhm, M.; Brooks, T.M.; Contreras-MacBeath, T.; Cox, N.A.; Harrison, I.; et al. One-quarter of freshwater fauna threatened with extinction. *Nature* **2025**, *638*, 138–145. [CrossRef] [PubMed]
2. Paerl, H.W.; Barnard, M.A. Mitigating the global expansion of harmful cyanobacterial blooms: Moving targets in a human- and climatically altered world. *Harmful Algae* **2020**, *96*, 101845. [CrossRef] [PubMed]
3. Igwaran, A.; Kayode, A.J.; Moloantoa, K.M.; Khetsha, Z.P.; Unuofin, J.O. Cyanobacteria Harmful Algae Blooms: Causes, Impacts, and Risk Management. *Water Air Soil Pollut.* **2024**, *235*, 71. [CrossRef]
4. IUCN World Commission on Protected Areas (WCPA); IUCN Species Survival Commission (SSC); Freshwater Conservation Sub-Committee. *Monographies Series 2024*; IUCN: Gland, Switzerland, 2024; Volume 8, p. 166. ISBN 978-2-8317-2314-3.
5. Valentim, H.I.L.; Feio, M.J.; Almeida, S.F.P. Assessing the effectiveness of terrestrial protected areas towards riverine ecosystems. *Sci. Total Environ.* **2025**, *958*, 177878. [CrossRef] [PubMed]
6. Azevedo-Santos, V.; Frederico, R.G.; Fagundes, C.K.; Pompeu, P.S.; Pelicice, F.M.; Padial, A.A.; Nogueira, M.G.; Fearnside, P.M.; Lima, L.B.; Daga, V.S.; et al. Protected areas: A focus on Brazilian freshwater biodiversity. *Divers Distrib.* **2018**, *25*, 442–448. [CrossRef]
7. Ostle, C.; Thompson, R.C.; Broughton, D.; Gregory, L.; Wootton, M.; Johns, D.G. The rise in ocean plastics evidenced from a 60-year time series. *Nat. Commun.* **2019**, *10*, 1622. [CrossRef] [PubMed]
8. Tekman, M.B.; Walther, B.A.; Peter, C.; Gutow, L.; Bergmann, M. Impacts of Plastic Pollution in the Oceans on Marine Species, Biodiversity and Ecosystems. In *Impacts of Plastic Pollution in the Oceans on Marine Species, Biodiversity and Ecosystems*; WWF Germany: Berlin, Germany, 2022; pp. 1–221. [CrossRef]
9. Oliver, S.; Braccini, M.; Newman, S.J.; Harvey, E.S. Global patterns in the bycatch of sharks and rays. *Mar. Policy.* **2015**, *54*, 86–97. [CrossRef]

10. Moore, A.B.M. Are guitarfishes the next sawfishes? Extinction risk and an urgent call for conservation action. *Endang. Species Res.* **2017**, *34*, 75–88. [CrossRef]
11. Pacoureau, N.; Rigby, C.L.; Kyne, P.M.; Sherley, R.B.; Winker, H.; Carlson, J.K.; Fordham, S.V.; Barreto, R.; Fernando, D.; Francis, M.P.; et al. Half a century of global decline in oceanic sharks and rays. *Nature* **2021**, *589*, 567–571. [CrossRef] [PubMed]

water

MDPI

Article

Testing the Remane Diagram: Occurrences of Benthic Macroinvertebrates in Oligohaline to Hyperhaline Salinities

Ruth Lim [1,2], Stephanie A. Fourie [1,2], Emily J. Stout [1,2], Ben J. Roots [1,2], Sorcha Cronin-O'Reilly [1,2], Essie M. Rodgers [1,2] and James R. Tweedley [1,2,*]

1 School of Environmental and Conservation Sciences, College of Environmental and Life Sciences, Murdoch University, 90 South Street, Murdoch, WA 6150, Australia

2 Centre for Sustainable Aquatic Ecosystems, Harry Butler Institute, Murdoch University, 90 South Street, Murdoch, WA 6150, Australia

* Correspondence: j.tweedley@murdoch.edu.au

Abstract: Starting with the Remane diagram, various conceptual models have been proposed to show how species richness varies along a salinity gradient. However, as relatively few estuaries experience extreme hypersalinity, quantitative data are lacking to evaluate the model. We used data for 1891 samples of benthic macroinvertebrates from 12 estuaries in southwestern Australia (salinity 0–122 ppt) to determine the salinities in which 257 taxa were recorded. The pattern of richness differed from the conceptual models, with relatively few species (≤20%) recorded in freshwater and oligohaline salinities. Richness peaked at 35 ppt (seawater, 44%) before declining precipitously, with 21% and 10% of taxa recorded in hyperhaline salinities of 40 and 48 ppt, respectively. Taxa were recorded across the full salinity range, and several holohaline annelids, crustaceans, and insects were identified. Descriptive statistics and the frequency distribution of each taxon along the salinity gradient are provided. These identify stenohaline taxa and those with different extents of euryhalinity and how the occurrence of these taxa changes with salinity. The results help predict how benthic macroinvertebrate species and assemblages in estuaries in southwestern Australia and other Mediterranean climatic regions may shift due to climate change, particularly increased incidences and magnitude of hypersalinity.

Keywords: euryhaline; holohaline; hypersalinity; lagoon; low in-flow; Mediterranean climate; osmoconformation; osmoregulation; salinity gradient; salt lake

1. Introduction

Being located at the interface of freshwater derived from land drainage and marine water from tidal movement, estuaries are dynamic, and their environmental conditions vary markedly over a range of temporal scales, e.g., from a tidal cycle to several years of drought [1,2]. Many definitions of estuaries include statements such as "free connection with the open sea" and "seawater is measurably diluted with fresh water", e.g., [3,4] and thus salinity values would typically range from 0 to 35 ppt. However, subsequent definitions have been broader and recognise that some estuaries become periodically disconnected from the sea and can become hypersaline [5,6]. These systems, which represent ~3% of all estuaries, have received increasing attention in recent years [7,8] and salinities > 300 ppt have been recorded in several [9,10].

The highly variable salinities present in estuaries pose osmoregulatory challenges compared to the relatively stable environments found in freshwater and marine ecosystems [11,12]. As a result, estuaries typically harbour a unique fauna, often comprising a smaller number of species, many of which are relatively euryhaline, meaning they can tolerate a wide range of salinities [13]. In contrast, stenohaline species, which can only tolerate a narrow range of salinities (<10 ppt), are less common in these environments [14]. Sampling conducted for the UK's National Marine Monitoring Programme recorded 784 species (14 phyla) from coastal areas and 296 species (11 phyla) from estuaries, resulting in a marked difference in faunal composition [15]. Typically, the benthic macroinvertebrate species in macrotidal estuaries (tidal range > 4 m) are annelids, crustaceans (excluding insects), and molluscs. For example, species in these three groups comprised 93.2% of the 296 species recorded in 16 UK estuaries [15]. Similarly, of the species recorded from intertidal areas from the mouth to the tidal freshwater reaches of the Scheldt Estuary (The Netherlands and Belgium), 45% were annelids, 29% were molluscs, and 23% were crustaceans [16]. However, insects are an additional group that can occur in microtidal estuaries (tidal range < 2 m) in Europe [17], South Africa [18], and Australia [19].

Numerous natural environmental variables influence the composition of the benthic macroinvertebrate fauna within estuaries. These include habitat, particularly the structural complexity provided by macrophyte beds and shellfish reefs, sediment composition (e.g., grain size and organic matter content), and the physicochemical properties of water [17–19]. Salinity, in particular, plays a crucial role as species have evolved physiological and life-cycle adaptations to specific salinity ranges [20]. Osmoregulation, the physiological process by which organisms maintain the balance of water and salts in their bodies, is essential for survival in estuarine environments. When animals are exposed to salinities outside their tolerance range, it can disrupt homeostasis, leading to physiological stress, reduced fitness, altered behaviour, and even mortality [21].

Elevated salinity levels can cause dehydration and ion imbalance as organisms lose water to their hypertonic environment, leading to cellular and systemic stress. Conversely, low salinity levels can result in water influx, causing cells to swell and potentially rupture, disrupting normal cellular functions. Osmoregulation can be energy-intensive, as it involves the active transport of ions across cell membranes, which demands a significant amount of energy [14]. There are two main strategies for dealing with salinity changes: osmoregulation or osmoconformation. Osmoregulators actively regulate their internal osmolarity to remain constant, regardless of the external environment. This strategy is common in many fish, mammals, and birds and involves specialised organs like kidneys, salt glands, or gills to remove excess water or ions [22]. In contrast, osmoconformers match their internal osmolarity to the osmolarity of their surrounding environment. This strategy is typical of many marine invertebrates and is less energetically demanding but limits their ability to survive in fluctuating salinity conditions. Few organisms can osmoregulate or complete their life cycles along the full salinity gradient from fresh to seawater, and many are adapted to a salinity range that is taxon-specific [23]. Consequently, biological communities are structured by the salinity gradient, e.g., [24,25].

The relationship between salinity and species richness has become a paradigm of estuarine ecology [26] and is illustrated by the Remane diagram [27]. This conceptual model, which was based on benthic macroinvertebrate fauna from the Baltic Sea, was designed to show how the number of species changes along a salinity gradient (0–35 ppt) and the salinity tolerance of those species, i.e., freshwater, brackish, and marine. Numerous authors have adapted the Remane diagram since its original publication in 1934; see [28] for a review. In particular, Hedgpeth [29], based on studies from bays and lagoons in Texas, most notably the Laguna Madre, extrapolated the salinity gradient to 310 ppt. He also developed two terms: (i) holoeuryhaline, i.e., species distributed from freshwater to extreme hypersalinity that may be absent from marine waters, and (ii) ultrahaline, i.e., marine species occurring in oligohaline to extreme hypersaline waters. Similarly, Khlebovich [30] also extended the salinity gradient to 100 ppt. The revised diagrams produced by both authors showed a decrease in the number of species after 35 ppt, albeit with a slightly different shape to the curve. Based on a review of the many modifications to the Remane diagram and examining patterns of richness for various estuarine faunas, Whitfield et al. [28] produced a conceptual model for estuarine biodiversity from freshwater to hyperhaline conditions (0–60 ppt). It shows that richness is moderately high in freshwater (0–0.5 ppt), declines to a low in the horohalinicum (5–8 ppt), and increases to a maximum at 35 ppt and then decreases to 60 ppt. These authors also termed holohaline (i.e., species found between 0 and 100+ ppt), which includes both holoeuryhaline and ultrahaline species.

Despite its wide usage in estuarine ecology literature, the Remane diagram provides a qualitative assessment of the biodiversity in different salinities [28]. It has seldom been tested, particularly in estuaries that can become hyperhaline. Therefore, the primary aim of this study is to collate data across a range of estuaries in the same climatic region that experience a salinity gradient from oligohaline to markedly hyperhaline conditions and determine how the number of species changes and compare this to the various modifications to the Remane diagram. A secondary aim is then to examine the occurrence distribution of species along the salinity gradient. The results are useful in assessing and predicting how climate change-induced changes in salinity might impact the fauna of estuaries.

2. Materials and Methods

2.1. Site Description and Sampling

Southwestern Australian estuaries are microtidal and experience a Mediterranean climate, with hot, dry summers and cool, wet winters [31]. Over 85% of rainfall occurs between May and October, and there is a gradient ranging from ~1400 mm per year in the southwestern corner to a minimum of ~400 mm north and eastwards, where it becomes more variable [32]. These conditions result in some of the estuaries becoming disconnected from the ocean for several years at a time and, if rainfall is low, evaporation can cause them to be amongst the most hyperhaline globally [33,34].

Data on the benthic macroinvertebrate fauna were collected from 12 estuaries distributed along the coast of southwestern Australia (Figure 1) and differed in their physical characteristics and environmental conditions (Table 1 and Table S1). The Swan–Canning and Peel–Harvey estuaries are large, permanently-open systems on the lower west coast. Oyster Harbour and Waychiniup Estuary are also always open to the ocean, albeit smaller and located on the south coast. The Vasse–Wonnerup (west coast) and Broke, Torbay, Taylor, Normans and Cheyne inlets and Cordinup River (all south coast) are open to the ocean for part of each year (annually-open), while Beaufort Inlet may be closed for several years at a time (normally-closed). Due to these differences in bar status, the size of their catchments and riverine flow, they experience marked variations in salinity ranging from 0 to 122 ppt (Figure A1).

Full details of the sampling protocols are provided in the relevant source publications and are summarised briefly in Table 1. Sampling was conducted along the longitudinal axis in all estuaries to incorporate the full range of salinities experienced during sampling. Moreover, the sites (12–60 per estuary) encompassed a range of sediment types and areas with and without submerged aquatic vegetation (e.g., seagrass and charophytes). Sampling was conducted between two and 15 times in each estuary to account for temporal variation.

This extensive sampling yielded 1891 samples, with between 48 and 480 samples from each of the 12 estuaries.

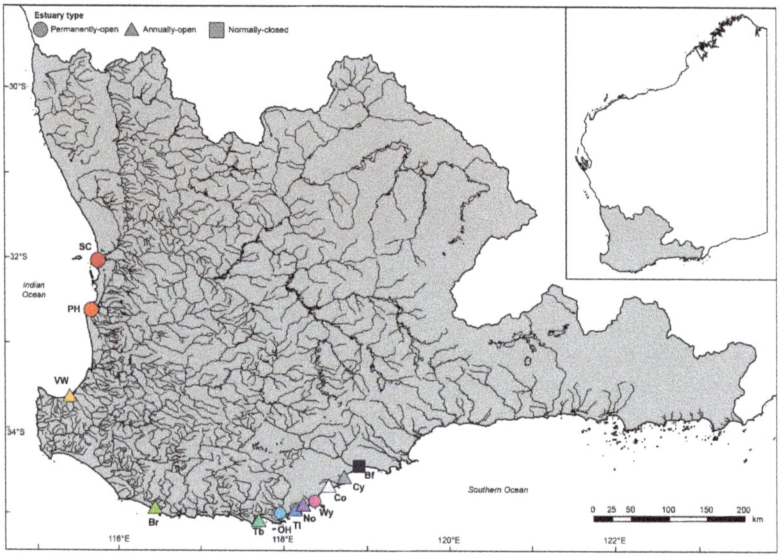

Figure 1. Map of southwestern Australia showing the rivers in the southwest drainage division (grey shading) and the location of the 12 estuaries where samples of benthic macroinvertebrates were collected. SC, Swan–Canning Estuary; PH, Peel–Harvey Estuary; VW, Vasse–Wonnerup Estuary; Br, Broke Inlet; Tb, Torbay Inlet; OH, Oyster Harbour; Tl, Taylor Inlet; No, Normans Inlet; Wy, Waychinicup Estuary; Co, Cordinup River; Cy, Cheyne Inlet; Bf, Beaufort Inlet. Insert shows the location of the southwest drainage division within Western Australia.

Two broad sampling methods were employed depending on the depth of the water body. Sediment from shallow waters (0.5–1 m deep) was collected using a cylindrical corer with a diameter of 11 cm (sampling area: 96 cm^2) that sampled to a depth of 15 cm. For sites where sediments are collected in waters deeper than 1 m, an Ekman grab (225 cm^2) that sampled to a depth of 15 cm was employed. Peel–Harvey Estuary was the only exception where samples were collected (regardless of water depth) using a smaller corer with a 6 cm diameter (sampling area: 28 cm^2). Once collected, sediment was wet sieved through a 500 µm mesh field sieve, and the residual material was transferred into a container and preserved with 4% formalin solution buffered in estuary water and/or 70% ethanol. At each site on each sampling occasion, water temperature (°C), salinity (ppt), and dissolved oxygen concentration (mgL^{-1}) were measured at the bottom of the water column using the YSI 556 or ProDSS Multiparameter Digital Water Quality Meter (Yellow Springs Instrument, Yellow Springs, OH, USA). As the instruments were only capable of accurately reading to a maximum salinity of 70 ppt, water samples were diluted using deionised water to allow accurate measurement [9]. Although the SI unit of salinity is the practical salinity scale (PSU), this is a ratio (i.e., has no units) and its calculation was validated only for values between 2 and 42 [35]. As such, salinities in this study are provided in parts per thousand (ppt), which is also the unit of measurement provided by the water quality probes.

Table 1. Summary of the characteristics of each of the 12 estuaries taken from [36], and details of the sampling regime and the mean and range of values recorded for a suite of water physicochemical parameters. n = number of samples; PO = permanently-open, AO = annually-open and NC = normally-closed.

	Estuary Characteristics				Sampling Regime					Water Physicochemical Parameters						
	Area (km^2)	Mean Catchment Rainfall (mm)	Median Annual Flow (ML)	Estuary Type	Region	Sites	Timing	n	Method	Salinity (ppt)		Temperature (°C)		DO (mgL^{-1})		Reference
										Mean	Range	Mean	Range	Mean	Range	
Swan-Canning Estuary	38.1	409	600,000	PO	Upper	5	11 months (2010/11)	309	Ekman grab (225 cm^2)	19.0	4.0–32.5	21.2	11.6–29.7	3.3	0.2–11.0	[37]
					Lower	4	4 seasons (2023/24)	80	Corer (96 cm^2)	34.5	19.1–39.5	22.4	14.3–30.9	8.3	6.4–10.5	[38]
					Lower	6	1 season (2023)	90	Corer (96 cm^2)	35.2	34.8–35.6	25.1	23.5–27.6	8.0	6.1–10.6	[39]
Peel-Harvey Estuary	133.6	662	810,000	PO	All	60	2 seasons (2017/18)	120	Corer (28 cm^2)	26.7	2.1–56.2	21.3	13.9–28.9	5.8	0.1–10.6	[40]
Vasse-Wonnerup Estuary	6.1	905	40,300	AO	All	30	15 seasons (2017–20)	344	Ekman grab (225 cm^2)	26.5	0.6–96.5	20.5	12.0–34.3	8.8	0.6–18.2	[41,42]
Broke Inlet	45.6	1331	162,000	AO	All	40	5 seasons (2007–08)	480	Corer (96 cm^2)	22.9	2.1–45.1	18.1	12.6–28.3	6.0	2.5–20.4	[43]
Torbay Inlet	0.9	943	75,000	AO	All	12	4 seasons (2020)	48	Corer (96 cm^2)	14.1	1.0–34.7	19.1	12.4–25.7	8.0	2.2–11.6	[44]
Oyster Harbour	17.7	949	97,200	PO	All	12	4 seasons (2020)	48	Corer (96 cm^2)	34.1	17.4–37.6	17.6	12.3–22.3	8.3	1.5–12.2	[44]
Taylor Inlet	0.5	800	1400	AO	All	12	4 seasons (2020)	48	Corer (96 cm^2)	20.0	16.1–24.2	17.5	12.5–22.5	9.6	6.1–16.3	[44]
Normans Inlet	0.2	810	1800	AO	All	12	4 seasons (2020)	48	Corer (96 cm^2)	4.2	2.4–6.2	16.9	12.5–22.0	8.4	4.3–10.4	[44]
Waychinicup Estuary	0.1	760	8000	PO	All	12	4 seasons (2020)	48	Corer (96 cm^2)	31.9	3.4–35.4	18.3	15.5–21.5	8.1	6.3–10.3	[44]
Cordinup River	0.1	610	1700	AO	All	12	4 seasons (2020)	48	Corer (96 cm^2)	12.6	7.9–19.3	17.6	14.3–23.2	8.4	4.4–13.9	[44]
Cheyne Inlet	0.2	610	1800	AO	All	12	4 seasons (2020)	48	Corer (96 cm^2)	42.5	33.0–54.3	18.0	13.4–25.7	7.6	3.7–10.7	[44]
Beaufort Inlet	6.5	410	36,000	NC	All	12	11 times (2020–23)	132	Corer (96 cm^2)	56.3	3.2–121.8	17.7	8.6–25.5	7.4	3.6–16.2	[45]

In the laboratory, invertebrates were removed from the residual sediment and stored in 100% ethanol. Each specimen was identified to the lowest possible taxon. Insects were identified to a higher taxonomic level than other taxa due to the differences in the resolution of identification keys for the various life stages (i.e., larva, pupa, adults). To ensure consistency in identification, the reference collections from the various studies were examined, and the taxonomy of all species was checked and updated where necessary using the World Register of Marine Species [46]. The abundance of each macroinvertebrate taxon in each replicate sample was converted into a density (individuals 100 cm^{-2}).

2.2. Data Analysis

The density and water physicochemical data were used to create a data matrix of the salinities of the samples in which each of the 257 taxa was present. These data were then used to calculate the minimum and maximum salinities in which a taxon occurred. As not all species were recorded in each salinity between 0 and 122 ppt, the values for the range were used to generate a presence/absence data matrix of each taxon in each salinity (1 ppt resolution). Species were assigned as stenohaline or euryhaline based on the range of salinities in which they were recorded, i.e., ≤10 or ≥10 ppt, respectively. The number of taxa in each salinity was counted and used to construct a quantitative version of the Remane diagram. Published modifications of the Remane diagram made by Whitfield et al. [28], Hedgpeth [29] and Khlebovich [30] were taken from the former paper, and the data for the total number of species in each salinity (1 ppt resolution) were extracted using https://www.graphreader.com/. As the *y*-axis in some versions of the Remane diagram has no quantitative unit of measure, the results from those papers and those generated in the current study were converted to a percentage of the maximum to compare how the values change along the salinity gradient. A stacked area chart was produced to show the number of stenohaline and euryhaline species in each salinity. Note that this plot was restricted to the 208 taxa recorded more than once.

To visualise how the proportional richness of various phyla and classes differed along the salinity gradient (groups of 10 ppt), the percentage contribution of taxa in each salinity category was calculated. Only phyla and classes comprising at least five taxa were included.

The minimum and maximum salinity in which each annelid, mollusc, and arthropod taxon was recorded were used to create a dumbbell plot. For taxa occurring in ≥10 of the 1891 samples, the distribution of those occurrences across the salinity gradient was plotted in a ridgeline plot using kernel density estimation. Dumbbell and ridgeline plots were generated using rStudio v4.3.1; [47], utilising the ggplot, ggridges, and ggdist packages [48–50].

3. Results

3.1. Faunal Description

A total of 257 benthic macroinvertebrate taxa, representing 11 phyla, were recorded across 1891 samples from the 12 estuaries in southwestern Australia (Table 2). Among these taxa, 94% were arthropods (118 taxa), annelids (82 taxa), and molluscs (42 taxa) and combined, they contributed 98% of all individuals. The 118 arthropod taxa included 58 malacostracans and 47 hexapods. Malacostracans were more numerous and frequently recorded than hexapods, representing 17% of all individuals and occurring in 58% of samples (vs. 5 and 32%, respectively). Although relatively few ostracod species (7 taxa) were recorded and not abundant, they were found in 7% of samples. The vast majority of the 82 annelid taxa were polychaetes (94%). Polychaetes represented 51% of the 53% of annelid individuals present and were found in 85% of samples. Oligochaetes were also relatively frequently recorded (14%). Of the 42 mollusc taxa, 24 were gastropods and 16 were bivalves. Bivalves were more abundant (19 vs. 3% of individuals) and more common (48 vs. 21% of samples; Table 2). Among the other phyla, nematodes and nemerteans were also relatively frequently recorded (~5% of samples).

Table 2. Richness (# taxa), percentage contribution to total abundance (%C), number of samples (*n*) and frequency of occurrence (%F) of taxa belonging to each phylum (boldface) and order or class.

Phylum and Order/Class	# taxa	%C	N	%F
Cnidaria	**2**	**0.05**	**12**	**0.63**
Hexacorallia	1	0.05	11	0.58
Scyphozoa	1	>0.01	1	0.05
Echinodermata	**5**	**0.01**	**7**	**0.37**
Ophiuroidea	2	0.01	4	0.21
Echinoidea	2	>0.01	2	0.11
Asteroidea	1	>0.01	1	0.05
Chordata	**1**	**>0.01**	**1**	**0.05**
Thaliacea	1	>0.01	1	0.05
Nematoda	**1**	**1.86**	**108**	**5.71**
Arthropoda	**118**	**22.80**	**1403**	**74.19**
Malacostraca	58	16.88	1091	57.69
Hexapoda	47	5.00	610	32.26
Ostracoda	7	0.84	130	6.87
Copepoda	4	0.08	23	1.22
Branchiopoda	2	0.01	8	0.42
Platyhelminthes	**3**	**0.03**	**24**	**1.27**
Mollusca	**42**	**21.43**	**1096**	**57.96**
Bivalvia	16	18.51	911	48.18
Gastropoda	23	2.92	397	20.99
Polyplacophora	2	>0.01	2	0.11
Scaphopoda	1	>0.01	2	0.11
Annelida	**82**	**53.67**	**1633**	**86.36**
Polychaeta	77	51.33	1610	85.14
Oligochaeta	3	2.31	256	13.54
Sipuncula	1	0.02	12	0.63
Hirudinea	1	0.01	5	0.26
Nemertea	**1**	**0.15**	**83**	**4.39**
Phoronida	**1**	**>0.01**	**1**	**0.05**
Foraminifera	**1**	**>0.01**	**1**	**0.05**

Five taxa, namely the polychaetes *Capitella* spp. (which includes the *Capitella capitata* species complex), *Simplisetia aequisetis* and *Pseudopolydora* cf. *kempi*, the bivalve *Arthritica semen*, and amphipod *Corophium minor* were the most abundant taxa, representing 24 to 5% of all individuals (54% combined) and were found in 51 to 22% of samples (Table S2). Other common species included the polychaetes *Scoloplos normalis*, *Prionospio cirrifera*, *Desdemona ornata*, and the amphipod *Grandidierella* spp. (i.e., *Grandidierella japonica* and *Grandidierella propodentata*).

3.2. Relationship Between Richness and Salinity

The number of taxa increased rapidly from 13 (5% of all taxa) in freshwater to 13–75 in oligohaline, 88–106 in mesohaline and 105–124 in euhaline waters, reaching a peak of 173 taxa (66%) at a salinity of 35 ppt (Figure 2a). After this, richness declined steeply to 83 by 40 ppt, 38 by 50 ppt, 21 by 60 ppt, and 11 by 90 ppt. Six taxa were present in salinities > 100 ppt, and a single species was found in the highest salinity recorded (122 ppt). This pattern of richness along the salinity gradient differs markedly from the conceptual models by Remane and the various modifications in several regards (Figure 2b). Firstly, the other models have a relatively high richness in freshwater, which declined rapidly through the horohalinicum before increasing. In contrast, in southwestern Australia, there was a rapid increase in richness in oligohaline salinities. Secondly, although the curves for all models peaked at a salinity of 35 ppt, the decline in richness in southwestern Australia was more rapid in salinities between 40 and 90 ppt. In the current study, 65% of the taxa were classified as euryhaline (i.e., found in a salinity range $\geq$ 10 ppt), including 55%, 70%, and 76% of the annelid, arthropod, and mollusc taxa, respectively. Stenohaline species were generally found in salinities between 32 and 38 ppt, representing 15–33% of all taxa (Figure 2c). While some stenohaline taxa were recorded in salinities < 5 ppt, they comprised 8% of the total number of taxa. Almost a third of all taxa were recorded across a salinity range greater than 35 ppt, and for five taxa, this exceeded 100 ppt (Figure 2d).

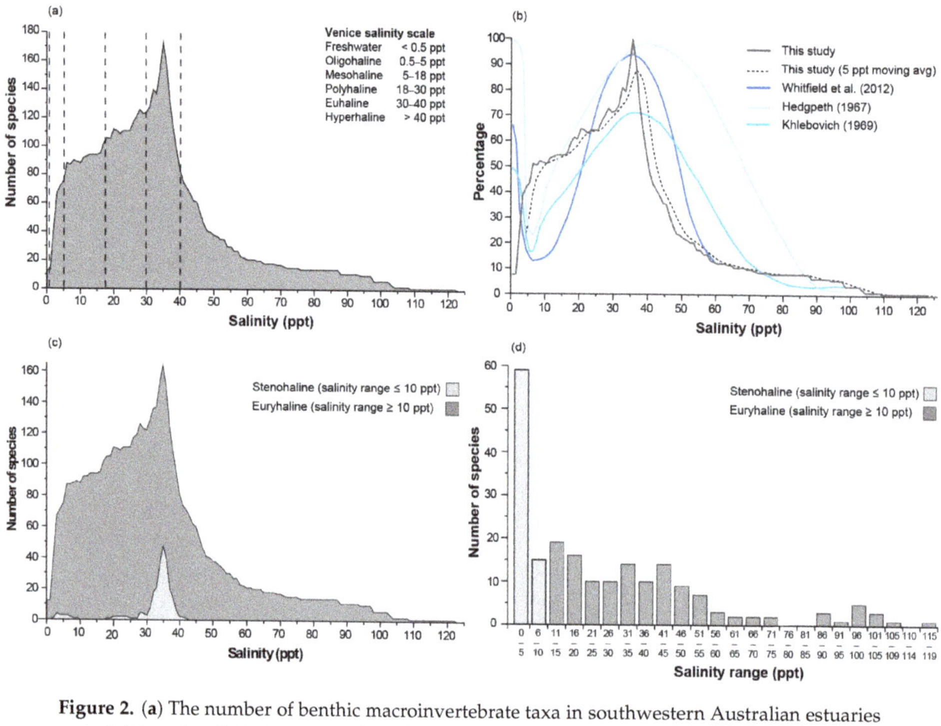

Figure 2. (**a**) The number of benthic macroinvertebrate taxa in southwestern Australian estuaries recorded in each salinity increment (1 ppt), (**b**) relationship between the number of taxa expressed as a percentage of the maximum value and salinity found in southwestern Australia and various authors' modifications to the Remane diagram [28–30], (**c**) the number of stenohaline and euryhaline species in each salinity increment, and (**d**) the number of taxa within each salinity range, defined by the difference between the minimum and maximum salinity that each taxon was recorded in. Dashed lines in (**a**) denote boundaries between salinity categories on the Venice salinity scale.

Among the phyla, arthropods were recorded in all salinity categories, with annelids only absent from salinities > 111 ppt and molluscs from those >100 ppt (Figure 3a). In contrast, echinoderms were restricted to euryhaline conditions (31–40 ppt). The percentage richness of arthropod, annelid, and mollusc taxa increased sequentially to a peak in euhaline conditions (31–40 ppt) before declining rapidly, with only 19, 11, and 13% of taxa recorded in salinities > 50 ppt. The pattern for these three phyla was mirrored by those for polychaetes, malacostracans, gastropods, and particularly bivalves, which were not recorded in salinities > 60 ppt (Figure 3b). However, that for hexapods was different, with the percentage richness being similar in salinities between 0 and 30 ppt (~67%) before declining more gradually and remaining > 10% until salinity exceeded 100 ppt.

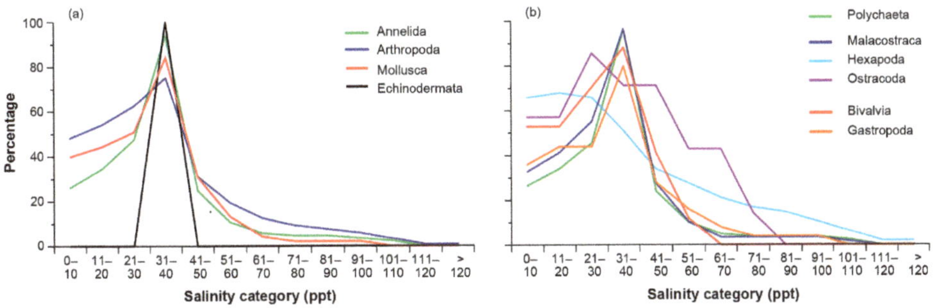

Figure 3. Percentage number of taxa belonging to each (**a**) phylum and (**b**) class in each salinity category (10 ppt).

3.3. Distribution of Annelids

Polychaetes were absent from freshwater, but 15 taxa were recorded in oligohaline conditions and 20 from hyperhaline conditions (Figure 4). Among the relatively common taxa, *Eusyllis* sp., *Aricidea* (*Acmira*) *lopezi*, *Spirorbis* sp., *Namanereis* sp., *Exogone* sp., and *Neanthes cricognatha* occurred over a narrow range of euhaline salinities. All taxa recorded in oligohaline conditions were also found in salinities > 35 ppt. Among these, some occurred over a wide range, including *D. ornata* (2–43 ppt), *P. cirrifera* (2–51 ppt), *Scoloplos normalis* (1–54 ppt), *S. aequisetis* (1–63 ppt), and particularly *Capitella* spp. (0–103 ppt) and *P.* cf. *kempi* (1–107 ppt). Among the other annelids, Hirudinea spp. and the oligochaetes Enchytraeidae sp. and Naididae sp. were recorded in salinities < ~30 ppt and *Sipuncula* spp. in 28–45 ppt. However, *Oligochaeta* spp. occurred over a far broader range of 1–86 ppt (Figure 4).

Visual interpretation of the ridgeline plots provides evidence for a shift in the distribution of common annelid taxa along the salinity gradient from oligohaline to hyperhaline conditions, after which occurrences are limited. However, due to the relatively broad range of salinity values in which many of these species were recorded, this trend is based more on their median salinity and relative frequency of occurrences (Figure 5). Although none of the common polychaete species were restricted to salinities < 35 ppt, species such as *Boccardiella limnicola* and *D. ornata* had the largest proportions of their occurrences in oligohaline to polyhaline salinities (median = 16 and 19 ppt, respectively). *Pseudopolydora* cf. *kempi*, *Ficopomatus enigmaticus*, and *P. cirrifera* were most often found in polyhaline salinities (median = 22–24 ppt), albeit they were present across a marked range of salinities. Similarly, a group of taxa comprising *S. normalis*, *S. aequisetis*, *Pseudopolydora* sp., *Capitella* spp., and *Armandia intermedia* were most often recorded in salinities of ~30 ppt (median = 26–32 ppt). Most of the remaining taxa, e.g., *Heteromastus filiformis*, *Spirorbis* sp., *Eusyllis* sp., and *Exogone* sp., were mainly found in a salinity of ~35 ppt and all except *H. filiformis* (range = 3–47 ppt) were recorded in a relatively restricted range of salinities (i.e., range = ~5 ppt).

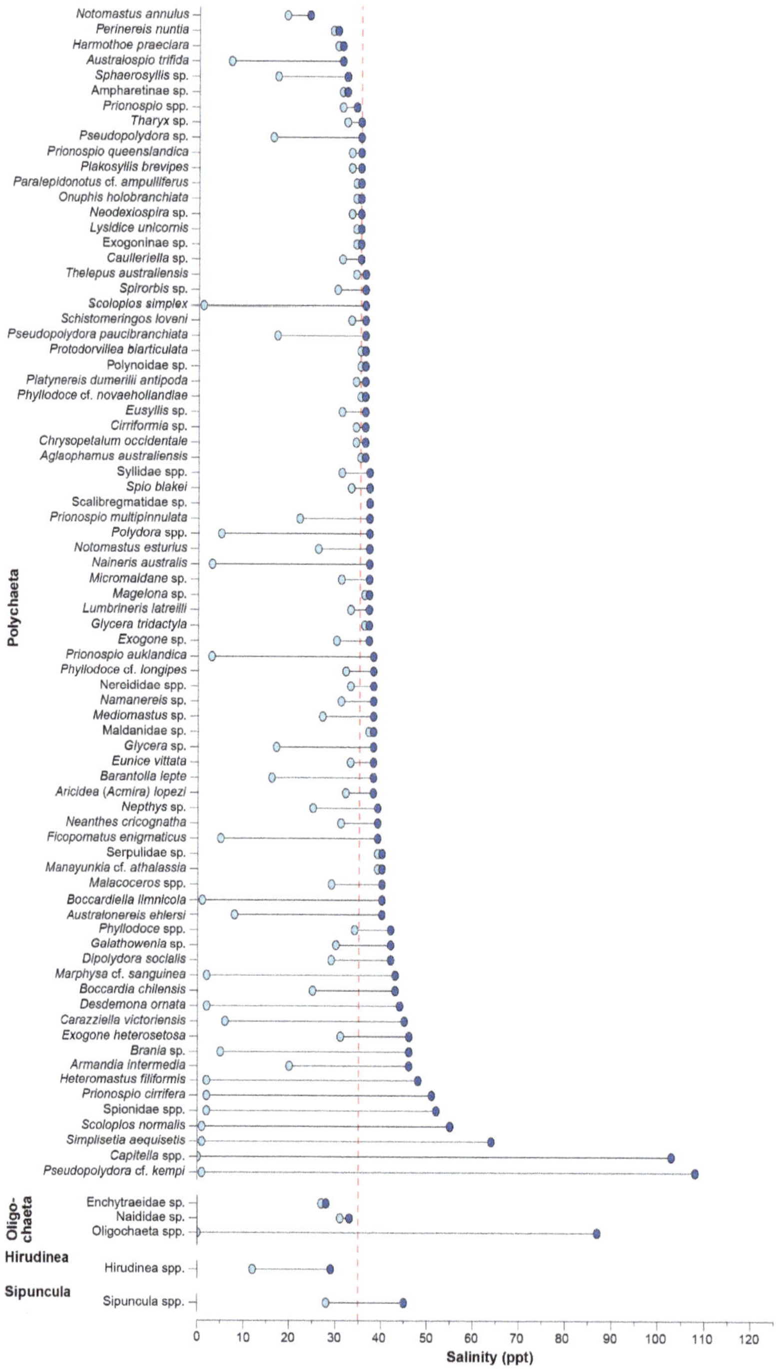

Figure 4. Dumbbell plot of the minimum and maximum salinity in which each annelid taxon was recorded. The dashed red line denotes salinity around full-strength seawater.

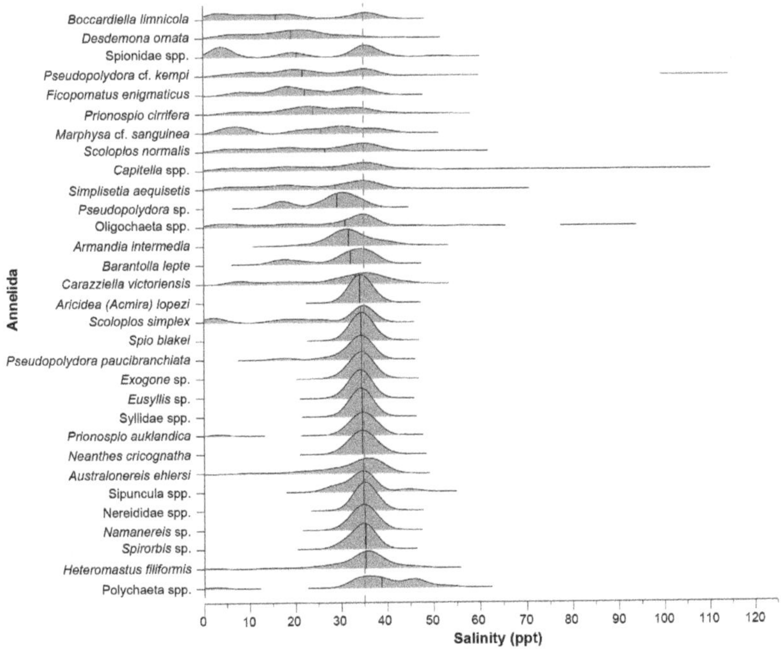

Figure 5. Ridgeline plot of the frequency distribution of each annelid taxon recorded in ≥10 samples across the salinity gradient. The dashed red line denotes salinity around full-strength seawater and the black vertical solid lines indicate the median salinity of each taxon generated by the kernel density estimation.

3.4. Distribution of Molluscs

As with the polychaetes, no bivalves were recorded in freshwater, and almost all taxa found in oligohaline conditions occurred over a salinity range of >35 ppt (Figure 6). Of these, *Arcuatula senhousia* and *Xenostrobus securis* were recorded in salinities of up to 36 and 39 ppt, respectively, *Hiatula biradiata* and *Fluviolanatus subtortus* up to ~46 ppt, and *Spisula trigonella* and *A. semen* to 51 and 54 ppt, respectively. Other species such as *Macomona deltoidalis*, *Venerupis galactites* and *Theora lubrica* occurred over ranges of 13–21 ppt, albeit in salinities that were polyhaline and euhaline. The gastropods *Potamopyrgus* spp. and *Coxiella striatula* occurred over the widest range of 0–97 ppt and 3–70 ppt, respectively. Among the other relatively common species, *Ascorhis occidua* and *Batillaria australis* occurred over a slightly narrow range of 5–40 ppt and 1–47 ppt, while *Tatea rufilabris* occurred in a more restricted range of lower salinities (1–24 ppt) than both *Tritia burchardi* and *Nassarius* spp. (16–46 ppt and 16–37 ppt, respectively). The scaphopod *Paradentalium hemileuron* and polyplacophorans *Leptochiton matthewsianus* and *Acanthochitonidae* sp. were essentially restricted to salinities ~35 ppt (Figure 6).

As with the annelids, there was a shift in the occurrences of species along the salinity gradient. The tateid (formerly hydrobiid) gastropod *A. occidua* was most frequently recorded in oligohaline salinities, while another species from this family, *T. rufilabris* and the bivalve *F. subtortus*, were mainly found in mesohaline waters (Figure 7). *Arthritica semen*, *X. securis*, *Potamopyrgus* spp., and *S. trigonella* were broadly distributed from fresh to hyperhaline waters but were most often recorded in polyhaline salinities. Other molluscs, such as *Tritia burchardi, B. australis,* and *M. deltoidalis,* had a more restricted distribution along the salinity gradient in euhaline waters.

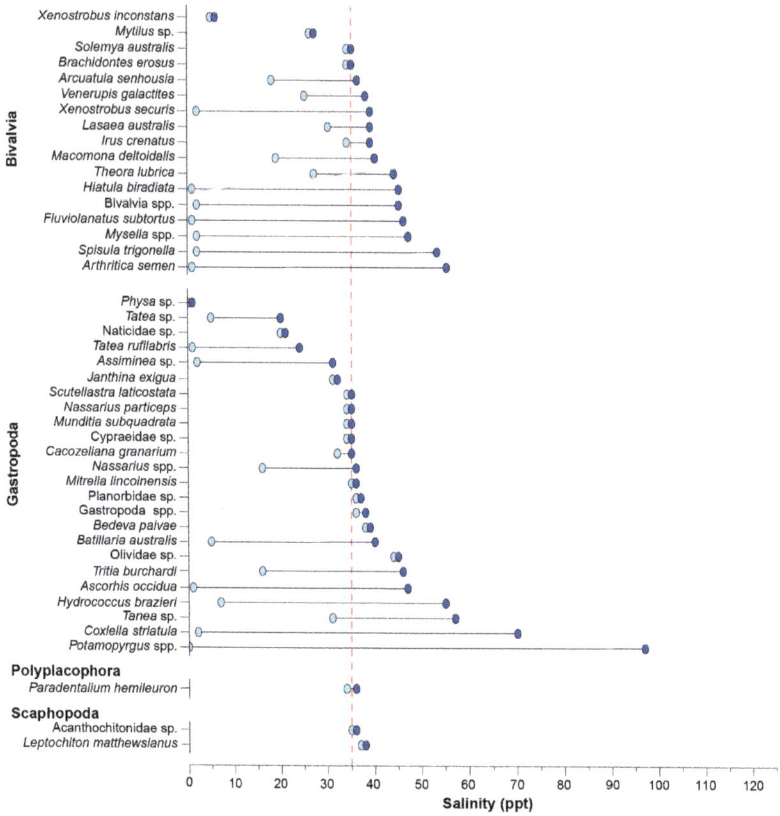

Figure 6. Dumbbell plot of the minimum and maximum salinity in which each molluscan taxon was recorded. The dashed red line denotes salinity around full-strength seawater.

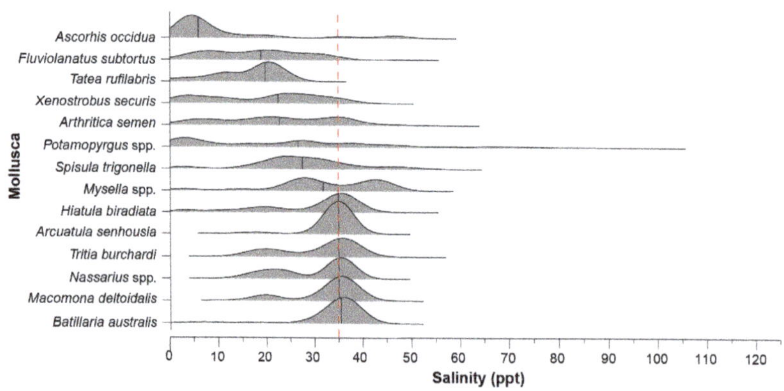

Figure 7. Ridgeline plot of the frequency distribution of each molluscan taxon recorded in $\geq$10 samples across the salinity gradient. The dashed red line denotes salinity around full-strength seawater and the black vertical solid lines indicate the median salinity of each taxon generated by the kernel density estimation.

3.5. *Distribution of Arthropods*

Most of the malacostracan taxa tended to be either recorded over a narrow range of salinities around full-strength seawater (euhaline) or recorded over a broad range of salinities from oligohaline to hyperhaline (Figure 8). Species in the first group tended to be uncommon and included the amphipods *Aora maculata*, *Birubius eake*, and *Ampithoe valida* and the hymenosomatid and portunid crabs *Halicarcinus ovatus* and *Ovalipes australiensis*. Twenty four of the 58 malacostracans were found in a range of ≥20 ppt. With the exception of species such as the isopod *Paracerceis sculpta*, mysid *Gastrosaccus sorrentoensis*, amphipod *Caprella scaura*, and hymenosomatid *Lucascinus bedfordi*, most of these taxa were recorded in salinities < 5 ppt (oligohaline). Because of this, the extent of their salinity range was determined by the highest salinity in which they were recorded. For example, the isopods *Syncassidina aestuaria* and *Cirolanidae* sp., the tanaidid *Tanais dulongii*, and *C. scaura* were recorded in salinities up to 39–45 ppt. Other species, such as the caridean shrimp *Palaemon australis* and the amphipods *Perthia* sp., *Barnardomelita matilda*, *Austrochiltonia subtenuis*, and *Paracorophium excavatum*, were recorded in salinities up to 57 ppt. The malacostracans found in the highest salinities, i.e., 93 and 102 ppt, were the amphipods *Corophium minor* and *Grandidierella* spp., respectively.

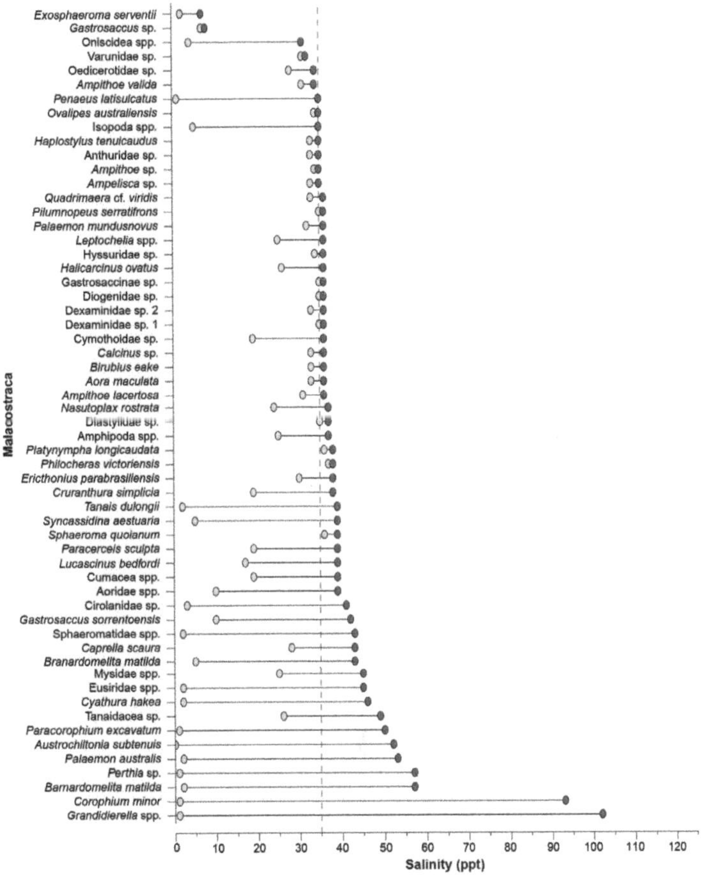

Figure 8. Dumbbell plot of the minimum and maximum salinity in which each malacostracan taxon was recorded. The dashed red line denotes salinity around full-strength seawater.

There was a clear sequential pattern of minimum and maximum salinities among the hexapods. A group including larval horse flies (*Tabanidae* sp.), mayflies (*Caenidae* spp.), beetles (*Dytiscidae* sp. 2 and *Coleoptera* spp.), and springtails (*Collembola* sp.) were restricted to freshwater and oligohaline salinities (Figure 9). Larvae of several mosquitoes (i.e., *Aedes* sp. and *Coquillettidia* spp.) and non-biting midges (chironomid; i.e., *Procladius villosimanus, Paratanytarsus grimmii, Chironomus alternans,* and *Chironomus occidentalis*) were recorded in mesohaline and polyhaline salinities (range up to 28 ppt). Almost all of the taxa recorded in hyperhaline salinities (≥40 ppt) were found over a wide range of salinities (range for a taxon of between 37 and 119 ppt). The taxa with the broadest ranges were also those recorded in the highest salinities, e.g., larvae of the water scavenger beetle (hydrophilid) *Berosus* spp. (97 ppt), *Diptera* spp. (102 ppt), biting midges *Ceratopogonidae* spp. (103 ppt), and the non-biting midge *Tanytarsus barbitarsis* (122 ppt). The latter species was also recorded over the widest range, i.e., 118 ppt (Figure 9).

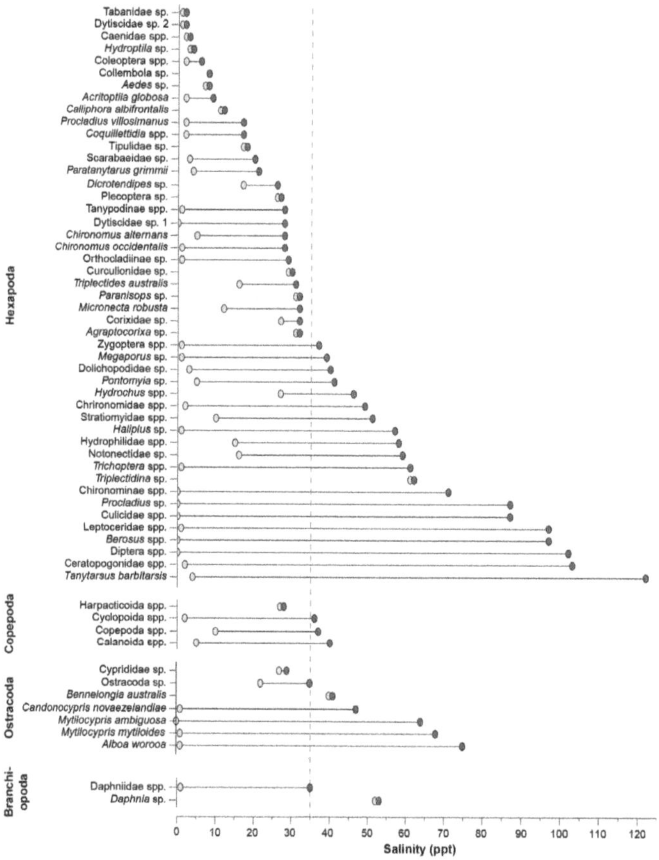

Figure 9. Dumbbell plot of the minimum and maximum salinity in which each hexapod, copepod, ostracod, and branchipod taxon was recorded. The dashed red line denotes salinity around full-strength seawater.

All ostracod taxa recorded on ten or more occasions were found in essentially freshwater (0 or 1 ppt) to hyperhaline conditions. For example, *Candonocypris novaezelandiae* was recorded in salinities up to 47 ppt, *Mytilocypris ambiguosa* and *Mytilocypris mytiloides* to 64 and 67 ppt, respectively, and *Alboa worooa* to 74 ppt (Figure 9). Copepod and branchiopod taxa were typically found in salinities of 0 to ~35 ppt, although individuals were only identified to a relatively low taxonomic resolution.

Despite the relatively broad range of salinities in which the most common taxa were recorded, there was a sequential pattern in the peak occurrences along the salinity gradient (Figure 10). Species most common in oligohaline waters included the ostracods *C. novaezelandiae* and *M. ambiguosa* and the non-biting midges *C. occidentalis* and *P. grimmii*. A similar suite of higher taxa was frequently recorded in mesohaline salinities, e.g., ostracods (*M. mytiloides* and *A. worooa*) and non-biting midges (*P. villosimanus*, *Chironominae* spp. and *Procladius* sp.). In contrast to oligohaline and mesohaline salinities, the taxa most common in polyhaline and euhaline salinities were a range of malacostracans, including amphipods (e.g., *Grandidierella* spp., *B. matilda*, and *C. minor*), isopods (e.g., *Cyathura hakea* and *S. aestuaria*) and decapods (*P. australis* and *L. bedfordi*). *Tanytarsus barbitarsis* was the only very common species with a median occurrence in hyperhaline conditions (92 ppt).

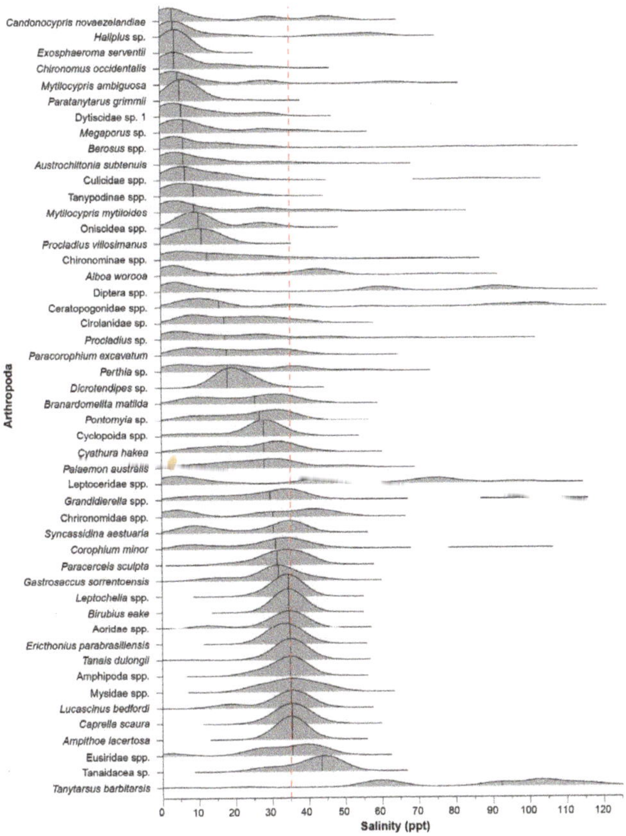

Figure 10. Ridgeline plot of the frequency distribution of each arthropod taxon recorded in ≥10 samples across the salinity gradient. The dashed red line denotes salinity around full-strength seawater and the black vertical solid lines indicate the median salinity of each taxon generated by the kernel density estimation.

4. Discussion

4.1. Comparisons to the Remane Diagram

The primary aim of this study was to determine, using quantitative data, whether the patterns of benthic macroinvertebrate species richness along a pronounced salinity gradient of freshwater to hyperhaline conditions matched the conceptual models in the literature. The pattern of richness in southwestern Australian estuaries peaked at a salinity of 35 ppt, as in all three models; however, the richness in (i) freshwater/oligohaline and (ii) hyperhaline waters differed. The Whitfield et al. [28], Khlebovich [30], and Hedgpeth [29] models all indicate, albeit to varying relative extents, that freshwaters are relatively speciose, and richness then declines through the horohalinicum before increasing. However, in southwestern Australia, taxa richness was lowest at a salinity of 0 and increased sequentially until 35 ppt. This could reflect the environmental characteristics of the region. Rainfall is highly seasonal, and thus most rivers are intermittent or ephemeral, and some are saline due to natural geology and/or secondary salinisation [32,51]. This would prohibit the occurrence of stenohaline freshwater species, which is supported by studies on wetlands on the Swan Coastal Plain in southwestern Australia, where salinity is one of the major drivers of invertebrate composition [52]. Moreover, the temporary nature of rivers is a driver of the relatively depauperate native fish fauna (8 species), all of which are small-bodied (typically < 5 cm total length) and endemic to the region [53]. The seasonal and/or saline riverine flow, together with the microtidal regime, is not conducive to the formation of tidal freshwater zones that are present in many macrotidal estuaries and support a unique faunal community [54,55]. From an evolutionary perspective, the aridification of the Australian climate, which began in the early Cenzoic era and accelerated in the last 33 ma, has resulted in increased salinity in rivers and lakes [56]. This process could have also led to the creation of selection pressures that favoured invertebrates with the ability to cope in saline water [57]. Similar to the current study, benthic macroinvertebrate species richness was lowest in the upstream reaches of the Gamtoos Estuary in South Africa [58]. The fauna in this region of the estuary comprised taxa typically associated with fluvial habitats (e.g., chironomids and oligochaetes) and a few estuarine taxa with tolerances to very low salinities (e.g., corophiid amphipods and some polychaetes) [58]. This mirrors the 18 taxa found in salinities of $\leq$1 ppt in southwestern Australia, which comprised the larvae of six hexapods (including several chironomids), nematodes, oligochaetes, three polychaetes, two gastropods, two bivalves, and a single ostracod and single amphipod species (Table S2). Sixteen of these 18 taxa were also recorded in salinities > 35 ppt, with seven found in salinities > 80 ppt. Thus, very few taxa are stenohaline. This supports the work of Cronin-O'Reilly et al. [42], who found that the benthic macroinvertebrates taxa recorded in the upper reaches of the Vasse–Wonnerup were mainly saline wetland-associated that can persist in a wide range of salinities.

The decline in richness between 40 and 90 ppt in southwestern Australia was greater than depicted in all three models. Moreover, whereas the models estimated a relatively equal richness on either side of a peak at 35 ppt, the decline in southwestern Australia was greater in hyperhaline than polyhaline salinities. For example, 173 taxa were recorded in 35 ppt, but 124 taxa occurred in a salinity of 30 ppt compared to 83 taxa at 40 ppt, a difference of 41 taxa. Furthermore, this differential increased to 50 taxa at $\pm$10 ppt and 74 taxa at $\pm$15 ppt. The proportionally fewer taxa able to survive in higher than lower salinities could reflect the low number of estuaries and the relative frequency with which these salinities occur [34] and also the smaller area they cover, leading to reduced selection pressure to develop behavioural strategies and physiological adaptations. Moreover, in permanently-open estuaries with a mesotidal or macrotidal regime, much of the estuary may become polyhaline or mesohaline each day, depending on the tidal cycle [59]. In contrast, hyperhaline conditions only occur in a subset of estuaries and typically over prolonged periods [8,9], which would reduce the effectiveness of some behavioural responses. However, these results highlight the vulnerability of benthic macroinvertebrates to hypersalinity [60]. Several studies have shown that pronounced hypersalinity, typically >60 ppt, simplifies the fauna and truncates food webs [61,62]. Extreme salinities can also result in the absence of benthic macroinvertebrates [10,63].

A far larger proportion of the species recorded in 35 ppt and euhaline salinities were stenohaline than in other salinities. Although reproductive information, e.g., the timing and location of spawning, is not known for many of these species, they are likely marine. An example of this is provided by the echinoderms, which all occurred in a narrow range of 33–37 ppt. This trend mirrors that in the UK, where 33 echinoderm taxa, which represented 4.2% of all invertebrates, were recorded in coastal waters, but only four such taxa comprising 1.4% of invertebrates were found in nearby estuaries [15]. Echinoderms are regarded as stenohaline as they lack an excretory organ, have highly permeable body walls, and their body fluids are ionically and osmotically similar to that of the surrounding water [64,65]. Although some species can tolerate short-term exposure to hyposaline conditions for several hours to days [66], echinoderm species in southwestern Australia are restricted to the lower reaches of permanently-open systems, which remain euhaline [44,67,68]. These species have also not been recorded in hypersaline conditions [34] and have been shown to be sensitive to discharges of desalination brine [69].

4.2. Holohaline Taxa

Globally, few species are regarded as holohaline [28,70,71], however, several holohaline taxa were identified in the current study, i.e., *T. barbitarsis*, *P.* cf. *kempi*, *Grandidierella* spp. (i.e., *Grandidierella japonica* and *Grandidierella propodentata*), *Capitella* spp. (which includes *C. capitata*), *Diptera* spp., and *Ceratopogonidae* spp. Furthermore, several other taxa, i.e., *Berosus* spp. and *Leptoceridae* spp., were found in salinities of ~0 to 97 ppt and in the future could be found in salinities $\geq$ 100 ppt. Among these eight taxa, *Capitella* spp., *P.* cf. *kempi* and *Grandidierella* spp., could be regarded as ultrahaline species as these polychaete and amphipod taxa are all known to occur and breed in coastal marine waters [72–74]. The remaining taxa, all of which are hexapods, are holoeuryhaline as they are typically absent from local marine waters [75] and more often associated with wetlands in Australia, some of which become markedly hypersaline [76].

While these taxa belong to several classes and exhibit a range of physiological adaptations to cope with the huge variation in salinities (see below), they all must obtain sufficient energy to survive, the need for which increases in more saline water [77]. At the same time, the diversity and abundance of faunal communities decreases [78,79], which has been found to telescope the food chain [80]. It is not surprising that many of the aforementioned holohaline taxa are deposit feeders (e.g., *P.* cf. *kempi* and *Capitella* spp. [81]) or suspension feed on phytoplankton and detritus or graze on microphytobenthos (e.g., *Grandidierella* spp. [82] and *Tanytarsini* larvae [83]). Moreover, *T. barbitarsis* is also known to ingest its faecal pellets [84]. Such a diet is beneficial as, in microtidal estuaries, most of the fine sediments and detritus that are transported into the estuary become 'trapped' and, in the Lagoa de Araruama (Brazil), benthic diatoms were present in salinities of 0–180 ppt [2,85]. The exceptions to this are larval ceratopogonids that suspension feed, graze, and predate on meiofauna [86], and larval *Berosus* spp., which are predators [87,88]. However, chironomids, capitellids, and ostracods were abundant and co-occurring. As a result, unlike the situation with the zoobenthivorous sparid in Beaufort Inlet [62], sufficient prey were always available for the beetle larvae.

4.3. Annelids

Among the three main phyla, annelids had the lowest proportion of taxa present in both freshwater/oligohaline and hyperhaline salinities, and this was particularly true for the sipunculans. This reflects the fact that annelids are mostly osmoconformers, although some taxa can osmoregulate and/or use behavioural adaptations to limit salinity exposure [89,90]. While sipunculans can tolerate considerable water loss and gain, they lack active regulation [91]. The range of salinities in which these taxa were recorded in the current study was relatively narrow, i.e., 28–45 ppt (median = 35 ppt), which matches that in the review by Murina [92]. The low number of polychaetes found in freshwater/oligohaline salinities can be attributed to their thin, permeable body walls and being osmoconformers [93]. However, several families of polychaetes have been documented to withstand low salinities, most notably nereids, spionids, sabellids, and fabriciids, which is consistent with the current study [93]. Many of the species recorded in oligohaline and mesohaline waters in southwestern Australia are also found in similar salinities in other estuaries. For example, *B. limnicola* was first described from lakes in New South Wales and was found in salinities between 0 and 12 ppt in the Hawkesbury Estuary (Australia) [94]. In addition to some of these species being able to osmoregulate [89,95], they often have thicker body walls, reside in tubes and use reproductive strategies such as brooding or embryo encapsulation to reduce the mortality of the more sensitive larval stage [93].

Some of these adaptations would also be effective in hyperhaline salinities and could potentially explain the broad salinity ranges in which some of the species, e.g., *P.* cf. *kempi*, *Capitella* spp., and *S. aequisetis*, were recorded in both southwestern Australia and elsewhere. For example, *Capitella* has been recorded in salinities of 1–138 ppt in the Coorong Lagoon (Australia) [96], the spionid *Polydora cornuta* and the nereid *Nereis pelagica* were recorded in salinities up to 75 ppt in the Laguna Madre (USA) [29], and several nereids were among the most abundant macrofaunal species in St Lucia (South Africa) during a prolonged drought [77]. The presence of polychaetes in salinities up to ~100 ppt is greater than that recorded in the Ría Lagartos (Mexico), where no such taxa were recorded in salinities of ≥60 ppt [97], and instead similar to hypersaline waterbodies in Crimea, where such species were recorded in salinities of up to 80–100 ppt [71].

It should also be noted that *C. capitata* (the main representative in *Capitella* spp.) and *P.* cf. *kempi* are thought to be species-complexes, and the holohaline nature of these 'species' could be an artefact of cryptic species. *Capitella capitata* was originally described in Greenland, and more detailed morphological and molecular studies are needed to confirm whether the species in southwestern Australia are the same as those elsewhere and if there are sibling species e.g., [98].

4.4. Molluscs

The richness of molluscan taxa was substantially greater than annelids in oligohaline to mesohaline waters before following a similar peak at 35 ppt and subsequent decline in hyperhaline salinities. The distribution of the common species largely conformed to the categories proposed by Chalmer et al. [99] and applied by Wells and Threlfall [100], i.e., freshwater (none in this study), estuarine (e.g., *A. occuidua* and *F. subortus*), marine affinity (e.g., *T. burchardi* and *S. trigonella*), and marine (e.g., *M. deltoidalis*). Moreover, the presence of *X. securis* in oligohaline and mesohaline waters and of *B. australis* in euhaline salinities aligns with previous surveys in the Swan–Canning Estuary [101,102]. As with annelids, most molluscs are osmoconformers [103], but they possess a range of behavioural adaptations. For example, *B. australis* was recorded infrequently in salinities as low as 6 ppt. Laboratory experiments conducted at 9, 18, and 36 ppt for two weeks resulted in no mortalities, which was due to individuals having a closed operculum preventing water exchange and being inactive, thus reducing metabolic needs [104]. Similar adaptations are also used by *Nassarius* spp. and *T. burchardi* [105] and some bivalves, e.g., *Brachidontes pharaoni* and *Cerastoderma endures*, in other parts of the world [106,107].

Species such as *A. semen* (1–74 ppt), *C. striatula* (3–70 ppt), and the hydrobiids *Hydrococcus brazieri* (7–55 ppt) and *Potamopyrgus* spp. (0–97 ppt) were all recorded over a wide range of salinities. *Arthritica semen* and an unidentified hydrobiid species were recorded in the Coorong Lagoon in salinities of 125 and 130 ppt, respectively [61]. Moreover, *A. semen* in the Peel–Harvey Estuary survived for 60 h in salinities of up to 54 ppt. The survival rate exceeded 90% in late summer, suggesting that individuals acclimatised as salinities increased over the dry summer period. The occurrence of *C. striatula* in hyperhaline conditions reflects the fact that, despite being an osmoconformer, *Coxiella* is the only gastropod genus where all species are halophilic and that individuals are active in salinities of 10 to 70 ppt, with a maximum recorded range of 0.3 to 130 ppt [108]. The hydrobiids *Hydrobia ulvae* and *Potamopyrgus jenkinsi* are capable of osmoregulating over a moderate range of salinities below 35 ppt [109], while *H. ulvae* and *Hydrobia glyca* were highly active in all salinities tested (i.e., 20–80 ppt) [110]. Wells and Threlfall [111] suggested that reproductive strategies such as embryo protection and rapid growth rates help *A. semen* and *H. brazieri* cope with salinity stress.

4.5. Arthropods

Arthropods were also found across the entire salinity gradient and, of the four main phyla, had the highest percentage of number of taxa in all salinity categories except 31–40 ppt. The pattern of relative richness varied among classes; hexapods were most rich in freshwater to mesohaline salinities, malacostracans in euhaline salinities, and ostracods and hexapods in moderately and extremely hyperhaline salinities, respectively.

Malacostracans typically occurred over a more restricted range of salinities than the other classes of arthropods. However, several species of amphipod were recorded in oligohaline and hyperhaline salinities. Among these, *A. subtenuis* was mainly recorded in salinities below 20 ppt, but was found in 52 ppt. This species is regarded as freshwater based on its pattern of occurrence but was recorded in wetlands in the Wheatbelt region of WA in salinities between 0.2 and 77 ppt [112]. The corophiid amphipods *C. minor* and *P. excavatum*, both of which are common in southwestern Australian estuaries and likely breed there [38,113], were found in salinities of 1–92 ppt and 1–49 ppt, respectively. Although the osmoregulation ability of these species is unknown, the estuarine-breeding *Corophium volutator* maintains hyperosmotic haemolymph concentration until 50 ppt [114]. In the current study, *Grandidierella* spp. (i.e., *G. japonica* and *G. propodentata*) was the malacostracan found in the broadest range of salinities (1–102 ppt). Species in this genus are known to occur in hypersaline waters, e.g., *Grandidierella lignorum* in South Africa (0–56 ppt) and *Grandidierella halophila* in Malaysia (7–80 ppt) [115,116]. Moreover, *G. propodentata* was first described from a solar salt farm in Australia [117]. The ability of *Grandidierella* spp. to occur in a broad range of salinities could be related to the presence of salt-excreting tissues, as is present in the coxal gills of *G. japonica* [118].

Several ostracod species belonging to the Mytilocypridinae were found across the gradient from oligohaline through to hypersaline conditions in this study. Both *M. ambiguosa* and *M. mytiloides* are common in Australian salt lakes and have been reported in salinities of

0.5–65 ppt and 1.3–173 ppt, respectively [119]. The congeneric *Mytilocypris praenuncia* is an osmoregulator with mitochondria-rich cells in the inner carapace layer, which help excrete salt in hypo-osmotic regulation [119]. Furthermore, the desiccation-resistant eggs produced by mytilocyprid ostracods would enable recolonisation from drying [108], which often occurs during dry periods and resulting hyperhaline conditions in southwestern Australian estuaries [120]. Copepods and daphniids were also found in a broad range of salinities, reflecting their occurrences in freshwater and marine environments. Although species-level identifications were not conducted in the source publications, species such as the copepod *Tigriopus californicus* have been recorded in salinities of 2 to at least 190 ppt and temperatures up to 40 °C [121]. Moreover, Pinder et al. [112] recorded 49 freshwater and 25 halophilic copepod species in wetlands in the Wheatbelt. The same authors also found that species of *Daphniopsis* (which some authors argue should be subsumed into *Daphnia* [108]) were most frequently recorded in <50 ppt but did occur in salinities > 150 ppt. Despite being common in salt lakes, no brine shrimp, i.e., *Artemia* and *Parartemia*, were recorded in hypersaline estuaries, although Tweedley and Krispyn [120] did observe an individual in Gordon Inlet in a salinity of 179 ppt.

The majority of the hexapods recorded in southwestern Australian estuaries, particularly those that are abundant, are characteristic of saline wetlands [52]. Thus, it is unsurprising that 30 of the 37 taxa recorded at least twice were classified as euryhaline and around 42% of all hexapods were present in salinities >35 ppt. Moreover, some of the taxa found in the highest and broadest ranges of salinities, i.e., *Berosus*, *Procladius*, *Tanytarsus*, and the ceratopogonid *Culicoides*, were identified as among the most salt-tolerant genera in Australia by Chessman [76] in his Invertebrate-Based Salinity Index designed to detect the ecological impact of salinisation in lentic and lotic inland waters. Species of *Berosus* have been recorded in salinities of up to 150 ppt in a salt pan in the Camargue (France) [122] and *Berosus munitipennis* and *Berosus discolor* in 0.8–91 ppt and <0.1–120 ppt, respectively, in the Wheatbelt [112]. Chironomids often dominate the benthic macroinvertebrate fauna of hypersaline waters [61,123] and, among these taxa in Australia, *T. barbitarsi* is regarded as the most salt-tolerant, being recorded in up to 210 ppt [112]. This species is known to osmoregulate; however, the high energetic cost has consequences, i.e., that it has a relatively long generation time and only occurs in eutrophic systems, suggesting a large food demand [124].

4.6. Limitations and Future Directions

In many Mediterranean climatic regions precipitation will decrease due to climate change, which can also have a disproportionate influence on freshwater discharge [125]. Combined with localised anthropogenic impacts, e.g., water diversion and secondary salinisation, these estuaries will generally become more saline, and the spatial extent of oligohaline waters will decrease [31,102]. Increasing salinity has varying effects on estuaries depending on their morphology [31]. On one hand, in estuaries that are permanently open to the ocean and those that experience regular bar breaches, marinisation can facilitate the recruitment of marine taxa, thereby increasing species richness [38,68], albeit the intrusion of more saline water may also lead to a decline in species adapted to lower salinities [102]. In contrast, in estuaries that are normally closed (i.e., Beaufort Inlet), a lack of tidal exchange, combined with reduced freshwater inflow and increased evaporation, can lead to hypersalinity and a dramatic decrease in richness and, in extreme cases, mass mortality events [45,62]. As such, there is a need to document the in situ environmental conditions in which species are recorded. Where possible, these observations should be paired with controlled laboratory experiments to determine the impacts of salinity alone and the synergistic effects of salinity with other variables (e.g., temperature). Most of the existing information details the effect of relatively low salinities on freshwater taxa rather than the effect of hyperhaline salinities on estuarine fauna [126,127]. This information is vital in helping to understand how estuaries may change in the future. The current study does have some limitations; as there were relatively few samples collected in salinities > 50 ppt, some of the taxa in the source publications were not identified to species level, and the analyses conducted only considered the frequency of occurrence and not the abundance of each taxon. Given the predicted continued drying climate and the shifts in species along the salinity gradient, more work should be conducted in hyperhaline estuaries. Finally, the accuracy of future studies would greatly benefit from improved taxonomy to resolve the status of cosmopolitan species such as *P. kempi* and *C. capitata* [128].

5. Conclusions

The pattern of benthic macroinvertebrate richness in southwestern Australia differs from that in the various modifications of the Remane diagram. This reflects a lack of stenohaline freshwater species and the proportionately fewer taxa that can survive in hyperhaline conditions. Most taxa were either stenohaline marine or euryhaline, with some euryhaline taxa also being holohaline. Although many of the euryhaline species of annelid, mollusc, and arthropod were recorded in oligohaline to euryhaline salinities, their frequency distribution showed zonation along the salinity gradient. Thus, although they can tolerate a broad range of salinities, there is a narrower suite of preferred salinities. This research highlights that future changes in the salinity regime of estuaries will have a marked impact on the suite of benthic macroinvertebrate species present.

Supplementary Materials: The following supporting information can be downloaded at: https://www.mdpi.com/article/10.3390/w17111642/s1, Table S1: Summary of the characteristics of the 12 estuaries sampled in southwestern Australia. Table S2: Mean density (100 cm^{-2}), percentage contribution (%C), number of samples (*n*), and frequency of occurrence (%F) of each benthic macroinvertebrate taxon and the minimum, maximum, mean and median salinity and salinity range (R) in which it was recorded in south-western Australian estuaries.

Author Contributions: Conceptualization, J.R.T. and R.L.; methodology, J.R.T., S.C.-O. and E.M.R.; software, J.R.T.; validation, R.L. and J.R.T.; formal analysis, R.L., J.R.T. and B.J.R.; investigation, R.L., E.J.S., S.C.-O., J.R.T. and S.A.F.; resources, J.R.T., S.C.-O. and E.M.R.; data curation, R.L., S.A.F., E.J.S., S.C.-O. and J.R.T.; writing—original draft preparation, R.L., J.R.T. and E.M.R.; writing—review and editing, all authors; visualization, J.R.T. and B.J.R.; supervision, J.R.T., S.C.-O. and E.M.R. All authors have read and agreed to the published version of the manuscript.

Funding: This research received no external funding.

Data Availability Statement: The data presented in this study are available on request from the corresponding author.

Acknowledgments: The authors acknowledge the Noongar people as the Traditional Custodians of the land on which this research took place and pay their respects to Elders past, present and emerging. We also thank the friends and colleagues who assisted with collecting the samples in the original studies, particularly Kurt Krispyn, Charles Maus, and Chris Hallett. Gratitude is also expressed to the organisations that funded the original studies, namely the Department of Biodiversity, Conservation and Attractions through the Swan River Trust, the Australian Research Council and its Linkage partners, the Department of Water and Environmental Regulation, and Murdoch University. Ben Roots is supported by an Australian Government Research Training Program (RTP) Scholarship.

Conflicts of Interest: The authors declare no conflicts of interest.

Appendix A

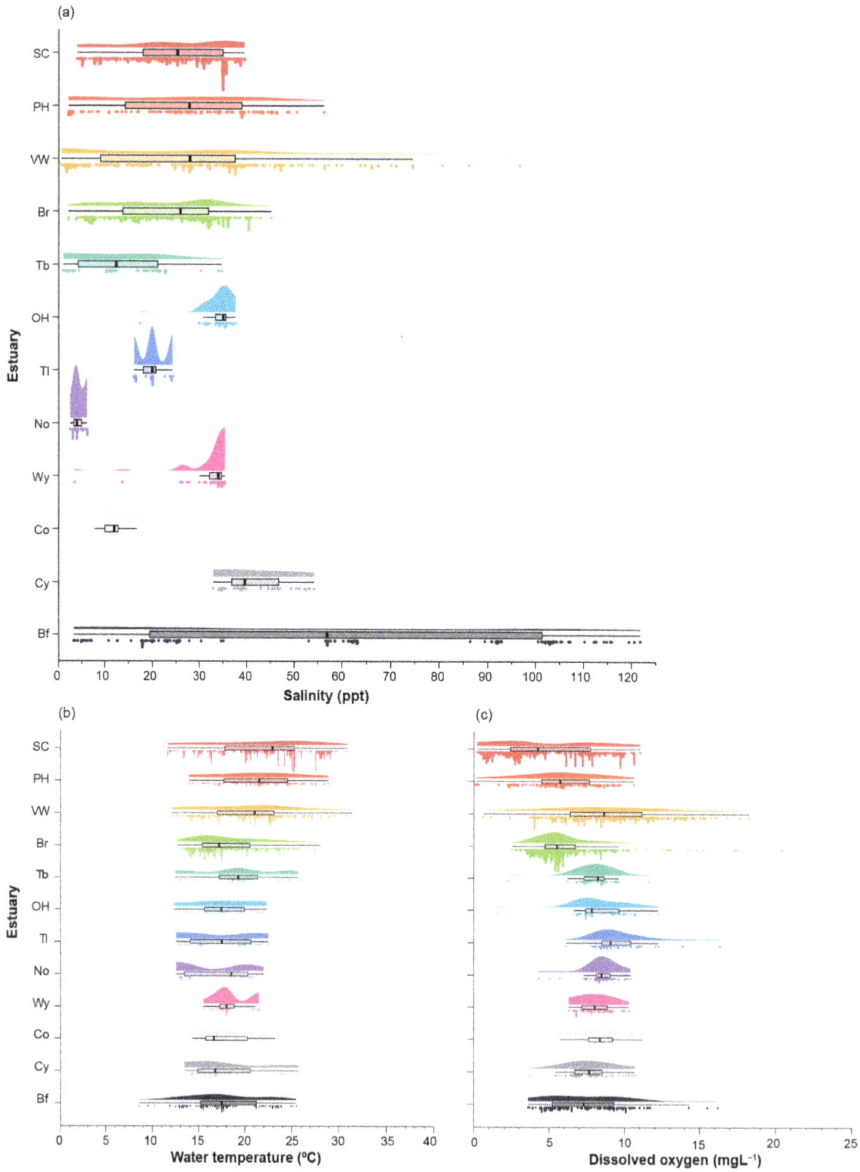

Figure A1. Raincloud plots, i.e., the combination of a scatter plot, probability density and a box plot of the (**a**) salinity (ppt), (**b**) water temperature (°C) and (**c**) dissolved oxygen concentration (mgL^{-1}) in each of the 12 estuaries in southwestern Australia. Data for each estuary is shown in a different colour; SC, Swan-Canning Estuary; PH, Peel-Harvey Estuary; VW, Vasse-Wonnerup Estuary; Br, Broke Inlet; Tb, Torbay Inlet; OH, Oyster Harbour; Tl, Taylor Inlet; No, Normans Inlet; Wy, Waychinicup Estuary; Co, Cordinup River; Cy, Cheyne Inlet; Bf, Beaufort Inlet.

References

1. McLusky, D.S.; Elliott, M. *The Estuarine Ecosystem: Ecology, Threats and Management*, 3rd ed.; Oxford University Press: Oxford, UK, 2004.
2. Tweedley, J.R.; Warwick, R.M.; Potter, I.C. The contrasting ecology of temperate macrotidal and microtidal estuaries. *Oceanogr. Mar. Biol. Annu. Rev.* **2016**, *54*, 73–172. [CrossRef]
3. Pritchard, D.W. What is an estuary: A physical viewpoint. *Am. Assoc. Adv. Sci.* **1967**, *83*, 3–5.
4. Tagliapietra, D.; Sigovini, M.; Ghirardini, A.V. A review of terms and definitions to categorise estuaries, lagoons and associated environments. *Mar. Freshw. Res.* **2009**, *60*, 497–509. [CrossRef]
5. Day, J.H. The nature, origin and classification of estuaries. In *Estuarine Ecology: With Particular Reference to Southern Africa*; Day, J.H., Ed.; A.A. Balkema: Cape Town, South Africa, 1981; pp. 1–6.
6. Potter, I.C.; Chuwen, B.M.; Hoeksema, S.D.; Elliott, M. The concept of an estuary: A definition that incorporates systems which can become closed to the ocean and hypersaline. *Estuar. Coast. Shelf Sci.* **2010**, *87*, 497–500. [CrossRef]
7. McSweeney, S.L.; Kennedy, D.M.; Rutherfurd, I.D.; Stout, J.C. Intermittently closed/open lakes and lagoons: Their global distribution and boundary conditions. *Geomorphology* **2017**, *292*, 142–152. [CrossRef]
8. Largier, J.L. Recognizing low-inflow estuaries as a common estuary paradigm. *Estuaries Coasts* **2023**, *46*, 1949–1970. [CrossRef]
9. Hoeksema, S.D.; Chuwen, B.M.; Tweedley, J.R.; Potter, I.C. Factors influencing marked variations in the frequency and timing of bar breaching and salinity and oxygen regimes among normally-closed estuaries. *Estuar. Coast. Shelf Sci.* **2018**, *208*, 205–218. [CrossRef]
10. Tweedley, J.R.; Krispyn, K.N. Protracted bar closure temporarily transforms an estuary into a salt lake. *Pac. Conserv. Biol.* **2024**, *30*, PC24007. [CrossRef]
11. Charmantier, G.U.Y. Ontogeny of osmoregulation in crustaceans: A review. *Invertebr. Reprod. Dev.* **1998**, *33*, 177–190. [CrossRef]
12. González-Ortegón, E.; Pascual, E.; Cuesta, J.A.; Drake, P. Field distribution and osmoregulatory capacity of shrimps in a temperate European estuary (SW Spain). *Estuar. Coast. Shelf Sci.* **2006**, *67*, 293–302. [CrossRef]
13. Whitfield, A.; Elliott, M. Ecosystem and biotic classifications of estuaries and coasts. In *Treatise on Estuarine and Coastal Science*; Wolanski, E., McLusky, D.S., Eds.; Academic Press: Waltham, MA, USA, 2011; pp. 99–124.
14. Willmar, P.; Stone, G.; Johnston, I. Animal water balance, osmoregulation and excretion. In *Environmental Physiology of Animals*, 2nd ed.; Willmar, P., Stone, G., Johnston, I., Eds.; Blackwell: Oxford, UK, 2005; pp. 76–110.
15. Tweedley, J.R.; Warwick, R.M.; Potter, I.C. Can biotic indicators distinguish between natural and anthropogenic environmental stress in estuaries? *J. Sea Res.* **2015**, *102*, 10–21. [CrossRef]
16. Ysebaert, T.; Meire, P.; Maes, D.; Buijs, J. The benthic macrofauna along the estuarine gradient of the Schelde Estuary. *Neth. J. Aquat. Ecol.* **1993**, *27*, 327–341. [CrossRef]
17. Edgar, G.J.; Shaw, C.; Watsona, G.F.; Hammond, L.S. Comparisons of species richness, size-structure and production of benthos in vegetated and unvegetated habitats in Western Port, Victoria. *J. Exp. Mar. Biol. Ecol.* **1994**, *176*, 201–226. [CrossRef]
18. Teske, P.R.; Wooldridge, T.H. What limits the distribution of subtidal macrobenthos in permanently open and temporarily open/closed South African estuaries? Salinity vs. sediment particle size. *Estuar. Coast. Shelf Sci.* **2003**, *57*, 225–238. [CrossRef]
19. Edgar, G.J.; Barrett, N.S. Benthic macrofauna in Tasmanian estuaries: Scales of distribution and relationships with environmental variables. *J. Exp. Mar. Biol. Ecol.* **2002**, *270*, 1–24. [CrossRef]
20. Whitfield, A.K. Estuaries—How challenging are these constantly changing aquatic environments for associated fish species? *Environ. Biol. Fishes* **2021**, *104*, 517–528. [CrossRef]
21. Thabet, R.; Ayadi, H.; Koken, M.; Leignel, V. Homeostatic responses of crustaceans to salinity changes. *Hydrobiologia* **2017**, *799*, 1–20. [CrossRef]
22. Marshall, W.S. Osmoregulation in estuarine and intertidal fishes. In *Fish Physiology*; McCormick, S.D., Farrell, A.P., Brauner, C.J., Eds.; Academic Press: Cambridge, MA, USA, 2012; Volume 32, pp. 395–434.
23. Cloern, J.E.; Jassby, A.D.; Schraga, T.S.; Nejad, E.; Martin, C. Ecosystem variability along the estuarine salinity gradient: Examples from long-term study of San Francisco Bay. *Limnol. Oceanogr.* **2017**, *62*, S272–S291. [CrossRef]
24. Ysebaert, T.; Herman, P.M.J.; Meire, P.; Craeymeersch, J.; Verbeek, H.; Heip, C.H.R. Large-scale spatial patterns in estuaries: Estuarine macrobenthic communities in the Schelde estuary, NW Europe. *Estuar. Coast. Shelf Sci.* **2003**, *57*, 335–355. [CrossRef]
25. Valesini, F.J.; Wildsmith, M.D.; Tweedley, J.R. Predicting estuarine faunal assemblages using enduring environmental surrogates, with applications in systematic conservation planning. *Ocean. Coast. Manag.* **2018**, *165*, 80–98. [CrossRef]
26. Elliott, M.; Whitfield, A.K. Challenging paradigms in estuarine ecology and management. *Estuar. Coast. Shelf Sci.* **2011**, *94*, 306–314. [CrossRef]
27. Remane, A. Die brackwasserfauna. *Verhandlungen Der Dtsch. Zool. Ges.* **1934**, *36*, 34–74.
28. Whitfield, A.K.; Elliott, M.; Basset, A.; Blaber, S.J.M.; West, R.J. Paradigms in estuarine ecology—A review of the Remane diagram with a suggested revised model for estuaries. *Estuar. Coast. Shelf Sci.* **2012**, *97*, 78–90. [CrossRef]

29. Hedgpeth, J.W. Ecological aspects of the Laguna Madre, a hypersaline estuary. In *Estuaries*; Lauff, G.H., Ed.; AAAS Publication 83, American Association for the Advancement of Science: Washington, DC, USA, 1967; pp. 408–419.
30. Khlebovich, V.V. Aspects of animal evolution related to critical salinity and internal state. *Mar. Biol.* **1969**, *2*, 338–345. [CrossRef]
31. Hallett, C.S.; Hobday, A.J.; Tweedley, J.R.; Thompson, P.A.; McMahon, K.; Valesini, F.J. Observed and predicted impacts of climate change on the estuaries of south-western Australia, a Mediterranean climate region. *Reg. Environ. Change* **2018**, *18*, 1357–1373. [CrossRef]
32. Brearley, A. *Ernest Hodgkin's Swanland*, 1st ed.; University of Western Australia Press: Crawley, UK, 2005; p. 550.
33. Hodgkin, E.P.; Hesp, P. Estuaries to salt lakes: Holocene transformation of the estuarine ecosystems of south-western Australia. *Mar. Freshw. Res.* **1998**, *49*, 183–201. [CrossRef]
34. Tweedley, J.R.; Dittmann, S.R.; Whitfield, A.K.; Withers, K.; Hoeksema, S.D.; Potter, I.C. Hypersalinity: Global distribution, causes, and present and future effects on the biota of estuaries and lagoons. In *Coasts and Estuaries*; Wolanski, E., Day, J.W., Elliott, M., Ramachandran, R., Eds.; Elsevier: Amsterdam, The Netherlands, 2019; pp. 523–546.
35. Lewis, E.L. The practical salinity scale 1978 and its antecedents. *IEEE J. Ocean. Eng.* **1980**, *5*, 3–8. [CrossRef]
36. Krispyn, K.N. *The Fish Faunas of Estuaries in the Albany Region of South-Western Australia*; Murdoch University: Perth, Australia, 2021.
37. Tweedley, J.R.; Hallett, C.S.; Warwick, R.M.; Clarke, K.R.; Potter, I.C. The hypoxia that developed in a microtidal estuary following an extreme storm produced dramatic changes in the benthos. *Mar. Freshw. Res.* **2016**, *67*, 327–341. [CrossRef]
38. Lim, R. *Multidecadal Changes in the Benthic Macroinvertebrates Assemblages of the Swan-Canning Estuary*; Murdoch University: Perth, Australia, 2025.
39. Stout, E. *Benthic Macroinvertebrate Communities of the Swan Estuary Marine Park*; Murdoch University: Perth, Australia, 2025.
40. Cronin-O'Reilly, S. *Benthic Community Structure, Health and Function of a Microtidal Estuary in South-Western Australia*; Murdoch University: Perth, Australia, 2021.
41. Tweedley, J.R.; Cronin-O'Reilly, S.; Cottingham, A.; Beatty, S.J. *Vasse-Wonnerup Integrated Monitoring Review of 2017-20: Benthic Macroinvertebrate Component*; Report for the Department of Water and Environmental Regulation; Murdoch University: Perth, Australia, 2021; p. 65.
42. Cronin-O'Reilly, S.; Cottingham, A.; Kalnejais, L.H.; Lynch, K.; Tweedley, J.R. Tidal exclusion barriers fragment an invertebrate community into taxonomically and functionally distinct estuarine and wetland assemblages. *J. Mar. Sci. Eng.* **2025**, *13*, 635. [CrossRef]
43. Tweedley, J.R. The Relationships Between Habitat Types and Faunal Community Structure in Broke Inlet, Western Australia. Ph.D. Thesis, Murdoch University, Perth, Australia, 2011.
44. Fourie, S.A. *Benthic Macroinvertebrate Faunas of Microtidal Estuaries in Albany, South-Western Australia*; Murdoch University: Perth, Australia, 2024.
45. Cronin-O'Reilly, S.; Krispyn, K.N.; Maus, C.; Standish, R.J.; Loneragan, N.R.; Tweedley, J.R. Empirical evidence of alternative stable states in an estuary. *Sci. Total Environ.* **2024**, *954*, 176356. [CrossRef]
46. WoRMS Editorial Board. World Register of Marine Species. Available online: https://www.marinespecies.org (accessed on 27 March 2025).
47. R Core Team. *R: A Language and Environment for Statistical Computing*; R Foundation for Statistical Computing: Vienna, Austria, 2024.
48. Wickham, H. *ggplot2: Elegant Graphics for Data Analysis*; Springer: New York, NY, USA, 2016; p. 260.
49. Wilke, C.O. ggridges: Ridgeline Plots in "ggplot2"; R Package Version 0.5.6. 2024. Available online: https://cran.r-project.org/web/packages/ggridges/index.html (accessed on 16 October 2024).
50. Kay, M. ggdist: Visualizations of distributions and uncertainty in the grammar of graphics. *IEEE Trans. Vis. Comput. Graph.* **2024**, *30*, 414–424. [CrossRef] [PubMed]
51. Nielsen, D.L.; Brock, M.A.; Rees, G.N.; Baldwin, D.S. Effects of increasing salinity on freshwater ecosystems in Australia. *Aust. J. Bot.* **2003**, *51*, 655–665. [CrossRef]
52. Davis, J.A.; Rosich, R.S.; Bradely, J.S.; Growns, J.E.; Schmidt, L.G.; Cheal, F. Wetland classification on the basis of water quality and invertebrate community data. In *Wetlands of the Swan Coastal Plain*; Water Authority of Western Australia and Environmental Protection Authority: Perth, Australia, 1993; Volume 6, p. 242.
53. Unmack, P.J. Biogeography of Australian freshwater fishes. *J. Biogeogr.* **2001**, *28*, 1053–1089. [CrossRef]
54. Barendregt, A. Tidal freshwater wetlands: The fresh dimension of the estuary. In *The Wetland Book: II: Distribution, Description, and Conservation*; Finlayson, C.M., Milton, G.R., Prentice, R.C., Davidson, N.C., Eds.; Springer: Dordrecht, The Netherlands, 2018; pp. 155–168.

55. Little, S.; Wood, P.J.; Elliott, M. Quantifying salinity-induced changes on estuarine benthic fauna: The potential implications of climate change. *Estuar. Coast. Shelf Sci.* **2017**, *198*, 610–625. [CrossRef]
56. Fujioka, T.; Chappell, J. History of Australian aridity: Chronology in the evolution of arid landscapes. In *Geological Society, London, Special Publications*; Bishop, P., Pillans, B., Eds.; Geological Society of London: London, UK, 2010; Volume 346, pp. 121–139.
57. Lawrie, A.D.A.; Chaplin, J.; Rahman, M.; Islam, M.A.; Webzell, K. Insights into the recent evolutionary history of salt lake gastropods (*Coxiella*) in Australia. *Hydrobiologia* **2025**, *852*, 3309–3325. [CrossRef]
58. Bate, G.C.; Whitfield, A.K.; Adams, J.B.; Huizinga, P.; Wooldridge, T.H. The importance of the river-estuary interface (REI) zone in estuaries. *Water SA* **2002**, *28*, 271–280. [CrossRef]
59. Valle-Levinson, A. Classification of estuarine circulation. In *Treatise on Estuarine and Coastal Science*; Wolanski, E., McLusky, D., Eds.; Academic Press: Waltham, MA, USA, 2011; pp. 75–86.
60. Warwick, R.M.; Tweedley, J.R.; Potter, I.C. Microtidal estuaries warrant special management measures that recognise their critical vulnerability to pollution and climate change. *Mar. Pollut. Bull.* **2018**, *135*, 41–46. [CrossRef]
61. Dittmann, S.; Baring, R.; Baggalley, S.; Cantin, A.; Earl, J.; Gannon, R.; Keuning, J.; Mayo, A.; Navong, N.; Nelson, M.; et al. Drought and flood effects on macrobenthic communities in the estuary of Australia's largest river system. *Estuar. Coast. Shelf Sci.* **2015**, *165*, 36–51. [CrossRef]
62. Krispyn, K.N.; Loneragan, N.R.; Whitfield, A.K.; Tweedley, J.R. Salted mullet: Protracted occurrence of *Mugil cephalus* under extreme hypersaline conditions. *Estuar. Coast. Shelf Sci.* **2021**, *261*, 107533. [CrossRef]
63. Wooldridge, T.H.; Adams, J.B.; Fernandes, M. Biotic responses to extreme hypersalinity in an arid zone estuary, South Africa. *S. Afr. J. Bot.* **2016**, *107*, 160–169. [CrossRef]
64. Diehl, W.J. Osmoregulation in echinoderms. *Comp. Biochem. Physiol. Part A Physiol.* **1986**, *84*, 199–205. [CrossRef]
65. Barrett, N.J.; Harper, E.M.; Peck, L.S. The impact of acute low salinity stress on Antarctic echinoderms. *Proc. R. Soc. B Biol. Sci.* **2024**, *291*, 20241038. [CrossRef] [PubMed]
66. Barker, M.F.; Russell, M.P. The distribution and behaviour of *Patiriella mortenseni* and *P. regularis* in the extreme hyposaline conditions of the southern New Zealand fiords. *J. Exp. Mar. Biol. Ecol.* **2008**, *355*, 76–84. [CrossRef]
67. Poh, B.; Tweedley, J.R.; Chaplin, J.A.; Trayler, K.M.; Crisp, J.A.; Loneragan, N.R. Influence of physico-chemical and biotic factors on the distribution of a penaeid in a temperate estuary. *Estuar. Coast. Shelf Sci.* **2019**, *218*, 70–85. [CrossRef]
68. Wildsmith, M.D.; Rose, T.H.; Potter, I.C.; Warwick, R.M.; Clarke, K.R.; Valesini, F.J. Changes in the benthic macroinvertebrate fauna of a large microtidal estuary following extreme modifications aimed at reducing eutrophication. *Mar. Pollut. Bull.* **2009**, *58*, 1250–1262. [CrossRef]
69. Fernández-Torquemada, Y.; González-Correa, J.M.; Sánchez-Lizaso, J.L. Echinoderms as indicators of brine discharge impacts. *Desalin. Water Treat.* **2013**, *51*, 567–573. [CrossRef]
70. Seale, A.P.; Cao, K.; Chang, R.J.A.; Goodearly, T.R.; Malintha, G.H.T.; Merlo, R.S.; Peterson, T.L.; Reighard, J.R. Salinity tolerance of fishes: Experimental approaches and implications for aquaculture production. *Rev. Aquac.* **2024**, *16*, 1351–1373. [CrossRef]
71. Anufriieva, E.V.; Shadrin, N.V. Diversity of fauna in Crimean hypersaline water bodies. *J. Sib. Fed. Univ. Biol.* **2018**, *11*, 294–305.
72. Blake, J.A.; Woodwick, K.H. Reproduction and larval development of *Pseudopolydora paucibranchiata* (Okuda) and *Pseudopolydora kempi* (Southern) (Polychaeta: Spionidae). *Biol. Bull.* **1975**, *149*, 109–127. [CrossRef]
73. Méndez, N.; Inez, L.-G.; Forbes, V.E. Variability in reproductive mode and larval development within the *Capitella capitata* species complex. *Invertebr. Reprod. Dev.* **2000**, *38*, 131–142. [CrossRef]
74. Munari, C.; Bocchi, N.; Mistri, M. *Grandidierella japonica* (Amphipoda: Aoridae): A non-indigenous species in a Po delta lagoon of the northern Adriatic (Mediterranean Sea). *Mar. Biodiv. Rec.* **2016**, *9*, 12. [CrossRef]
75. Wildsmith, M.D.; Potter, I.C.; Valesini, F.J.; Platell, M.E. Do the assemblages of the benthic macroinvertebrates in nearshore waters of Western Australia vary among habitat types, zones and seasons? *J. Mar. Biol. Assoc. UK* **2005**, *85*, 217–232. [CrossRef]
76. Chessman, B.C. A new salinity index for the invertebrate fauna of Australian inland waters. *Hydrobiologia* **2023**, *850*, 3539–3550. [CrossRef]
77. Anufriieva, E.V.; Shadrin, N.V. General patterns of salinity influence on the energy balance of aquatic animals in hypersaline environment. *Biol. Bull. Rev.* **2023**, *13*, 420–430. [CrossRef]
78. Roots, B.J.; Lim, R.; Cronin-O'Reilly, S.; Fourie, S.A.; Rodgers, E.M.; Stout, E.J.; Tweedley, J.R. Hypersalinity leads to dramatic shifts in the invertebrate fauna of estuaries. *Animals* **2025**.
79. Pillay, D.; Perissinotto, R. The benthic macrofauna of the St. Lucia Estuary during the 2005 drought year. *Estuar. Coast. Shelf Sci.* **2008**, *77*, 35–46. [CrossRef]
80. Deegan, B.M.; Lamontagne, S.; Aldridge, K.T.; Brookes, J.D. *Trophodynamics of the Coorong: Spatial Variability in Food Web Structure Along a Hypersaline Coastal Lagoon*; CSIRO: Water for a Healthy Country National Research Flagship: Canberra, Australia, 2010; p. 52.

81. Jumars, P.A.; Dorgan, K.M.; Lindsay, S.M. Diet of worms emended: An update of polychaete feeding guilds. *Annu. Rev. Mar. Sci.* **2015**, *7*, 497–520. [CrossRef]
82. Aikins, S.; Kikuchi, E. Grazing pressure by amphipods on microalgae in Gamo Lagoon, Japan. *Mar. Ecol. Prog. Ser.* **2002**, *245*, 171–179. [CrossRef]
83. Ingvason, H.R.; Ólafsson, J.S.; Gardarsson, A. Food selection of *Tanytarsus gracilentus* larvae (Diptera: Chironomidae): An analysis of instars and cohorts. *Aquat. Ecol.* **2004**, *38*, 231–237. [CrossRef]
84. Paterson, C.; Walker, K. Recent history of *Tanytarsus barbitarsis* Freeman (Diptera: Chironomidae) in the sediments of a shallow, saline lake. *Mar. Freshw. Res.* **1974**, *25*, 315–325. [CrossRef]
85. Sylvestre, F.; Beck-Eichler, B.; Duleba, W.; Debenay, J.-P. Modern benthic diatom distribution in a hypersaline coastal lagoon: The Lagoa de Araruama (R.J.), Brazil. *Hydrobiologia* **2001**, *443*, 213–231. [CrossRef]
86. Hribar, L.J. Mouthpart morphology and feeding behavior of biting midge larvae (Diptera: Ceratopogonidae). In *Functional Morphology of Insect Feeding*; Schaefer, C.W., Leschen, R.A.B., Eds.; Thomas Say Publications in Entomology: Proceedings of the Entomological Society of America: Lanham, Maryland; Entomological Society of America: Annapolis, MD, USA, 1993; pp. 43–58.
87. Rodriguez, G.; Fikáček, M.; Minoshima, Y.N.; Archangelsky, M.; Torres, P.L.M. Going underwater: Multiple origins and functional morphology of piercing-sucking feeding and tracheal system adaptations in water scavenger beetle larvae (Coleoptera: Hydrophiloidea). *Zool. J. Linn. Soc.* **2020**, *193*, 1–30. [CrossRef]
88. Yee, D.A.; Kehl, S. Order Coleoptera. In *Thorp and Covich's Freshwater Invertebrates*, 4th ed.; Thorp, J.H., Rogers, D.C., Eds.; Academic Press: Boston, MA, USA, 2015; pp. 1003–1042.
89. Preston, R.L. Osmoregulation in annelids: Cells and animals. In *Osmotic and Ionic Regulation*; Evans, D.H., Ed.; CRC Press: Boca Raton, FL, USA, 2008; pp. 135–165.
90. Castellano, G.C.; Lana, P.d.C.; Freire, C.A. Euryhalinity of subtropical marine and estuarine polychaetes evaluated through carbonic anhydrase activity and cell volume regulation. *J. Exp. Zool. Part A Ecol. Integr. Physiol.* **2020**, *333*, 316–324. [CrossRef] [PubMed]
91. Oglesby, L.C. Salinity-stress and desiccation in intertidal worms. *Am. Zool.* **2015**, *9*, 319–331. [CrossRef]
92. Murina, G.-V.V. Ecology of Sipuncula. *Mar. Ecol. Prog. Ser.* **1984**, *17*, 1–7. [CrossRef]
93. Glasby, C.J.; Erséus, C.; Martin, P. Annelids in extreme aquatic environments: Diversity, adaptations and evolution. *Diversity* **2021**, *13*, 98. [CrossRef]
94. Hutchings, P.A.; Murray, A. Taxonomy of polychaetes from the Hawkesbury River and the southern estuaries of New South Wales, Australia. *Rec. Aust. Mus. Suppl.* **1984**, *3*, 1–118. [CrossRef]
95. Mucciolo, S.; Desiderato, A.; Leal, S.M.; Mastrodonato, M.; Lana, P.; Freire, C.A. Variability in the degree of euryhalinity of neotropical estuarine annelids. *J. Exp. Mar. Biol. Ecol.* **2021**, *544*, 151617. [CrossRef]
96. Rolston, A.; Gannon, R.; Dittmann, S. *Macrobenthic Invertebrates of the Coorong, Lower Lakes and Murray Mouth Ramsar Site: A Literature Review of Responses to Changing Environmental Conditions*; Report to the Department for Environment and Heritage; Flinders University: Adelaide, Australia, 2010; p. 36.
97. Dávila-Jiménez, Y.; Papiol, V.; Hernández-Alcántara, P.; Enriquez, C.; Sauma-Castillo, L.; Chiappa-Carrara, X. Polychaete assemblages in a tropical hypersaline coastal lagoon of the southeastern Gulf of Mexico during the rainy season. *Rev. Biol. Trop.* **2019**, *67*, 136–156. [CrossRef]
98. Silva, C.F.; Seixas, V.C.; Barroso, R.; Di Domenico, M.; Amaral, A.C.Z.; Paiva, P.C. Demystifying the *Capitella capitata* complex (Annelida, Capitellidae) diversity by morphological and molecular data along the Brazilian coast. *PLoS ONE* **2017**, *12*, e0177760. [CrossRef]
99. Chalmer, P.N.; Hodgkin, E.P.; Kendrick, G.W. Benthic faunal changes in a seasonal estuary of south-western Australia. *Rec. West. Aust. Mus.* **1976**, *4*, 383–410.
100. Wells, F.E.; Threlfall, T.J. Molluscs of the Peel-Harvey Estuary with a comparision with other south-western Australian estuaries. *J. Malacol. Soc. Aust.* **1981**, *5*, 101–111.
101. Wildsmith, M.D.; Valesini, F.J.; Robinson, S.F. The value of enduring environmental surrogates as predictors of estuarine benthic macroinvertebrate assemblages. *Estuar. Coast. Shelf Sci.* **2017**, *197*, 159–172. [CrossRef]
102. Cottingham, A.; Bossie, A.; Valesini, F.; Maus, C.; Tweedley, J.R. Habitat compression of an estuarine mytilid following half a century of streamflow decline. *Estuar. Coast. Shelf Sci.* **2023**, *282*, 108253. [CrossRef]
103. Medeiros, I.P.M.; Faria, S.C.; Souza, M.M. Osmoionic homeostasis in bivalve mollusks from different osmotic niches: Physiological patterns and evolutionary perspectives. *Comp. Biochem. Physiol. Part A Mol. Integr. Physiol.* **2020**, *240*, 110582. [CrossRef]

104. Thomsen, M.S.; Wernberg, T.; Tuya, F.; Silliman, B.R. Ecological performance and possible origin of a ubiquitous but under-studied gastropod. *Estuar. Coast. Shelf Sci.* **2010**, *87*, 501–509. [CrossRef]
105. Wells, F.E.; Threlfall, T.J.; Wilson, B.R. *The Peel–Harvey Estuarine System Study 1976–1980, Technical Report, Biology of Molluscs*; Department of Conservation and Environment: Perth, WA, Australia, 1980; p. 121.
106. Sarà, G.; Romano, C.; Widdows, J.; Staff, F.J. Effect of salinity and temperature on feeding physiology and scope for growth of an invasive species (*Brachidontes pharaonis*—MOLLUSCA: BIVALVIA) within the Mediterranean Sea. *J. Exp. Mar. Biol. Ecol.* **2008**, *363*, 130–136. [CrossRef]
107. Verdelhos, T.; Marques, J.C.; Anastácio, P. The impact of estuarine salinity changes on the bivalves *Scrobicularia plana* and *Cerastoderma edule*, illustrated by behavioral and mortality responses on a laboratory assay. *Ecol. Indic.* **2015**, *52*, 96–104. [CrossRef]
108. Lawrie, A.D.A.; Chaplin, J.; Pinder, A. Biology and conservation of the unique and diverse halophilic macroinvertebrates of Australian salt lakes. *Mar. Freshw. Res.* **2021**, *72*, 1553–1576. [CrossRef]
109. Todd, M.E. Osmotic balance in *Hydrobia ulvae* and *Potamopyrgus jenkinsi* (Gastropoda: Hydrobiidae). *J. Exp. Biol.* **1964**, *41*, 665–677. [CrossRef]
110. Pascual, E.; Drake, P. Physiological and behavioral responses of the mud snails *Hydrobia glyca* and *Hydrobia ulvae* to extreme water temperatures and salinities: Implications for their spatial distribution within a system of temperate lagoons. *Physiol. Biochem. Zool. Ecol. Evol. Approaches* **2008**, *81*, 594–604. [CrossRef] [PubMed]
111. Wells, F.E.; Threlfall, T.J. Reproductive strategies of *Hydrococcus brazeri* (Tenison Woods, 1876) and *Arthritica semen* (Menke, 1843) in Peel Inlet Western Australia. *J. Malacol. Soc. Aust.* **1982**, *5*, 157–166.
112. Pinder, A.M.; Halse, S.A.; McRae, J.M.; Shiel, R.J. Occurrence of aquatic invertebrates of the wheatbelt region of Western Australia in relation to salinity. *Hydrobiologia* **2005**, *543*, 1–24. [CrossRef]
113. Tweedley, J.R.; Warwick, R.M.; Clarke, K.R.; Potter, I.C. Family-level AMBI is valid for use in the north-eastern Atlantic but not for assessing the health of microtidal Australian estuaries. *Estuar. Coast. Shelf Sci.* **2014**, *141*, 85–96. [CrossRef]
114. Dobrzycka, A.; Szaniawska, A. The effect of salinity on osmoregulation in *Corophium volutator* (Pallas) and *Saduria entomon* (Linnaeus) from the Gulf of Gdańsk. *Oceanologia* **1995**, *37*, 111–122.
115. Masikane, N.F.; Newman, B.K.; Scharler, U.M. Salinity tolerance of the South African endemic amphipod *Grandidierella lignorum* (Amphipoda: Aoridae). *Afr. J. Aquat. Sci.* **2014**, *39*, 151–156. [CrossRef]
116. Shahin, S.; Okomoda, V.T.; Ishak, S.D.; Waiho, K.; Fazhan, H.; Azra, M.N.; Azman, A.R.; Wongkamhaeng, K.; Abualreesh, M.H.; Rasdi, N.W.; et al. Lagoon amphipods as a new feed resource for aquaculture: A life history assessment of *Grandidierella halophila*. *J. Sea Res.* **2023**, *192*, 102360. [CrossRef]
117. Moore, P.G. A new species in the genus *Grandidierella* Coutière (Crustacea: Amphipoda) from an Australian solar salt-works. *J. Nat. Hist.* **1986**, *20*, 1393–1399. [CrossRef]
118. Kikuchi, S.; Matsumasa, M. The osmoregulatory tissue around the afferent blood vessels of the coxal gills in the estuarine amphipods, *Grandidierella japonica* and *Melita setiflagella*. *Tissue Cell* **1993**, *25*, 627–638. [CrossRef]
119. Rahman, M.; Chaplin, J.; Pinder, A. The biology of giant ostracods (Crustacea, Cyprididae), a review focusing on the Mytilocypridinae from Australian inland waters. *Mar. Freshw. Res.* **2023**, *74*, 1–19. [CrossRef]
120. Tweedley, J.R.; Krispyn, K.N. Fishes and salinities of low-inflow estuaries in the Fitzgerald Biosphere. *FiSHMED Fishes Mediterr. Environ.* **2025**, *001*, 1–14.
121. Terry, C.E.; Liebzeit, J.A.; Purvis, E.M.; Dowd, W.W. Interactive effects of temperature and salinity on metabolism and activity of the copepod *Tigriopus californicus*. *J. Exp. Biol.* **2024**, *227*, jeb248040. [CrossRef] [PubMed]
122. Britton, R.H.; Johnson, A.R. An ecological account of a Mediterranean salina: The Salin de Giraud, Camargue (S. France). *Biol. Conserv.* **1987**, *42*, 185–230. [CrossRef]
123. Shadrin, N.V.; Anufriieva, E.V.; Belyakov, V.P.; Bazhora, A.I. Chironomidae larvae in hypersaline waters of the Crimea: Diversity, distribution, abundance and production. *Eur. Zool. J.* **2017**, *84*, 61–72. [CrossRef]
124. Kokkinn, M. Osmoregulation, salinity tolerance and the site of ion excretion in the halobiont chironomid, *Tanytarsus barbitarsis* Freeman. *Mar. Freshw. Res.* **1986**, *37*, 243–250. [CrossRef]
125. Cowley, P.D.; Tweedley, J.R.; Whitfield, A.K. Conservation of estuarine fishes. In *Fish and Fisheries in Estuaries*; Wiley: Hoboken, NJ, USA, 2022; pp. 617–683.
126. Kefford, B.J.; Fields, E.J.; Clay, C.; Nugegoda, D. Salinity tolerance of riverine microinvertebrates from the southern Murray Darling Basin. *Mar. Freshw. Res.* **2007**, *58*, 1019–1031. [CrossRef]

127. Kefford, B.J.; Papas, P.J.; Metzeling, L.; Nugegoda, D. Do laboratory salinity tolerances of freshwater animals correspond with their field salinity? *Environ. Pollut.* **2004**, *129*, 355–362. [CrossRef]
128. Hutchings, P.; Kupriyanova, E. Cosmopolitan polychaetes—Fact or fiction? Personal and historical perspectives. *Invertebr. Syst.* **2018**, *32*, 1–9. [CrossRef]

water

MDPI

Article

Relationships Between Land Use and Stream Macroinvertebrate Biotic Integrity in Central Ohio, USA

Douglas Spieles *, Zoe Krashes, Khiem Nguyen, Summer Rodgers, Lillian Ruiz and Marco Vigilante

McPhail Center for Sustainability and Environmental Studies, Denison University, Granville, OH 43023, USA; nguyen_k5@denison.edu (K.N.); rodger_s1@denison.edu (S.R.); vigila_m1@denison.edu (M.V.)
* Correspondence: spielesd@denison.edu

Abstract: Land use is known to be an important factor in the composition and function of adjacent freshwater lotic ecosystems. However, the relative effects of land use type, extent, intensity, and proximity on aquatic ecosystem quality are not fully understood. We evaluate these questions in low-order streams within 30 watersheds in developed, agricultural, and less developed landscapes of central Ohio, USA. We assess the relationships of land use cover percentage and spatial scale with stream macroinvertebrate community diversity and biotic integrity. We also investigate the importance of impervious cover and subsurface tile drainage within each watershed and Active River Area (ARA). We find that the percentage of coverage of developed land at the watershed scale is the strongest predictor of stream macroinvertebrate community diversity and integrity. High-intensity development is a stronger negative correlate than low-intensity development or agriculture. There is a significant decline in stream macroinvertebrate diversity and biotic integrity at the watershed and ARA scales when undeveloped land coverage falls below 20–30%. We do not find a significant relationship between stream macroinvertebrate metrics and land use at the 1 km^2 scale or in comparison with any instream habitat attributes except sinuosity. Impervious cover has a significant negative relationship with both macroinvertebrate taxon richness and biotic integrity at the watershed and ARA scales. However, subsurface tile-drained land does not have a significant relationship with the stream macroinvertebrate community at any scale. We conclude that impervious land cover at the watershed and ARA scales is a critical factor for the biotic integrity of small streams in this region. Collectively, our conclusions provide evidence to support practices of ecologically sensitive land use planning.

Keywords: stream biotic integrity; land use; land cover; impervious; subsurface tile drainage; watershed; active river area

1. Introduction

Aquatic ecosystem structure and function are influenced by adjacent terrestrial conditions in human-dominated landscapes, meaning that aquatic systems can be neither protected nor restored in isolation [1]. Aquatic–terrestrial linkages are both essential components for ecological conservation and challenging aspects of land use planning [2]. Many scholars have attempted to elucidate the relative effects of land use type, extent, proximity, intensity, and duration on various indicators of aquatic ecosystem biotic integrity, e.g., [3–5]. While much has been learned, these linkages are still not fully understood in terms of the aquatic community response to land use change. Continued research has important implications for regional land management, as we attempt to maintain ecosystem services amidst a growing human environmental footprint. For example, understanding the spatial thresholds of watershed land use may help prevent deleterious functional shifts in adjacent aquatic ecosystems [6].

Within this framework, a great deal of attention has been paid to land use change and the biotic integrity of lotic ecosystems. Biotic integrity is understood here to be "the capacity of a land to support characteristic functional and structural communities in the context of normal variability, to resist loss of this function and structure due to a disturbance, and to recover from such disturbance" [7]. In lotic systems, biotic integrity is typically assessed with the resident fish and aquatic macroinvertebrate communities; we focus on macroinvertebrates here. Macroinvertebrates are critical components of the faunal biomass, abundance, and diversity in freshwater ecosystems [8,9]. As such they play key roles in ecological processes such as nutrient cycling [10], decomposition [11], and bioturbation [12], as well as many provisional and cultural ecosystem services [13]. At the same time, the wide sensitivity range and relatively rapid response to changing biological and physicochemical conditions make macroinvertebrate communities useful indicators of biotic integrity [14].

Several recent studies of large data sets show that land use intensification can have deleterious effects on the biotic integrity of riverine systems. In a meta-analysis of 37 global studies, Petsch et al. [15] find that while any type of land use alteration from pristine native vegetation can reduce freshwater stream diversity, urbanization is associated with the strongest negative effect. Hughes et al. [16] use the National Rivers and Streams Assessment (NRSA) to analyze macroinvertebrate taxon richness (i.e., the abundance of different macroinvertebrate types as identified to the lowest practicable taxon) of 3475 streams across the conterminous Unites States. According to this study, the percentages of forested land (positive correlation) and agricultural land (negative correlation) within the watershed are the best predictors of macroinvertebrate taxon diversity. This is consistent with other studies, e.g., [17,18], which find greater taxon richness in forested streams compared with agricultural watersheds, though Sandefur et al. [19] report no significant correlation between macroinvertebrate indices and agricultural land use in a similar national-scale data set. In a meta-analysis of 62 studies from around the world, Camana et al. [20] find that a macroinvertebrate community response is most discernable in landscapes with a greater percentage of watershed land use change. The type of land use change, land use history, physical geography, and climate are comparatively unimportant. In temperate landscapes, 50% land use change from natural cover is an approximate threshold for a stream macroinvertebrate community response, while this threshold varies widely in tropical and subtropical regions [20].

The conditions of the watershed, riparian corridor, and reach-scale instream habitat are hierarchically nested and interconnected. Many regional studies have attempted to discern the importance of the different scales on stream biotic integrity. In a single-stream ecosystem in Greece, Karaouzas et al. [21] find land development at both the floodplain and watershed scales to be negatively correlated with stream macroinvertebrate diversity and quality (i.e., the relative abundance of sensitive taxa). Robinson et al. [22] study the spatial land use effects of two watersheds in Switzerland and report different primary drivers of the stream macroinvertebrate community in the two regions: a linear relationship with the percentage of land development in one, and a more complex amalgamation of watershed and small-scale water quality effects in the other. In a study of 1017 streams in Germany, Palt et al. [23] find watershed-level land use factors to be more strongly related to macroinvertebrate biotic integrity than small-scale metrics of the riparian corridor. Both Sulikowski [24] and Hoover and Wood [25] find watershed land use to be a strong predictor of stream macroinvertebrate biotic integrity, along with significant local correlation of riparian forest cover [24] and instream water quality [25].

Fewer studies have considered the importance of instream habitat relative to watershed-scale land use factors, and those that have are contradictory. For example, Hrodey et al. [26] find instream features like channel width, bank stability, and the percentage of fine substrates to be better predictors of macroinvertebrate community biotic integrity than watershed-scale factors. Karaouzas et al. [21] also report instream microhabitat, such as woody debris, artificial substrate, water chemistry, and stream morphology, to be more important than watershed-scale factors. However, Henderson and Christian [27] consider reach-scale stream habitat but find no significant correlation with the instream macroinvertebrate community, similar to the findings of Stepenuck et al. [28]. There is a need for additional studies on this question.

A second issue in need of further exploration is the nuanced character of developed land. Urbanized land is not monolithic, and we need a better understanding of the types of urban development and their relative effects on stream structure and function. Wang et al. [29] classify urban development as the percentage of impervious land in the watershed and find a significant decline in stream macroinvertebrate biotic integrity at a mere 3.5–5.5% impervious cover. Stepenuck et al. [28] report steep declines in stream quality metrics at 8–12% watershed impervious cover, while Huang and Gergel [30] find the best macroinvertebrate biotic integrity in streams with 16% or less watershed impervious cover, and the lowest integrity in streams with greater than 60% watershed impervious cover. Agricultural land may similarly be classified and analyzed based on agricultural practices, such as artificial drainage. Some watersheds are more than 80% drained by surface ditches and subsurface tiles, causing profound changes in the hydrologic and chemical load of receiving streams [31]. Our analysis of the relative relationship of impervious cover and tile-drained land provides an assessment of specific land development attributes and stream biotic integrity.

In this study, we explore three hypotheses. First, we hypothesize that the type and degree of watershed land use, broadly considered to be developed land, agricultural land, and less developed land, influences stream macroinvertebrate community diversity and biotic integrity. Second, we hypothesize that physical land features at various scales—watershed, river corridor, small catchment, and instream attributes—have differential effects on stream macroinvertebrate communities. Third, we hypothesize that watershed impervious cover and subsurface tile drainage affect in-stream macroinvertebrate communities in similar ways.

2. Materials and Methods

2.1. Study Area

We analyzed watersheds and low-order streams of Central Ohio, USA, in Delaware, Franklin, Hocking, Knox, and Licking Counties (Figure 1). Central Ohio features a variety of land uses, including urban and suburban sectors of Columbus and the surrounding cities, intensive row-crop agriculture, pasture, rural residential areas, forest, and surface water. Within this region, we selected Hydrologic Unit Code (HUC) level 12-scale watersheds based on their land use characteristics according to the United States Geological Survey National Land Cover Database (USGS NLCD) [32,33]. The HUC system delineates nested watersheds of the United States, with the HUC12 level denoting "subwatersheds", which averaged approximately 80 km^2 in our data set (Table 1) [32,33]. We used data from the 2019 USGS NLCD [34] to estimate impervious cover and data from Valayamkunnath et al. [35] to estimate the area of subsurface tile drainage within each HUC12 watershed. From a population of over 70 HUC12 watersheds, we selected 30 for analysis (Table 1): 10 with a high percentage of impervious cover (developed); 10 with a high percentage of subsurface tile drainage (agriculture); and 10 with comparatively low impervious cover and tile drainage (less developed). Within each watershed, we selected a low-order stream, and at least one reach within that stream, for spatial and field analyses. Stream reaches and access points were chosen by consultation with and with permission from local landowners. In three watersheds, we sampled multiple stream reaches; thus, for 30 watersheds, we had 33 field sampling sites.

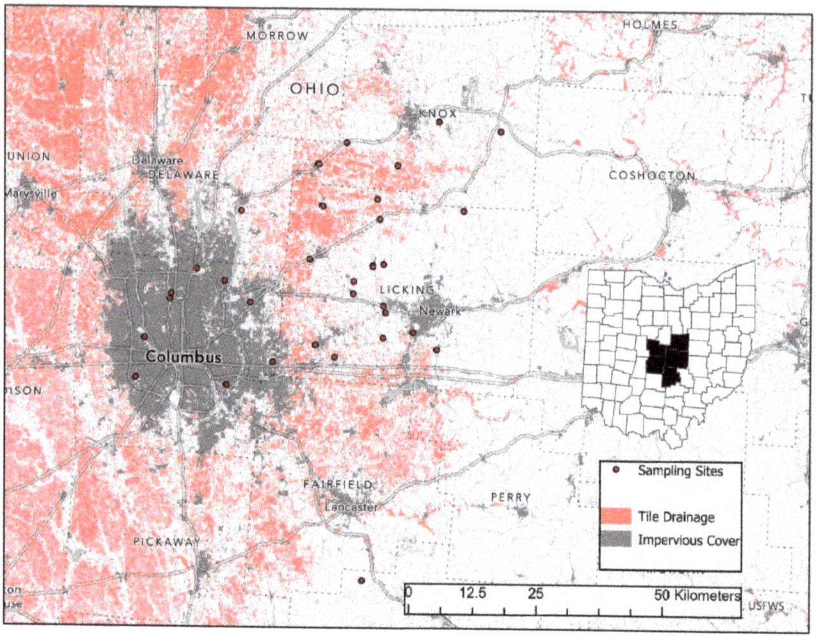

Figure 1. Stream sampling locations, impervious cover, and estimated subsurface tile drainage in central Ohio, USA. Impervious cover was estimated from the 2019 USGS National Land Cover Database [34], and data from Valayamkunnath et al. were used to estimate the area of subsurface tile drainage [35].

Table 1. Descriptive statistics (mean ± SE) of 71 HUC12 watersheds in central Ohio and their Active River Area (ARA), compared with the 30 watersheds and associated ARA sampled for this study. Impervious cover was estimated from the 2019 USGS National Land Cover Database [34], and data from Valayamkunnath et al. [35] were used to estimate the area of subsurface tile drainage.

	HUC12 Watershed			Active River Area (ARA)		
	Area (km^2)	Imperv%	Tiled%	Area (km^2)	Imperv%	Tiled%
Developed						
Total (27)	79 ± 5.7	61 ± 4.6	11 ± 2.5	21 ± 2.6	9.1 ± 1.0	1.8 ± 0.5
Sampled (10)	82 ± 9.1	64 ± 9.0	5.7 ± 3.0	37 ± 4.7	57 ± 9.1	3.2 ± 1.2
Agriculture						
Total (20)	86 ± 8.4	8.6 ± 1.4	55 ± 1.5	23 ± 4.1	3.2 ± 1.2	12 ± 1.4
Sampled (10)	76 ± 9.6	15 ± 6.3	30 ± 1.8	37 ± 5.1	17 ± 5.7	18 ± 3.1
Less Developed						
Total (24)	79 ± 5.2	8.3 ± 0.5	6.6 ± 1.0	19 ± 1.8	1.8 ± 0.2	1.4 ± 0.2
Sampled (10)	71 ± 6.9	12 ± 1.8	11 ± 2.6	26 ± 4.0	17 ± 3.5	7.1 ± 1.2

2.2. Spatial Analysis

For each of the 30 watersheds, we used USGS NLCD 2019 data to quantify the HUC12 area and the Active River Area (ARA) [36]. The ARA consists of the main river channel plus the base zone (parafluvial and riparian corridors), the 100 year floodplain, and material contribution areas [36]. We used Model My Watershed [37] to determine the NLCD 2019 developed area, agricultural area, and undeveloped area of each HUC12 watershed, its ARA, and a 1 km^2 square centered on each field sampling site (Figure 2). Developed land, by our definition, includes low-density residential as well as medium-density commercial and high-intensity industrial land uses. Agricultural land includes cultivated cropland and pastures, while undeveloped land includes forests, wetlands, grasslands, shrublands, and open water. We also used Model My Watershed to obtain the simulated hydrologic load, nitrogen load, phosphorus load, and sediment load for each HUC12 and its ARA. These simulations were based on a hypothetical 24 h storm in each land use context, as generated by a hybrid of the Source Loading and Management Model (SLAMM), the TR-55 model, and the Spreadsheet Tool for the Estimation of Pollutant Load [38–40]. Since these are simulated values, they serve as indicators of watershed runoff load rather than actual hydrologic conditions. We estimated impervious cover and subsurface tile drainage for each HUC12, ARA, and 1 km^2 by extracting USGS NLCD data and the most-likely tile-drained area based on soil drainage class and topographic slope [35] using ArcPro 3.3.

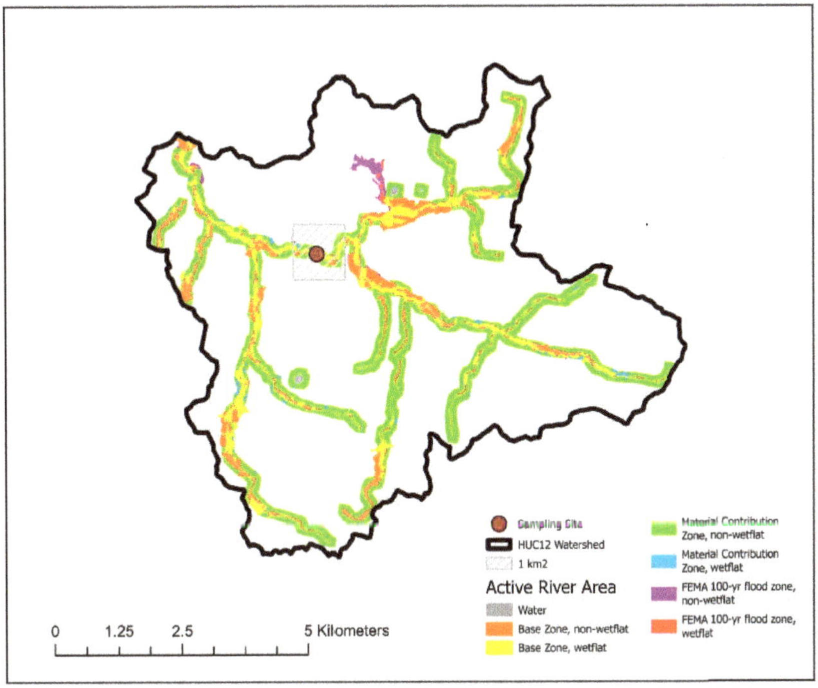

Figure 2. Sampling scheme for spatial and field data in each HUC12 watershed and its Active River Area (ARA; [36]), and 1 km^2 zone centered on stream sampling location.

2.3. Field Analysis

We visited 33 second- or third-order streams within the 30 HUC12 watersheds in June and July 2024. At each location, we demarcated a 100 m reach along the thalweg, with perpendicular cross sectional transects at 10 m intervals. Within every reach, we measured the sinuosity, dissolved oxygen, water and air temperature, electrical conductivity, and total suspended solids. We also rated every reach according to the Ohio Environmental Protection Agency Qualitative Headwater Habitat Evaluation Index (QHEI) and the United States Environmental Protection Agency (USEPA) HabitatAssessment for Low-Gradient Streams [41,42]. Along each transect, we established 2 to 5 sampling locations, depending on stream width, where we measured the depth, flow rate, substrate size, instream vegetation cover, canopy cover, bank stability, riparian vegetation cover, character of woody debris, and instream habitat type (riffle, run, glide, pool). We collected and tallied aquatic macroinvertebrates from the Phyla Annelida, Arthropoda, and Mollusca within every 100 m reach by hand-picking rocks and woody debris and with fine-mesh seines. At each location, we collected macroinvertebrates for 120 person-minutes (typically 6 individuals for 20 min). Voucher specimens were preserved in alcohol for later identification to the lowest taxon practicable. For each macroinvertebrate community, we determined the total number of individuals collected; the total number of taxa; the total number of Ephemeroptera, Plecoptera, and Trichoptera (EPT) taxa; the Shannon diversity index [43]; and the Headwater Macroinvertebrate Field Evaluation Index (HMFEI) [44]. The HMFEI is a biotic integrity index in which benthic macroinvertebrates, identified to the order or family level, are assigned points based on their tolerance and fidelity to habitat conditions [44].

2.4. Statistical Analysis

We used ANOVA to compare mean macroinvertebrate community statistics between locations according to the following land use characteristics: percentage of developed land; percentage of agricultural land; percentage of undeveloped land; and simulated hydrologic, nitrogen, phosphorus, and sediment loads. These comparisons were conducted at the HUC12, ARA, and 1-km^2 scales. Similarly, we used ANOVA to compare macroinvertebrate community characteristics at the lowest, middle, and highest terciles of instream habitat attributes according to the QHEI and the USEPA Habitat Assessment for Low-Gradient Streams. To evaluate the relative effects of the impervious area and estimated tile-drained area, we performed a linear regression of the land use quintile means and macroinvertebrate community scores. Finally, we used a correlation matrix (Pearson's r) and multiple linear regression to identify the best predictors of stream macroinvertebrate biotic integrity. All statistical analyses were conducted with JMP 17, with significance evaluated at $\alpha = 0.05$ and $\alpha = 0.1$.

3. Results

3.1. Descriptive Statistics

The HUC12 watersheds in our analysis are representative of the watersheds in the greater central Ohio region (Table 1). Those that we classify as developed have impervious areal coverage greater than 60%, while agricultural regions are estimated to have more than 30% subsurface tile drainage. Watersheds that we classify as less developed have less than 25% combined coverage of impervious and tile-drained lands. The ARA zones generally have statistically similar or marginally less developed and agricultural land compared to the corresponding HUC12 watersheds (Table 1). The intensity of HUC12 land use ranges from highly developed (85% residential, commercial, and industrial land use; 95% impervious cover) to highly agricultural (75% crop and pastureland; 40% subsurface tile drainage) to predominantly forested (70% forest; 6% impervious cover; 3% tile-drained). HUC12 land use is strongly correlated with simulation model-generated hydrologic load (Pearson's r = 0.94), but not significantly correlated with nitrogen, phosphorus, or sediment load.

Instream habitat features vary throughout the 33 stream reaches of this study. The QHEI ranges from 70 to 32 and correlates with the USEPA Stream Habitat Assessment Score (r = 0.72). The habitat variables with the greatest range include sinuosity; canopy cover; instream vegetation and periphyton; substrate material; and the relative percentage of riffles, runs, and pools. Within these streams we cataloged 3957 individual macroinvertebrate organisms of 96 taxa (Table 2). The Shannon diversity ranges from 2.7 to 1.4, EPT taxa from 11 to 2, and HMFEI from 48 to 12. The total taxa, EPT taxa, Shannon Diversity, and HMFEI are all positively correlated with each other; thus, we use total taxa as our primary measure of diversity and the HMFEI as our primary measure of biotic integrity.

Table 2. Taxonomic classification of aquatic macroinvertebrates collected in low-order streams of central Ohio, USA, in June and July 2024 by the authors of this study. Abundance indicates the total number of each taxon collected, and Locations indicates the number of sites in which each taxon was collected out of a possible 33 stream reaches.

Phylum	Class	Order	Family	Abundance	Locations
Annelida	Clitellata	Rhynchobdellida	Glossiphoniidae	187	22
	Clitellata	Other		4	3
Arthropoda	Arachnida	Araneae	Pisauridae	2	1
	Insecta	Coleoptera	Dryopidae	10	6
			Dytiscidae	4	4
			Elmidae	74	24
			Gyrinidae	28	4
			Haliplidae	4	3
			Hydrophilidae	1	1
			Psephenidae	241	17
			Ptilodactylidae	1	1
		Diptera	Athericidae	1	1
			Ceratopogonidae	1	1
			Chironomidae	59	20
			Ephydridae	1	1
			Limoniidae	2	2
			Simuliidae	257	21
			Tipulidae	24	11
		Ephemeroptera	Ameletidae	11	3
			Baetidae	213	26
			Caenidae	1	1
			Ephemerellidae	17	6
			Heptageniidae	279	26
			Leptophlebiidae	10	6
			Potamanthidae	1	1
			Siphlonuridae	3	3
		Hemiptera	Gerridae	17	8
			Veliidae	12	6
		Megaloptera	Corydalidae	2	2
		Odonata	Aeshnidae	6	6
			Calopterygidae	3	2
			Coenagrionidae	10	6
		Plecoptera	Capniidae	1	1
			Chloroperlidae	33	7
			Leuctridae	3	1
			Nemouridae	10	5
			Peltoperlidae	1	1
			Perlidae	27	7
			Perlodidae	15	5
			Taeniopterygidae	2	2
		Trichoptera	Apataniidae	1	1
			Brachycentridae	32	8
			Dipseudopsidae	19	6
			Glossosomatidae	27	2
			Helicopsychidae	226	12
			Hydropsychidae	814	32
			Hydroptilidae	13	5
			Lepidostoma	4	1
			Leptoceridae	3	2
			Limnephilidae	33	4
			Philopotamidae	275	27
			Polycentropodidae	18	4

Table 2. *Cont.*

Phylum	Class	Order	Family	Abundance	Locations
			Psychomyiidae	9	4
			Rhyacophilidae	23	5
			Thremmatidae	2	1
			other	97	23
	Malacostraca	Amphipoda	Gammaridae	7	5
		Decapoda	Cambaridae	93	43
Mollusca	Bivalvia	Unionida	Unionidae	43	13
	Gastropoda	Architaenioglossa	Viviparidae	3	2
		Basommatophora	Ancylidae	16	4
			Lymnaeidae	15	4
			Physidae	179	25
			Planorbidae	32	14
		Caenogastropoda	Pleuroceridae	15	1
		Gyraulus	Planorbidae	2	1
		Heterostropha	Valvatidae	1	1
		Littorinimorpha	Amnicolidae	2	1
			Hydrobiidae	2	2
		Neotaenioglossa	Pleuroceridae	334	15
		334	Pleurocera	2	1
		Sorbeoconcha	Arionidae	4	3
		Stylommatophora		7	1
		Other		26	4

3.2. *Analysis of Hypotheses*

Our first hypothesis suggests that land use type and extent within the HUC12 watershed is associated with macroinvertebrate community diversity and biotic integrity. To test this, we calculated the correlation coefficients of the HUC12 land use variables with the macroinvertebrate community variables (Table 3). The percentage of developed land within the watershed and the associated hydrologic load has a significant negative correlation with every measure of macroinvertebrate diversity and biotic integrity. Conversely, the percentage of undeveloped land has a positive correlation with every variable except total individuals. We therefore accept this hypothesis. The relative amount of developed and undeveloped land at the HUC12 scale is the strongest predictor of stream macroinvertebrate community diversity and biotic integrity in our study (Table 3; Figure 3).

To evaluate our second hypothesis, we compared the mean number of macroinvertebrate taxa and the HMFEI by category of land use at the HUC12, ARA, 1 km^2, and instream scales (Figure 3). Using the percentage of undeveloped land (i.e., forest, shrubland, grassland, and surface water) as the indicator of land use, we find a significant decline in both total taxa and the HMFEI at the HUC12 and ARA scales (Figure 3A,B). At both scales, macroinvertebrate communities are significantly less diverse and of lower integrity when undeveloped land drops below 20–30%. We do not find a significant decline in total taxa or the HMFEI at the 1 km^2 scale (Figure 3C). Similarly, the tercile comparison of stream habitat scores (Figure 3D) indicates no significant change in macroinvertebrate community metrics. Based on this evidence, we accept the hypothesis and conclude that land use at the watershed and river corridor scales has a greater influence on macroinvertebrates than smaller-scale land use or instream attributes in this data set.

In our third hypothesis, we propose that impervious cover and subsurface tile drainage will have similar effects on the stream macroinvertebrate community. To test this hypothesis, we compare macroinvertebrate taxa and the HMFEI to the quantile means of percentage of impervious cover and percentage of tiled land at the HUC12 and ARA scales (Figure 4). At both scales, impervious cover has a significant negative relationship with both taxa and the HMFEI. However, the percentage of tile-drained land does not have a significant relationship with the stream macroinvertebrate community at either scale. Thus, we reject the hypothesis and conclude that impervious cover has a greater association with the stream macroinvertebrate community than the percentage of tile-drained land. Multiple linear regression models (Figure 5) using the HUC12 percentage of development and stream sinuosity as predictors can explain about 43% of the variation in macroinvertebrate taxa (y = 23.2 − 0.2*%HucDev + 0.9*sinuosity) and about 52% of the variation in the HMFEI (y = 12.6 − 0.1*%HucDev + 0.4*sinuosity).

Table 3. Correlation (Pearson's r) of stream macroinvertebrate community metrics with HUC12 watershed characteristics and reach-level instream habitat characteristics. Bold values indicate significance at the 0.05 ** and 0.1 * levels.

	Macroinvertebrate Community Metrics				
	Total Individuals	**Total Taxa**	**Shannon Diversity**	**HMFEI Score**	**EPT Taxa**
HUC12 Level					
% Developed	**−0.40 ****	**−0.54 ****	**−0.37 ****	**−0.57 ****	**−0.50 ****
% Agriculture	**0.56 ****	**0.41 ****	0.20	0.38	0.25
% Undeveloped	−0.06	**0.45 ****	**0.42 ****	**0.57 ****	**0.63 ****
Hydrologic Load	−0.13	**−0.51 ****	**−0.39 ****	**−0.60 ****	**−0.60 ****
Nitrogen Load	0.25	−0.28	−0.26	**−0.40 ****	**−0.47 ****
Phosphorus Load	**0.36 ****	−0.20	−0.22	**−0.34 ****	**−0.42 ****
Sediment Load	−0.20	−0.29 *	−0.21	−0.28 *	−0.26
Reach Level					
QHEI Score	−0.05	0.15	**0.35 ****	0.23	0.22
Habitat Score	0.09	0.28	**0.35 ****	0.30 *	0.28
Sinuosity	0.02	**0.37 ****	0.32 *	**0.42 ****	**0.43 ****
Canopy Cover	−0.01	0.19	0.23	0.23	0.26
Substrate Score	−0.02	−0.09	−0.10	−0.08	−0.13
% Riffle	−0.24	−0.12	−0.01	−0.05	0.05
Periphyton Cover	**0.43 ****	0.15	0.08	0.15	0.04

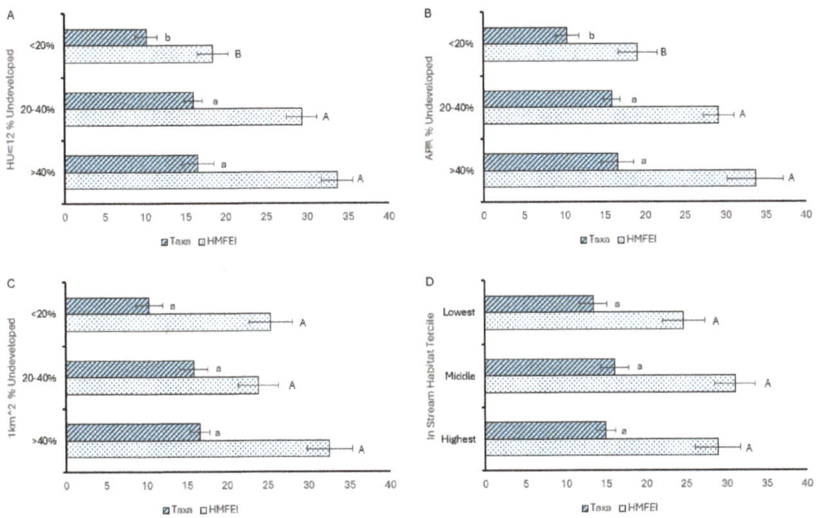

Figure 3. Stream macroinvertebrate taxa and HMFEI (mean ± SE) compared with respect to percentage of undeveloped land at (**A**) HUC12 watershed scale, (**B**) Active River Area (ARA) scale, and (**C**) 1 km^2 scale, and (**D**) with respect to tercile index of instream habitat metrics. Lowercase letters indicate significant differences among taxa in each group, and capital letters indicate significant differences among HMFEIs; α = 0.05.

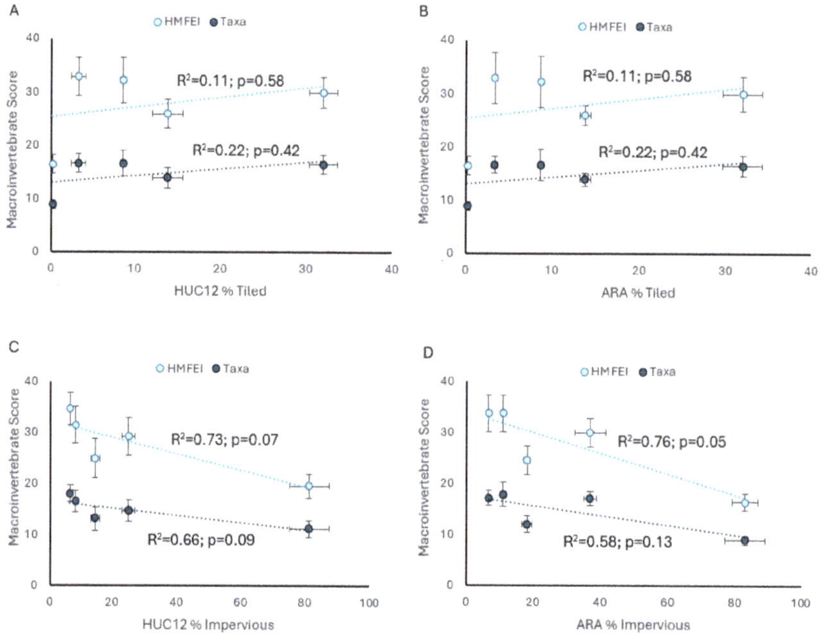

Figure 4. Linear regression of stream macroinvertebrate taxa and HMFEI (mean ± SE) with (**A**) mean ± SE HUC12 watershed percentage of subsurface tile-drained land; (**B**) Active River Area (ARA) percentage of subsurface tile-drained land; (**C**) HUC12 watershed percentage of impervious cover; and (**D**) ARA percentage of impervious cover. Impervious cover is estimated from 2019 USGS National Land Cover Database [34], and data from Valayamkunnath et al. [35] are used to estimate area of subsurface tile drainage.

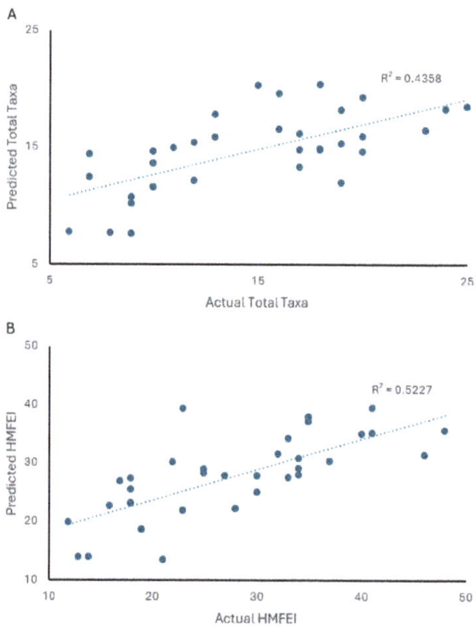

Figure 5. Multiple linear regression of actual vs. model-predicted macroinvertebrate taxa (**A**) and HMFEI (**B**) using HUC12 percentage of development and stream sinuosity as predictors.

4. Discussion

Other studies illustrate the importance of land use variation for the detection of lotic macroinvertebrate community effects. In our sample, developed HUC12 watersheds range from 60 to 84% combined residential, commercial, and industrial land use, spanning 31 to 95 percentage of impervious cover. By comparison, the global data set of Petsch et al. [15] considers developed sites that range from 20 to 96% urbanized land, with impervious cover over 40%. Agriculture-dominated watersheds in our study range from 58 to 76 percentage of row-crop or pasture lands, while Petsch et al. [15] sample studies ranging from 53 to 100% agriculture. The percentage of tile-drained land in HUC12 watersheds is not well represented in the literature, but our selection of 25–35% tile-drained land falls in the medium range of 3.5–60% for central Ohio HUC12 watersheds estimated by Valayamkunnath et al. [35]. Our less developed watersheds range from 40 to 68% forest cover, while undeveloped reference sites in Petsch et al. [15] average 77% $\pm$ 12% (mean $\pm$ SD) of forest cover. Other large data sets [16,20] analyze a similar range of land use coverage. Thus, while a study of this nature in a comparatively small region can restrict the variety of land use coverage [22], our data set appears to represent an adequate degree of land use heterogeneity.

Regarding the relationship of land use type and extent with the stream macroinvertebrate community, our study is largely in agreement with the literature. We accept our first hypothesis and find the HUC12 percentage of developed area and percentage of undeveloped area (primarily forest, with small areas of shrublands, grasslands, and surface water) to be the strongest predictors of macroinvertebrate community metrics. The percentage of agricultural land in the HUC12 watershed is the best positive predictor only for the total number of individual organisms collected at sampling locations; it is not a significant correlate with any other macroinvertebrate community metrics. Comparable studies [15,28,45,46] similarly find the percentage of urbanization to be the strongest predictor of low macroinvertebrate biotic integrity, while others report a strong positive correlation with the percentage of watershed forested land [4,16,36]. Interestingly, our analysis of NLCD developed land subcategories suggests that high-intensity development (i.e., industrial) has the strongest negative correlation with total macroinvertebrate taxa, while low-density residential development has the smallest negative correlation with HM-FEI scores. Within agricultural subcategories, the amount of pasture- and hay-production land has a much higher positive correlation with macroinvertebrate indices than the amount of row-crop agriculture land. These subtleties make it difficult to identify a threshold for deleterious land use change as Camana et al. have [20]. In our study, low-intensity development and low-density pastoral agriculture may allow for a high-integrity-stream macroinvertebrate community with only 20–30% forested cover in the watershed, while high-intensity development and row-crop agriculture have a disproportionately greater negative effect.

Our second hypothesis, proposing differential spatial scale effects of physical land features on the stream macroinvertebrate community, is confirmed by our analysis. We find significant and indistinguishable correlations of land use with stream biotic integrity at the HUC12 and ARA scales, and we find no significant correlation at the 1 km^2 or reach-level scales; this agrees with similar studies [27,28]. In our study the only significant reach-level factor is sinuosity. Our results are not necessarily in conflict with those that find greater predictive power at smaller spatial scales [21,26]. In our data set, the variance of landscape-level factors is greater than that of instream factors; regions with the inverse relative variance may well show the opposite spatial importance. In any case, the significance of ARA land use and sinuosity in our study, along with numerous studies in the literature, make it clear that suitable land cover at all spatial scales is necessary to maintain stream biotic integrity.

The analysis of our third hypothesis provides evidence that impervious cover in both the watershed and river corridor has a strong negative relationship with stream biotic integrity. Some studies suggest that a small percentage of impervious cover can have a strong negative effect on stream quality [28,29]. Wang and Kanehl [47] similarly report a significant decline in stream macroinvertebrate biotic integrity with watershed impervious cover of more than 7 percent. In the region of our study, stream macroinvertebrate biotic integrity declines when impervious cover in the watershed or river corridor reaches 30–60% (Figure 4), which is more in line with Huang and Gergel [30]. According to our data, the percentage of tile-drained land in the watershed or river corridor does not have the same negative effect. In fact, we find no direct relationship of the percentage of tile-drained land with stream biotic indices. Indirectly, we do find negative correlations of runoff volume and nutrient load—both enhanced by impervious cover and tile drainage—with the stream macroinvertebrate community. Our finding of a greater importance of imperviousness than tile drainage may be an artifact of site selection: some watersheds exceeded 95% impervious cover, but our greatest percentage of tile-drained coverage was only 38%. While our study provides a partial analysis of the effects of tile drainage on stream macroinvertebrates, we can find no other published studies for comparison. This question deserves more attention.

5. Conclusions

We preface our conclusions with three caveats. First, analyses of this type are context-specific. While our methods are transferable to other regions, local factors may yield nuanced results. Second, the assessment of ecosystem biotic integrity depends upon the metrics selected to quantify integrity. We select the total macroinvertebrate taxa, EPT taxa, Shannon diversity, and HMFEI as metrics for their general comparability with other studies. But these are certainly not the only, and maybe not the best, metrics for all situations. Different metrics may yield different conclusions. Indeed, our own analyses of individual Ephemeroptera, Coleoptera, Trichoptera, Mollusca, and Decapoda data did not show the same land use relationships as aggregate community indices. Finally, biotic indices of any collection of indicator organisms may not necessarily equate with ecosystem function, nor do they always indicate the status of desired ecosystem services.

With these qualifications, we offer three clear conclusions. First, watershed-scale and ARA land use are critical factors for the biotic integrity of small streams in our study region. Given suitable land use in the watershed and river corridor, a high-integrity macroinvertebrate community may exist in a variety of instream habitat conditions. Second, a greater degree of undeveloped land, or, conversely, a lesser degree of high-intensity development and row-crop agriculture, is associated with greater stream macroinvertebrate biotic diversity and integrity. Third, in both the river corridor and the watershed, impervious cover has a greater negative association with the stream macroinvertebrate community than the percentage of tile-drained land. These conclusions provide evidence to support practices of ecologically sensitive land use planning. Specifically, the protection of undeveloped land and minimization of impervious cover in the watershed and in the river corridor are associated with greater macroinvertebrate diversity and biotic integrity.

Author Contributions: All authors contributed to conceptualization, methodology, investigation, data curation, formal analysis, funding acquisition, and review and editing. The original draft preparation, supervision, and project administration were conducted by D.S. All authors have read and agreed to the published version of the manuscript.

Funding: This research was funded by the Denison University Anderson Endowment and the Denison University Research Foundation.

Data Availability Statement: The raw data supporting the conclusions of this article will be made available by the authors on request.

Acknowledgments: We thank the Licking County Park District, the Franklin County Soil and Water Conservation District, the Knox County Park District, Kenyon College, and Columbus Metroparks for granting us site access. The assistance of Sarah Lim and Alex Sket is gratefully acknowledged.

Conflicts of Interest: The authors declare no conflicts of interest.

References

1. Cheng, C.; Zhang, F.; Shi, J.; Kung, H.T. What is the relationship between land use and surface water quality? A review and prospects from remote sensing perspective. *Environ. Sci. Pollut. Res.* **2022**, *29*, 56887–56907. [CrossRef] [PubMed]
2. Hansen, A.J.; DeFries, R. Ecological mechanisms linking protected areas to surrounding lands. *Ecol. Appl.* **2007**, *17*, 974–988. [CrossRef]
3. Lin, J.; Huang, J.; Prell, C.; Bryan, B.A. Changes in supply and demand mediate the effects of land-use change on freshwater ecosystem services flows. *Sci. Total Environ.* **2021**, *763*, 143012. [CrossRef]
4. Fierro, P.; Bertrán, C.; Tapia, J.; Hauenstein, E.; Peña-Cortés, F.; Vergara, C.; Cerna, C.; Vargas-Chacoff, L. Effects of local land-use on riparian vegetation, water quality, and the functional organization of macroinvertebrate assemblages. *Sci. Total Environ.* **2017**, *609*, 724–734. [CrossRef] [PubMed]
5. Carpenter, S.R.; Stanley, E.H.; Vander Zanden, M.J. State of the world's freshwater ecosystems: Physical, chemical, and biological changes. *Ann. Rev. Environ. Res.* **2011**, *36*, 75–99. [CrossRef]
6. Scheffer, M.; Carpenter, S.; Foley, J.A.; Folke, C.; Walker, B. Catastrophic shifts in ecosystems. *Nature* **2001**, *413*, 591–596. [CrossRef]
7. Deng, X.; Du, J. Land Quality: Environmental and Human Health Effects. In *Encyclopedia of Environmental Health*; Nriagu, J.O., Ed.; Elsevier: Amsterdam, The Netherlands, 2011; pp. 362–365. [CrossRef]
8. Lin, X.W.; Zhang, X.G.; Tan, L.; Tian, Z.; Li, J.T.; Luo, Q.Y. Temporal ecological processes have different seasonal influences on multiple dimensions of riverine insect diversity in China. *Insect Conserv. Div.* **2024**, *18*, 246–259. [CrossRef]
9. Clarke, A.; Mac Nally, R.; Bond, N.; Lake, P.S. Macroinvertebrate diversity in headwater streams: A review. *Freshw. Biol.* **2008**, *53*, 1707–1721. [CrossRef]
10. Atkinson, C.L.; Capps, K.A.; Rugenski, A.T.; Vanni, M.J. Consumer-driven nutrient dynamics in freshwater ecosystems: From individuals to ecosystems. *Biol. Rev.* **2017**, *92*, 2003–2023. [CrossRef]
11. Graça, M.A. The role of invertebrates on leaf litter decomposition in streams–a review. *Int. Rev. Hydrobiol.* **2001**, *86*, 383–393. [CrossRef]
12. Chakraborty, A.; Saha, G.K.; Aditya, G. Macroinvertebrates as engineers for bioturbation in freshwater ecosystem. *Environ. Sci. Pollut. Res.* **2022**, *29*, 64447–64468. [CrossRef] [PubMed]
13. Macadam, C.R.; Stockan, J.A. More than just fish food. Ecosystem services provided by freshwater insects. *Ecol. Entomol.* **2015**, *40*, 113–123. [CrossRef]
14. Carter, J.L.; Resh, V.H.; Hannaford, M.J. Macroinvertebrates as biotic indicators of environmental quality. In *Methods in Stream Ecology*, 3rd ed.; Lamberti, G.A., Hauer, F.R., Eds.; Academic Press: Cambridge, MA, USA, 2017; pp. 293–318. [CrossRef]
15. Petsch, D.K.; Blowes, S.A.; Melo, A.S.; Chase, J.M. A synthesis of land use impacts on stream biodiversity across metrics and scales. *Ecology* **2021**, *102*, e03498. [CrossRef] [PubMed]
16. Hughes, R.M.; Herlihy, A.T.; Comeleo, R.; Peck, D.; Mitchell, R.M.; Paulsen, S.G. Patterns in and predictors of stream and river macroinvertebrate genera and fish species richness across the conterminous USA. *Knowl. Manag. Aquat. Ecosyst.* **2023**, *424*, 19. [CrossRef]
17. Rumschlag, S.L.; Mahon, M.B.; Jones, D.K.; Battaglin, W.; Behrens, J.; Bernhardt, E.S.; Bradley, P.; Brown, E.; de Laender, F.; Hill, R.; et al. Density declines, richness increases, and composition shifts in stream macroinvertebrates. *Sci. Adv.* **2023**, *9*, 18. [CrossRef]
18. Vinson, M.R.; Hawkins, C.P. Broad-scale geographical patterns in local stream insect genera richness. *Ecography* **2003**, *26*, 751–767. [CrossRef]
19. Sandefur, H.; Cummings, E.; Johnston, R.; Matlock, M.; Gustafson, D. Analyzing correlations between stream benthic macroinvertebrate community structure, water quality, and watershed land use in the continental United States. *Biol. Eng. Trans.* **2014**, *7*, 169–182. [CrossRef]
20. Camana, M.; Ortega, J.C.; Brejão, G.L.; Melo, A.S.; Dias, M.S.; Becker, F.G. A global meta-analysis of the effects of land use on the diversity of stream fish and macroinvertebrates. *Aqua. Sci.* **2024**, *86*, 86. [CrossRef]

21. Karaouzas, I.; Gritzalis, K.C.; Skoulikidis, N. Land use effects on macroinvertebrate assemblages and stream quality along an agricultural river basin. *Fres. Environ. Bull.* **2007**, *16*, 645–653.
22. Robinson, C.T.; Schuwirth, N.; Baumgartner, S.; Stamm, C. Spatial relationships between land-use, habitat, water quality and lotic macroinvertebrates in two Swiss catchments. *Aqua. Sci.* **2014**, *76*, 375–392. [CrossRef]
23. Palt, M.; Le Gall, M.; Piffady, J.; Hering, D.; Kail, J. A metric-based analysis on the effects of riparian and catchment land use on macroinvertebrates. *Sci. Total Environ.* **2022**, *816*, 151590. [CrossRef]
24. Sulikowski, T. An Examination of the Effects of Land Use and land Cover on Macroinvertebrate Communities at Two Landscape Scales. Master's Thesis, The Faculty of Montclair State University, Montclair, NJ, USA, August 2013.
25. Hoover, G.W.; Wood, J. Analysis of localized water chemistry and land use drivers of macroinvertebrate community structure in the Upper Ohio Watershed using multivariate modelling and ordination techniques. *Proc. West Va. Acad. Sci.* **2021**, *93*, 1. [CrossRef]
26. Hrodey, P.J.; Sutton, T.M.; Frimpong, E.A.; Simon, T.P. Land-use impacts on watershed health and integrity in Indiana warmwater streams. *Am. Mid. Nat.* **2009**, *161*, 76–95. [CrossRef]
27. Henderson, N.D.; Christian, A.D. Freshwater invertebrate assemblage composition and water quality assessment of an urban coastal watershed in the context of land-use land-cover and reach-scale physical habitat. *Ecologies* **2022**, *3*, 376–394. [CrossRef]
28. Stepenuck, K.F.; Crunkilton, R.L.; Wang, L. Impacts of urban landuse on macroinvertebrate communities in southeastern Wisconsin streams. *J. Am. Water Res. Assoc.* **2007**, *38*, 1041–1051. [CrossRef]
29. Wang, B.; Liu, D.; Liu, S.; Zhang, Y.; Lu, D.; Wang, L. Impacts of urbanization on stream habitats and macroinvertebrate communities in the tributaries of Qiangtang River, China. *Hydrobiologia* **2012**, *680*, 39–51. [CrossRef]
30. Huang, J.; Gergel, S.E. Effectiveness of landscape indicators for explaining the variability of benthic macro-invertebrates in urban streams. *J. Environ. Manag.* **2023**, *342*, 118130. [CrossRef]
31. Blann, K.L.; Anderson, J.L.; Sands, G.R.; Vondracek, B. Effects of agricultural drainage on aquatic ecosystems: A review. *Crit. Rev. Environ. Sci. Technol.* **2009**, *39*, 909–1001. [CrossRef]
32. Seaber, P.R.; Kapinos, F.P.; Knapp, G.L. *Hydrologic Unit Maps*; Water Supply Paper 2294; U.S. Department of the Interior, U.S. Geological Survey US Government Printing Office: Washington, DC, USA, 1987. [CrossRef]
33. Dewitz, J. *United States National Land Cover Database (NLCD) 2019 Products, v. 3.0.*; U.S. Department of the Interior, U.S. Geological Survey: Washington, DC, USA, 2024. [CrossRef]
34. Wickham, J.; Stehman, S.V.; Sorenson, D.G.; Gass, L.; Dewitz, J.A. Thematic accuracy assessment of the NLCD 2019 land cover for the conterminous United States. *GISci. Remote Sens.* **2023**, *60*, 2181143. [CrossRef]
35. Valayamkunnath, P.; Barlage, M.; Chen, F.; Gochis, D.J.; Franz, K.J. Mapping of 30-meter resolution tile-drained croplands using a geospatial modeling approach. *Sci. Data* **2020**, *7*, 257. [CrossRef]
36. Smith, M.P.; Schiff, R.; Olivero, A.; MacBroom, J. The Active River Area. In *A Conservation Framework for Protecting Rivers and Streams*; The Nature Conservancy: Boston, MA, USA, 2008; Eastern U.S. Freshwater Program 11 Avenue De Lafayette, 5th Floor; Available online: https://www.conservationgateway.org/Files/Pages/active-river-area-conserv.aspx (accessed on 1 June 2024).
37. Stroud Water Research Center. Model My Watershed [Software], v.1.35.0. 2024. Available online: https://wikiwatershed.org/ (accessed on 1 June 2024).
38. Pitt, R.; Voorhees, J. SLAMM, the source loading and management model. In *Wet-Weather Flow in the Urban Watershed*; CRC Press: Boca Raton, FL, USA, 2002; pp. 103–139. [CrossRef]
39. United States Soil Conservation Service. *Urban Hydrology for Small Watersheds*; Tech. Rep. No. 55 (TR-55); United States Soil Conservation Service: Washington, DC, USA, 1986. Available online: https://www.nrc.gov/docs/ML1421/ML14219A437.pdf (accessed on 1 June 2024).
40. United States Environmental Protection Agency. *Spreadsheet Tool for the Estimation of Pollutant Load (STEPL)*; Tetra Tech, Inc.: Fairfax, VA, USA, 2001. Available online: https://19january2021snapshot.epa.gov/nps/spreadsheet-tool-estimating-pollutant-loads-stepl (accessed on 1 June 2024).
41. Ohio Environmental Protection Agency. *Methods for Assessing Habitat in Flowing Waters: Using the Qualitative Habitat Evaluation Index (QHEI)*; Division of Surface Water, Ecological Assessment Section: Columbus, OH, USA, 2006. Available online: https://dam.assets.ohio.gov/image/upload/epa.ohio.gov/Portals/35/documents/QHEIManualJune2006.pdf (accessed on 1 June 2024).
42. Barbour, M.T. Rapid Bioassessment Protocols for Use in Wadable Streams and Rivers. In *Periphyton, Benthic Macroinvertebrates and Fish*; US Environmental Protection Agency, Office of Water: Washington, DC, USA, 1999. Available online: https://www3.epa.gov/region1/npdes/merrimackstation/pdfs/ar/AR-1164.pdf (accessed on 1 June 2024).
43. Shannon, C.E. A mathematical theory of communication. *Bell Syst. Tech. J.* **1948**, *27*, 379–423. [CrossRef]
44. Ohio Environmental Protection Agency. *Field Methods for Evaluating Primary Headwater Streams in Ohio, Version 4.1*; Ohio EPA Division of Surface Water: Columbus, OH, USA, 2020. Available online: https://dam.assets.ohio.gov/image/upload/epa.ohio.gov/Portals/35/wqs/headwaters/PHWHManual_2020_Ver_4_1_May_2020_Final.pdf (accessed on 1 June 2024).

45. Sterling, J.L.; Rosemond, A.D.; Wenger, S.J. Watershed urbanization affects macroinvertebrate community structure and reduces biomass through similar pathways in Piedmont streams, Georgia, USA. *Freshw. Sci.* **2016**, *35*, 676–688. [CrossRef]
46. Weigel, B.M.; Lyons, J.; Paine, L.K.; Dodson, S.I.; Undersander, D.J. Using stream macroinvertebrates to compare riparian land use practices on cattle farms in southwestern Wisconsin. *J. Freshw. Ecol.* **2000**, *15*, 93–106. [CrossRef]
47. Wang, L.; Kanehl, P. Influences of watershed urbanization and instream habitat on macroinvertebrates in cold water streams. *J. Am. Water Res. Assoc.* **2003**, *39*, 1181–1196. [CrossRef]

water

MDPI

Article

Predicting Harmful Algal Blooms Using Explainable Deep Learning Models: A Comparative Study

Bekir Zahit Demiray [1], Omer Mermer [1,*], Özlem Baydaroğlu [2,3] and Ibrahim Demir [4,5]

1 IIHR Hydroscience and Engineering, University of Iowa, Iowa City, IA 52242, USA; bekirzahit-demiray@uiowa.edu
2 NOAA, Global Systems Laboratory, Boulder, CO 80305, USA; ozlem.baydaroglu@noaa.gov
3 National Academies of Sciences, Engineering, and Medicine, Washington, DC 20001, USA
4 Department of River-Coastal Science and Engineering, Tulane University, New Orleans, LA 70118, USA; idemir@tulane.edu
5 ByWater Institute, Tulane University, New Orleans, LA 70118, USA
* Correspondence: omer-mermer@uiowa.edu

Abstract: Harmful algal blooms (HABs) have emerged as a significant environmental challenge, impacting aquatic ecosystems, drinking water supply systems, and human health due to the combined effects of human activities and climate change. This study investigates the performance of deep learning models, particularly the Transformer model, as there are limited studies exploring its effectiveness in HAB prediction. The chlorophyll-a (Chl-a) concentration, a commonly used indicator of phytoplankton biomass and a proxy for HAB occurrences, is used as the target variable. We consider multiple influencing parameters—including physical, chemical, and biological water quality monitoring data from multiple stations located west of Lake Erie—and employ SHapley Additive exPlanations (SHAP) values as an explainable artificial intelligence (XAI) tool to identify key input features affecting HABs. Our findings highlight the superiority of deep learning models, especially the Transformer, in capturing the complex dynamics of water quality parameters and providing actionable insights for ecological management. The SHAP analysis identifies Particulate Organic Carbon, Particulate Organic Nitrogen, and total phosphorus as critical factors influencing HAB predictions. This study contributes to the development of advanced predictive models for HABs, aiding in early detection and proactive management strategies.

Keywords: chlorophyll-a; deep learning (DL); explainable AI (XAI); harmful algal bloom (HAB); prediction; SHAP value; transformer; water quality

1. Introduction

The emergence and proliferation of harmful algal blooms (HABs), particularly characterized by cyanobacteria, also known as blue-green algae, have escalated into a pressing environmental issue affecting water bodies worldwide. These blooms significantly impact water quality, public health, and the balance of aquatic ecosystems [1]. The primary culprits behind the surge in HAB incidents include nutrient pollution, primarily from agricultural production and industrial waste, coupled with climatic variables such as rising water temperatures and altered water quality characteristics [2,3]. HABs are notorious for producing dangerous toxins, undermining the aesthetic and recreational value of waterways, and challenging the provision of clean drinking water [4]. The situation is further exacerbated by the dramatic rise in HAB events in recent years, attributed to the rapid expansion of human populations, extensive agricultural practices, escalating pollution levels, and the overarching effects of climate change [5,6].

This disturbing trend underscores the urgent need for the development and refinement of HAB monitoring, estimation, modeling, and prediction techniques to protect water resources and public health [7–10]. Given these challenges, it becomes imperative to monitor and develop forecasting tools for HABs to ensure public health and safety. However, direct measurement of algal concentrations is labor-intensive and time-consuming, where the chlorophyll-a concentration often serves as a proxy for algal blooms [11] and poor water quality [12]. Therefore, it would be desirable to predict the occurrence of HABs from other readily measured variables that are primary factors contributing to the proliferation of HABs.

HAB formation typically results from a combination of factors that create a favorable growth environment [13]. Key contributors to HABs include eutrophication, deteriorating water quality—especially high levels of nitrogen and phosphorus—and climate change [14,15]. While numerous studies have examined the effects of individual factors like nutrient levels, land use, or climate drivers, many rely on single data points related to the individual actors, leading to oversimplified prediction and potential errors [16–18]. Developing accurate predictive models requires a comprehensive understanding of these interrelated factors, along with specific water quality parameters that influence HAB dynamics. Variables such as particulate organic matter, total phosphorus, dissolved phosphorus, and turbidity play crucial roles in determining chlorophyll-a concentrations, a key indicator of algal growth and toxicity. However, while increased chlorophyll-a levels often correlate with cyanobacterial growth, their toxicity varies depending on site-specific environmental conditions. This highlights the complexity of HAB formation, underscoring the need for integrated research and strict monitoring to mitigate their harmful effects on ecosystems, the economy, and public health.

Effective monitoring of HABs often requires laboratory analysis of water samples to assess a variety of indicators such as chlorophyll-a [19], cyanobacteria [20], and various algal toxins [21]. The laboratory techniques used for these analyses encompass microscopy, spectrophotometry, liquid chromatography, and biochemical assays [22]. Such techniques are vital for identifying algae at the species level and require specialized technical expertise. Additionally, the use of remote sensing-derived data from satellites and UAVs provides valuable spatial information on HAB spread [23–26], making it easier to observe bloom conditions across different water bodies. Addressing the issue of algal blooms and their harmful effects necessitates a thorough understanding of their causes and the implementation of effective management strategies, including the timely prediction of HAB. Consequently, ensuring the accuracy of HAB predictions is crucial for effectively responding to this global environmental challenge and safeguarding both the ecological environment and human health [27].

There are various prediction models in the literature addressing HABs. Physical process-based models [28–31] specify physical and biochemical processes that influence HABs. Typical physical process-based models include Environmental Fluid Dynamic Code (EFDC), Water Quality Analysis Simulation (WASP), and QUAL2K [32–34]. While these models have high prediction accuracy with complete data, constructing perfect data, particularly with spatial resolution, is costly and involves practical limitations. In addition to process-based models, statistical models, which use formulas or fits to correlate various variables (physicochemical, meteorological, and hydrodynamic) to HAB indicators, sometimes struggle to capture complex non-linear relationships and lack predictive accuracy [35–37]. Both approaches have their contributions to studying HAB dynamics but often encounter limitations in accurately capturing complex interactions and maintaining predictive accuracy under varied conditions [38].

Due to the limitations of these traditional physical and statistical models, data-driven models have emerged as ideal alternatives and have recently gained popularity in being applied to solve various problems in hydrology and atmospheric studies [39,40], environmental monitoring [41], and environmental management [42–44]. Similarly, recent applications of these models have focused on predicting HABs by analyzing the relationship between various algal-related factors and water quality parameters [45–47]. Various machine learning (ML) models have been developed and are in practice for determining drivers of algal dynamics and multistep prediction in various lakes, rivers, and reservoirs [48–53]. Among them, Ai et al. [49] presented a comprehensive study to compare the performance of widely used ML algorithms, including bootstrap aggregating (BA), gradient boosting (GB), Gaussian processes (GPs), k-nearest neighbors (KNNs), Random Forrest (RF), support vector machine (SVM), and XGBoost (XGB), for predicting HABs. In addition to these algorithms, Yu et al. [54] used Adaptive Boosting (AdaBoost) and Gradient Boosting Decision Tree (GBDT) algorithms to predict HAB occurrence. However, limited information is available on criteria for selecting, validating, and evaluating the performance of ML models for HAB prediction [55].

Besides these ML models, researchers have also applied deep learning (DL) models which have recently gained popularity due to their significant advantages such as the ability to capture temporal dynamics and handle complex interactions among features effectively [56,57]. For example, several studies performed HAB prediction using different Long Short-Term Memory (LSTM)-based models [58–61]. Lee et al. [61] used and compared the integrated CNN model and LSTM to predict HABs for multiple streams from 29 stations in four South Korean rivers and showed that prediction performance for all monitoring stations of the integrated CNN model was better than LSTM-based models. In addition, a few recent studies have used state-of-the-art DL algorithms such as Gated Recurrent Unit (GRU) and the Transformer for water quality prediction [62]. Despite studies to predict HABs via ML/DL models increasing, making optimal HAB predictions by reflecting various parameters, time-series characteristics, and performance evaluation remains challenging [63–68].

Despite their predictive capabilities, deep learning models such as LSTM, CNN, and RNN often function as black-box systems, making it difficult to interpret the reasoning behind predictions [69]. This means that users can only access input and output data without understanding the underlying decision-making process [70]. This lack of interpretability poses challenges, especially when AI-driven predictions impact critical fields like medicine, law, and environmental science [71]. To address these concerns, explainable AI methodologies, such as SHAP analysis, have been developed to enhance transparency and interpretability by identifying the contribution of individual variables to model predictions [72,73]. In the context of HABs, predicting their occurrence remains complex due to the multitude of influencing factors and non-linear growth patterns. While previous studies have primarily focused on HAB prediction using field measurements, statistical methods, and machine learning techniques, few have leveraged SHAP to interpret model outputs. Recent research has demonstrated the effectiveness of explainable AI in identifying key environmental drivers of HABs [45,53,61]. For example, SHAP analysis has been applied to deep learning models to highlight the influence of water temperature and nutrient concentrations [61] or precipitation and turbidity [53] on chlorophyll-a levels. Another study used SHAP to explore spatiotemporal patterns of algal cell density in Lake Dianchi [45]. In contrast, this study employs SHAP values to interpret advanced deep learning models, including the Transformer, for HAB prediction in Lake Erie. By leveraging data from multiple monitoring stations over an eight-year period, it provides a comprehensive evaluation of HAB drivers, encompassing physical, chemical, and biological water quality parameters while capturing both spatial and temporal dynamics.

To address these gaps, we investigated the performance of deep learning models, particularly the Transformer model, as there are limited studies exploring their effectiveness in HAB prediction, considering multiple influencing parameters including physical, chemical, and biological water quality monitoring data from multiple stations located west of Lake Erie, and used SHAP values as an explainable AI tool to identify key input features affecting HABs. Addressing gaps in prior research, in summary, this study aims to (1) utilize comprehensive datasets that encompass physical, chemical, and biological water quality parameters from multiple monitoring stations; (2) evaluate and compare the performance of various deep learning models, with a particular emphasis on the Transformer model, for predicting Chl-a concentrations, which indicate HAB occurrences; and (3) uncover significant features influencing HAB formation by applying SHAP.

By combining deep learning methods with SHAP analysis, our study (model comparison with indices) will aid scientists and stakeholders in understanding HAB formation mechanisms and serve as a practical reference, theoretical guide, and potential prediction standard for protecting ecosystems globally. This paper is organized as follows: Section 2 introduces the dataset utilized in this study. Section 3 details the methods tested, their specifications, and the role of explainable artificial intelligence (XAI). Section 4 presents the results and discussion, emphasizing the role of SHAP in interpreting the models. Finally, Section 5 concludes with insights into the implications of our findings and offers suggestions for future research directions.

2. Study Area and Dataset

Lake Erie, part of the Great Lakes system, serves as a critical case study for examining HABs. The Great Lakes, the largest and most biodiverse freshwater reserve on Earth, encompass both industrial and agricultural regions [74,75]. Among them, Lake Erie is the shallowest and smallest by volume but the fourth largest by area. It holds ecological, cultural, and economic significance for approximately 12.5 million people in its watershed, supporting nearly 14,000 tons of commercial and traditional fisheries, over 33 million tons of freight, and generating more than USD 1.5 million in recreation and tourism annually [76]. However, due to its geographical location, particularly in the western basin, it experiences severe nutrient overload, making it highly susceptible to HABs [77]. Since 2002, chlorophyll-a concentration, an indicator of eutrophication and HABs, has risen dramatically in Lake Erie, reaching unprecedented levels in recent years [77,78]. Humans can be exposed to harmful algae through ingestion of contaminated fish and drinking water, as well as inhalation and dermal contact during recreational activities such as swimming and boating [79,80]. Given the risks HABs pose to human health, the economy, and the environment, accurate prediction and identification of key factors are crucial for implementing preventive measures to mitigate potential losses [81].

As nutrient load and HAB occurrence predominantly affect the western part of Lake Erie, this study focused on the western basin of Lake Erie, which includes the area from the western part of the lake to Point Pelee, ON, Canada, and Cedar Point, OH, USA. Water quality data, collected from seven monitoring stations on the U.S. side of the western part of Lake Erie (shown in Figure 1) and operated by the National Oceanic and Atmospheric Administration (NOAA) Great Lakes Environmental Research Laboratory (GLERL), were used. This dataset encompasses water quality measurements sampled from 2013 to 2020 [77].

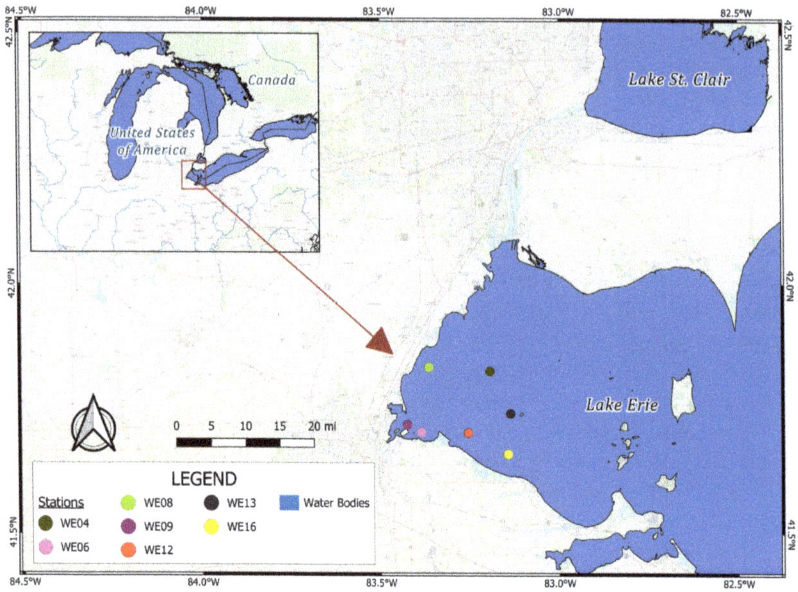

Figure 1. Location and description of western Lake Erie water quality monitoring stations managed by NOAA's Great Lakes Environmental Research Laboratory (original source available at NOAA GLERL's web page: https://www.glerl.noaa.gov/res/HABs_and_Hypoxia/rtMonSQL.php, accessed on 23 February 2025).

These stations are selected to represent the various nutrient and sediment inputs into the western basin of Lake Erie and highlight areas consistently susceptible to HABs. The selected water quality parameters measured at these stations included Secchi Depth (m), CTD Temperature (°C), CTD Specific Conductivity (µS/cm), CTD Dissolved Oxygen (mg/L), turbidity (NTU), total phosphorus (µg P/L), Total Dissolved Phosphorus (µg P/L), Ammonia (µg N/L), Nitrate + Nitrite (mg N/L), Particulate Organic Carbon (mg/L), Particulate Organic Nitrogen (mg/L), Total Suspended Solids (mg/L), and chlorophyll-a (µg/L). The primary objective was to predict chlorophyll-a concentration as a key indicator of HABs.

Table 1 presents a summary of the descriptive statistics of the complete dataset after pre-processing. Particulate Organic Carbon and Particulate Organic Nitrogen levels exhibited considerable variation, with maximum values far exceeding the averages, indicating the presence of organic matter in differing concentrations throughout the samples. Similarly, Total Suspended Solids and chlorophyll-a showed large ranges in value, reflecting varied conditions in the sampled water bodies. The standard deviation for each variable highlights the extent of variability in the measurements, with certain variables such as turbidity and total phosphorus displaying significant variability.

To prepare the dataset for modeling, the year 2020 was designated as the test set, while the remaining years were used for training purposes. From the training data, 20% was allocated as a validation set. The validation data were selected using a random stratified sampling approach to preserve the temporal and spatial characteristics of the dataset. Specifically, for each year in the training period (2013–2019), 20% of the samples were randomly selected for validation. This validation approach ensures that the model is tested against data that represent different years, seasons, and stations within the study area, reducing the risk of overfitting and improving generalizability. Standard Scaler was applied to standardize the features, ensuring each feature had a mean of zero and a standard deviation of one. The dataset was converted into a time-series format with a sequence length of three, meaning that each input sequence consisted of three consecutive time steps. One-hot encoding was used to specify each monitoring station, allowing the model to differentiate between data from different locations. To address the missing features in each input sequence provided to the models, the persistence method was employed [82], which involves carrying forward the last known value.

Table 1. Descriptive statistical summary of used dataset.

Variables	Min	Max	Mean	Median	Std. Dev
Secchi Depth	0.00	6.50	1.15	0.90	0.917
CTD Temperature	2.90	29.70	22.02	22.90	3.880
CTD Specific Conductivity	19.90	583.30	297.76	280.40	67.028
CTD Dissolved Oxygen	4.20	13.04	7.58	7.60	1.148
Turbidity	0.68	1148.00	19.76	9.30	55.561
Total Phosphorus	4.00	2482.24	77.83	48.14	132.919
Total Dissolved Phosphorus	0.16	273.58	20.42	9.27	28.248
Ammonia	0.04	2108.70	33.61	12.25	87.502
Nitrate + Nitrite	0.00	9.45	0.88	0.34	1.341
Particulate Organic Carbon	0.14	219.34	2.53	1.30	10.576
Particulate Organic Nitrogen	0.01	40.93	0.43	0.21	1.895
Total Suspended Solids	0.82	540.80	17.98	10.15	33.212
Chlorophyll-a	0.71	678.40	37.86	16.14	237.576

3. Methods

This study evaluates the effectiveness of several machine models—Lasso, Ridge, LSTM, GRU, and Transformer—in forecasting harmful algal blooms (HABs) in Lake Erie. This section describes the setup and implementation of these models, providing insights into their application for HAB prediction.

3.1. Lasso and Ridge Regression Models

In predictive modeling, particularly when dealing with high-dimensional data, regularization techniques such as Lasso and Ridge regression are essential for managing multicollinearity and preventing overfitting. Both methods modify the least squares loss function by adding a penalty term, aiding in stabilizing the coefficients, though they do so in distinct ways. Lasso regression (Least Absolute Shrinkage and Selection Operator) applies a penalty equal to the absolute value of the coefficients, promoting sparsity within the model parameters. This L1 regularization helps in performing variable selection by reducing some coefficients to zero, which is particularly beneficial for enhancing model interpretability [83]. The formulation of the Lasso regression cost function (Equation (1)) is as follows:

$$L(w) = \frac{1}{2n}\sum_{i=1}^{n}(y_i - X_i w)^2 + \alpha\|w\|_1, \tag{1}$$

where y_i denotes the observation values, w is the vector of model coefficients, α is the regularization parameter, and $\|\mathrm{w}\|_1$ is the L1 norm of the coefficients.

Conversely, Ridge regression incorporates L2 regularization by adding a penalty equal to the square of the magnitude of the coefficients [84]. This approach does not lead to zero coefficients but reduces their size uniformly, helping to manage multicollinearity without excluding any variables. This method is formulated through the following cost function (Equation (2)):

$$R(w) = \frac{1}{2n}\sum_{i=1}^{n}(y_i - X_i w)^2 + \alpha\|w\|_2^2, \quad (2)$$

where y_i denotes the observation values, X_i denotes the prediction values, w is the vector of model coefficients, α is the regularization parameter, and $\|w\|_2^2$ is the L2 norm of the coefficients.

3.2. LSTM and GRU Models

Long Short-Term Memory (LSTM) and Gated Recurrent Units (GRUs) are foundational to modern predictive modeling and are especially known for their efficiency in handling sequences and time-series data. These models are designed to solve the problems inherent in traditional recurrent neural networks (RNNs), specifically the difficulty in learning long-range temporal dependencies.

LSTMs were introduced by Hochreiter and Schmidhuber in 1997 [85] and have since become one of the most popular architectures for sequential data modeling due to their ability to maintain long-term dependencies. LSTM networks are composed of a series of units, each containing a cell state and various gates that regulate the flow of information. The key components of an LSTM unit are the cell state, forget gate, input gate, and output gate. The cell state is responsible for carrying information across different time steps and can be thought of as the memory of the network. The forget gate decides what information from the cell state should be discarded. It takes the previous hidden state and the current input and passes them through a sigmoid function. The input gate determines which new information should be added to the cell state. It comprises two parts: a sigmoid layer that decides which values to update and a *tanh* layer that creates a vector of new candidate values. The output gate decides what the next hidden state should be, which will be used in the next time step. It takes into account the current input, the previous hidden state, and the cell state. LSTMs have been widely applied in various domains, including natural language processing, speech recognition, and time-series forecasting. Their ability to capture long-term dependencies makes them particularly suitable for tasks where context over extended sequences is important.

Gated Recurrent Units (GRUs) were introduced by Cho et al. in 2014 [86] as a simplified variant of LSTMs. GRUs combine the forget and input gates into a single update gate and merge the cell state and hidden state, resulting in a more streamlined architecture. This simplification makes GRUs computationally more efficient while still retaining the capability to capture long-term dependencies. In a GRU, the update gate determines the amount of the previous hidden state that needs to be passed along to the future. This gate is computed using a sigmoid function that combines the previous hidden state and the current input. The reset gate, another key component of GRUs, decides how much of the past information to forget. It is also computed using a sigmoid function applied to the previous hidden state and the current input. The candidate hidden state is then calculated by applying a *tanh* function to the combination of the reset-modified previous hidden state and the current input. The final hidden state is obtained by linearly interpolating between the previous hidden state and the candidate hidden state using the update gate. GRUs have been found to perform comparably to LSTMs on many tasks, often with fewer parameters and less computational complexity. This makes them an attractive option for applications where computational resources are limited or where faster training times are desired.

Both LSTM and GRU architectures are adept at modeling environmental data, which is often non-linear and subject to sudden shifts due to underlying ecological dynamics. These models are versatile and have been employed in a variety of environmental studies [87], including streamflow forecasting [88], air quality prediction [89], and water quality prediction [90]. Each application showcases the ability of LSTM and GRU models to analyze and predict complex dynamic systems, making them invaluable tools in the field of environmental science.

These models have also been specifically applied in the context of harmful algal bloom (HAB) prediction [48,61], where their ability to process and make predictions based on diverse and dynamic environmental data inputs proves invaluable. By leveraging historical and current environmental data, they have demonstrated significant accuracy in predicting HABs, thereby supporting proactive management and mitigation strategies in water bodies. This practical application underlines the models' crucial role in not only understanding but also anticipating environmental changes that impact water quality and ecosystem health.

3.3. Transformer Models

The Transformer model, originally introduced for natural language processing tasks [91], has demonstrated a remarkable ability to capture complex dependencies in sequential data. It represents a paradigm shift in sequential data modeling by leveraging self-attention mechanisms instead of recurrence, enabling them to handle long-range dependencies more efficiently. The core concept of the Transformer model is the attention mechanism, which allows the model to weigh the importance of different elements within the input sequence dynamically.

The Transformer architecture consists of an encoder–decoder structure. However, in this study, we employed an encoder-based Transformer, which focuses on encoding the input sequence into a rich representation that captures temporal dependencies and interactions between various environmental factors. Typically, Transformer models use an embedding layer to convert input elements into continuous vector representations. However, in our implementation, we used a linear layer as the embedding layer to convert each element of the input sequence into continuous embeddings. These embeddings are then processed by multiple layers of self-attention and feedforward neural networks.

The self-attention mechanism is the cornerstone of the Transformer's architecture. It allows the model to focus on different parts of the sequence simultaneously by computing attention weights over the entire sequence. The self-attention mechanism (Equation (3)) calculates a weighted sum of the input values, where the weights are determined by the relevance of each input element to others. This process is defined by the following equations:

$$Attention(Q,\ K,\ V) = softmax\left(\frac{QK^T}{\sqrt{d_k}}\right)V, \tag{3}$$

where Q (queries), K (keys), and V (values) are the input embeddings, and d_k is the dimension of the keys. The attention mechanism computes the dot product of the queries and keys, scales it by the square root of the key dimension, and applies a softmax function to obtain the attention weights. These weights are then used to compute a weighted sum of the values, effectively allowing the model to focus on the most relevant parts of the sequence.

One of the innovative features of the Transformer model is the multi-head attention mechanism, which enhances the self-attention mechanism by enabling the model to attend to information from different representation subspaces at different positions. Instead of performing a single attention function, the multi-head attention mechanism projects the queries, keys, and values into multiple lower-dimensional spaces and performs the attention function in parallel. By attending to different parts of the sequence in parallel, the multi-head attention mechanism allows the model to capture a richer set of dependencies and interactions within the data.

Since the self-attention mechanism treats input sequences as sets without inherent order, positional encodings are added to the input embeddings to retain positional information, which is critical for sequence modeling tasks. These encodings use sine and cosine functions (Equation (4)) of different frequencies:

$$\begin{aligned} PE(pos,\ 2i) &= \sin\left(\frac{pos}{10{,}000^{\frac{2i}{d_{model}}}}\right) \\ PE(pos, 2i+1) &= \cos\left(\frac{pos}{10{,}000^{\frac{2i}{d_{model}}}}\right) \\ pos &= position,\ i = dimension \end{aligned} \quad (4)$$

After the self-attention layers, the output is passed through a position-wise feedforward network, which applies two linear transformations with a ReLU activation in between. This network helps in learning complex patterns by introducing non-linearity and improving the model's expressiveness. The output of the feedforward network is then added to the input of that layer through a residual connection, followed by layer normalization, ensuring stable and efficient training.

Transformers have achieved state-of-the-art results in various domains, including natural language processing, machine translation, and more recently, time-series forecasting [92–96]. Their ability to process entire sequences simultaneously rather than step by step as in RNNs or LSTMs makes them highly efficient, especially when dealing with long sequences. The Transformer model's architecture, leveraging self-attention mechanisms and positional encodings, can be a powerful tool for forecasting harmful algal blooms in Lake Erie. Its ability to capture complex temporal dependencies, handle large datasets, and offer explainable predictions makes it an invaluable asset in environmental monitoring and management. By enabling accurate and proactive HAB predictions, Transformers support efforts to protect water quality and ecosystem health, ultimately contributing to the sustainability of aquatic environments.

3.4. Specifications of Tested Models

The implementation of the models in this study was conducted using Python 3.10.12 frameworks and libraries to ensure optimal performance and accuracy in forecasting HABs in Lake Erie. Each model's training environment was tailored through grid search to find the best hyperparameters, which varied across different models. For the Lasso and Ridge regression models, the scikit-learn library in Python was utilized. During the experiments, the maximum iteration parameter was set to 100,000. The optimal alpha and tolerance values were found through grid search. Based on the experiments, for Lasso, the best solution was achieved with an alpha of 0.01 and a tolerance of 0.1, while for Ridge, the optimal alpha was 100 and the tolerance was 0.1.

The Long Short-Term Memory (LSTM), Gated Recurrent Unit (GRU), and Transformer models were developed and tested using the PyTorch 2.2.2 library. For all these models, the L1Loss function and Adam optimizer were used. During the training process, ReduceLROnPlateau and early stopping were employed. ReduceLROnPlateau is a learning rate scheduler that reduces the learning rate when a metric has stopped improving, helping to avoid plateaus during training. Early stopping is a technique used to stop training when the model's performance on a validation set stops improving, preventing overfitting. For the LSTM-based model, a linear layer was used at the beginning and end of the LSTM model, with dropout applied at the end. The training was set with 400 epochs for warm-up, 400 epochs for patience in ReduceLROnPlateau, and 1000 epochs for early stopping, with a learning rate of 0.001. The optimal solution was reached with a model consisting of two layers and a hidden dimension of 16, with a dropout rate of 0.

Similarly, for the GRU-based model, a linear layer was used at the beginning and end of the GRU, with a dropout layer at the end. The training configuration included 1000 epochs for warm-up, 100 epochs for patience, and 250 epochs for early stopping, with a learning rate of 1×10^{-5}. The optimal model had a hidden dimension of 32, four layers, and a dropout rate of 0.0. For the Transformer model, an encoder-based Transformer was used. The training setup included a learning rate of 0.001, 200 epochs for patience, 500 epochs for warm-up, and 500 epochs for early stopping. The model architecture included a linear layer as input embeddings. The Transformer used two attention heads, one layer, a model dimension (dmodel) of 128, a feedforward dimension of 64, and a dropout rate of 0.2. Positional encoding was tested but not used in the final model as it provided slightly lower performance. The results were very close, but since we used the models with the best MAE in other models, we followed the same criterion here as well.

The training process was performed on a machine equipped with NVIDIA GPUs to accelerate computation, particularly for the deep learning models. By utilizing grid search to tune hyperparameters, each model was optimized individually to achieve the best possible performance for HAB prediction. This approach acknowledges the unique requirements and strengths of each model, ensuring a fair and comprehensive comparison of their capabilities.

3.5. Model Interpretation Using XAI

Explainable artificial intelligence (XAI) aims to address one of the significant challenges in modern AI systems, particularly deep learning models: the lack of transparency and interpretability. As AI models, especially those based on deep learning, become more complex and widely used in critical applications such as environmental monitoring and public health, the ability to understand and trust the decisions made by these models becomes crucial.

In the context of predicting HABs, where decisions can significantly impact ecological management and public safety, the interpretability of models is not just a luxury but a necessity. Explainable models help stakeholders validate and trust the predictions, understand the contributing factors leading to an event, and facilitate informed decision-making. This is particularly important in environmental science, where the interactions between numerous biotic and abiotic factors are complex and often non-linear.

To interpret the effects of water quality parameters on the predicted chlorophyll-a concentrations via the deep learning models, SHAP (SHapley Additive exPlanations) was utilized. SHAP is a post hoc, model-agnostic technique that unifies multiple interpretability methods under the concept of Shapley values [97,98]. It establishes a new explanatory model for a given black-box system by uncovering associations between feature values and the model's output [99]. This method is based on coalitional game theory and Shapley values, which represent a unique distribution of the total surplus generated by a coalition of all players in a cooperative game [100].

The GradientExplainer from the SHAP library was employed to estimate SHAP values. GradientExplainer is specifically designed for deep learning models and leverages the gradients of the model's output with respect to its inputs to compute SHAP values. These gradients help determine the importance of each input feature in predicting the model's output, providing insights into the decision-making process.

In the Results Section, feature importance is visualized using bar graphs representing the average absolute contribution of each feature, with larger values indicating higher importance. To provide both global and local interpretability, summary plots were created, showing the SHAP values of each feature for individual samples. By averaging these values over multiple predictions, it was possible to identify which features were generally important across the entire dataset. For individual predictions, the SHAP values explained why a specific outcome was reached, illustrating the contribution of each feature to that prediction. The relationship between the magnitude of a feature's value and the predicted outcomes was indicated by color coding in the summary plots, which also displayed the distribution of feature values. Positive SHAP values suggested that a feature contributed to an increase in chlorophyll-a concentration, while negative SHAP values indicated a decrease.

By incorporating this XAI technique, the study aimed to bridge the gap between model performance and interpretability, ensuring that the predictive models for HABs were not only accurate but also understandable and actionable. This approach supports proactive management strategies for HAB mitigation, contributing to the sustainability of Lake Erie's ecosystem.

3.6. Performance Metrics

To evaluate the performance of HAB prediction models in Lake Erie, this study utilizes four key metrics: R-squared (R^2), Mean Absolute Error (MAE), Mean Absolute Percentage Error (MAPE), and Normalized Root Mean Square Error (NRMSE). These metrics were selected due to their effectiveness in capturing different aspects of model performance in environmental data modeling and are widely recognized in the field. Mean Absolute Error (MAE), Equation (5), provides an intuitive and easily interpretable measure of prediction model performance. The MAE between the measured and predicted chlorophyll-a concentrations is computed without considering their direction. A lower MAE indicates a better model fit, showing that the model's predictions are closer to the true values. This metric is beneficial when comparing different models on the same dataset, as it helps identify the model with the most accurate predictions.

R^2, Equation (6), indicates how well the predictive value explains the measured value. In this work, measured and predicted chlorophyll-a concentrations were taken as the dependent and independent variables, respectively, and R^2 was determined by applying linear regression analysis. The R^2 ranges from 0 to 1; the closer the value is to 1, the better the independent variable explains the dependent variable, meaning higher prediction accuracy. Mean Absolute Percentage Error (MAPE), Equation (7), is another widely used performance metric that provides an intuitive and easily interpretable measure of prediction model performance. MAPE emphasizes relative errors and remains unaffected by overall scaling adjustments of the target variable. Smaller MAPE values indicate enhanced accuracy and stronger predictive capabilities of the model.

$$MAE = \frac{\sum_{i=1}^{n}|y_i - \hat{y}_i|}{n}, \quad (5)$$

$$R^2 = 1 - \frac{\sum_{i=1}^{n}(y_i - \hat{y}_i)^2}{\sum_{i=1}^{n}(y_i - \overline{y})^2} \quad (6)$$

$$MAPE = \frac{1}{n}\sum_{i=1}^{n}\left|\frac{y_i - \hat{y}_i}{y_i}\right| \quad (7)$$

$$NRMSE = \frac{\sqrt{\frac{\sum_{i=1}^{n}(y_i - \hat{y}_i)^2}{n}}}{\overline{y}} \quad (8)$$

$$y_i = actual\ value\ at\ time\ i,$$
$$\hat{y}_i = predicted\ value\ at\ time\ i$$
$$n = number\ of\ sample$$
$$\overline{y} = mean\ of\ actual\ values$$

The final metric used in this study is the Normalized Root Mean Square Error (NRMSE). NRMSE (Equation (8)) measures the average magnitude of the errors between the predicted and observed Chl-a concentrations, normalizing these errors against the mean of the observed data. This normalization provides a relative assessment of the model's accuracy, facilitating comparisons across different datasets. Due to the varying conditions and environmental factors in this study, NRMSE is an appropriate choice for evaluation. The metric ranges from 0 to 1, with lower values indicating better model performance by denoting smaller errors relative to the mean of the observed Chl-a concentrations. Higher NRMSE values indicate larger errors and reduced predictive accuracy, highlighting areas where the model may need improvement.

4. Results and Discussion

This section presents the findings from our comprehensive experiments aimed at developing predictive models for HABs using deep learning algorithms, with a focus on the relationship between the HAB indicator, chlorophyll-a, and various water quality parameters. The primary goal was to evaluate the predictive performance of five distinct models: Lasso, Ridge, LSTM, GRU, and Transformer. The Transformer model, with its innovative self-attention mechanism, represents a relatively recent advancement in this field and warrants a thorough examination for its potential to capture complex temporal dependencies in environmental data.

We assessed each model's ability to predict chlorophyll-a concentrations using four key metrics: R-squared, MAE, MAPE, and NRMSE. These metrics provide a comprehensive assessment of each model's predictive accuracy and effectiveness. Furthermore, we applied the explainable AI (XAI) technique, specifically SHAP value analysis, to enhance the interpretability of the models by identifying the contribution of different water quality parameters to the predictions. This combination of quantitative metrics and qualitative insights ensures a thorough evaluation of the models' capabilities in predicting HAB events and supports informed decision-making for ecological management and public safety.

We analyzed chlorophyll-a concentrations measured at seven stations from 2013 to 2020 on a yearly (Figure 2a), monthly (Figure 2b), and station (Figure 2c) scale. These plots provide insight into the temporal and spatial variability of chlorophyll-a over the study period. The median chlorophyll-a concentration appears to fluctuate slightly over the years, with the highest median values observed in the earlier years, particularly between 2013 and 2015. However, significant outliers are present in nearly all years, with the highest concentrations recorded in 2015, indicating a substantial outbreak of HABs during this year. Notable peaks also occurred in 2019. In contrast, lower chlorophyll-a concentrations were recorded in years like 2016 and 2018. Based on month scale analysis, as shown in Figure 2b, the data indicate that high occurrences of HABs may occur in the period from late spring to early fall.

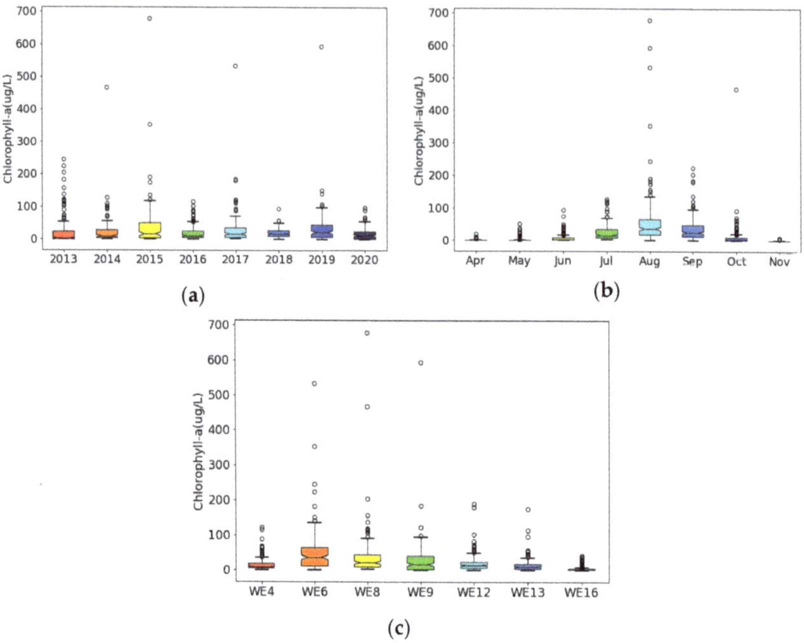

Figure 2. Box plot of chlorophyll-a concentrations in Lake Erie from 2013 to 2020: (**a**) yearly scale, (**b**) monthly scale, and (**c**) station scale.

Specifically, from May to September, chlorophyll-a concentrations were substantially increased. This period often aligns with increased rainfall, which can lead to higher nutrient runoff from surrounding areas into the lake, thereby further elevating HABs. Finally, the spatial distribution of chlorophyll-a concentrations is shown in the station scale (Figure 2c), where data from seven monitoring stations (WE4, WE6, WE8, WE9, WE12, WE13, and WE16) are compared. Stations WE6, WE8, and WE9 (closest to the Maumee River inflow) consistently exhibit higher median chlorophyll-a concentrations, suggesting that these areas may be more prone to algal blooms, possibly due to localized nutrient inputs from the river [77] compared to stations further out in the western basin of the lake (WE13 and WE16).

The results of our comprehensive experiments are summarized in Table 2. The comparative analysis of the models reveals distinct differences in their predictive performance. The Transformer model consistently outperformed the other models across all metrics, achieving the highest R^2 value and the lowest values for MAE, MAPE, and NRMSE. This indicates that the Transformer model is the most accurate and reliable for predicting HABs in Lake Erie among the tested models.

Table 2. Performance metrics for HAB prediction models.

Model	R^2	MAE	MAPE	NRMSE
Ridge	0.26	12.84	1.14	0.67
Lasso	0.38	11.56	1.00	0.61
LSTM	0.763	6.099	0.399	0.38
GRU	0.849	5.31	0.412	0.31
Transformer	0.861	5.06	0.348	0.30

To provide a comprehensive evaluation, we also implemented two traditional machine learning models, Ridge and Lasso regression, as baseline comparisons. These models are widely used in predictive modeling due to their ability to handle multicollinearity and feature selection [82,95], yet they inherently struggle with capturing complex, non-linear relationships that are characteristic of environmental systems such as HAB dynamics. As expected, the results indicate that while Ridge and Lasso regression models offer reasonable predictive capabilities, their performance is notably inferior to deep learning models, particularly in capturing temporal dependencies and intricate interactions among multiple water quality parameters.

Among the DL models, the GRU model also demonstrated strong performance, closely following the Transformer in performance. The GRU's simplified gating mechanism allows it to efficiently capture long-term dependencies, making it suitable for real-time monitoring and forecasting applications. This balance between computational efficiency and predictive power makes it a robust choice for HAB prediction.

Figure 3 presents a detailed comparison of the performance of three different models—LSTM, GRU, and Transformer—across different data subsets. The figure is organized into three rows, each dedicated to one of these models, and within each row, three scatter plots are displayed. The columns represent the different data subsets used for evaluation: the first column corresponds to the model's performance on the training data, the second column corresponds to the model's performance on the test data, and the third column corresponds to the model's performance on the combined dataset (training and test data together). In each plot, the x-axis represents the observed values, while the y-axis represents the predicted values from the model. The solid red diagonal line indicates the line of perfect agreement (1:1 line) where the model's predictions perfectly match the observed values. The blue dots (scatter points) show the actual data point, while the dotted blue line represents the best-fit linear regression line between the observed and predicted values. Across all nine plots, the alignment of the data points relative to the 1:1 line provides insight into the accuracy and bias of the models. Points that fall above the 1:1 line indicate overestimation by the model, while those below the line suggest underestimation. The closer the scatter points cluster around the perfect fit line, the better the model's predictive accuracy.

The scatter plots for the LSTM model (top row) reveal that while the LSTM can capture the overall trend of the data, there is a noticeable degree of scatter, particularly at higher concentration values for all subsets of data. This suggests that the LSTM model may struggle with predicting extreme values accurately, leading to underestimation in those cases. The GRU model (middle row) shows a slightly improved performance compared to the LSTM, with a tighter clustering of data points around the 1:1 line, particularly at moderate concentration levels and using test data.

However, similar to the LSTM, the GRU model also exhibits challenges in accurately predicting the highest concentrations, as indicated by the scatter and deviation of the regression line from the 1:1 line. The Transformer model (bottom row) demonstrates a distinct performance pattern compared to LSTM and GRU models. The scatter plots indicate that the Transformer may better handle the variability in the data, as evidenced by a closer alignment of the points with the 1:1 line, particularly at higher concentrations. This suggests that the Transformer model might be more robust in capturing the complex relationships in the dataset, leading to more accurate predictions across the full range of observed chlorophyll-a concentrations.

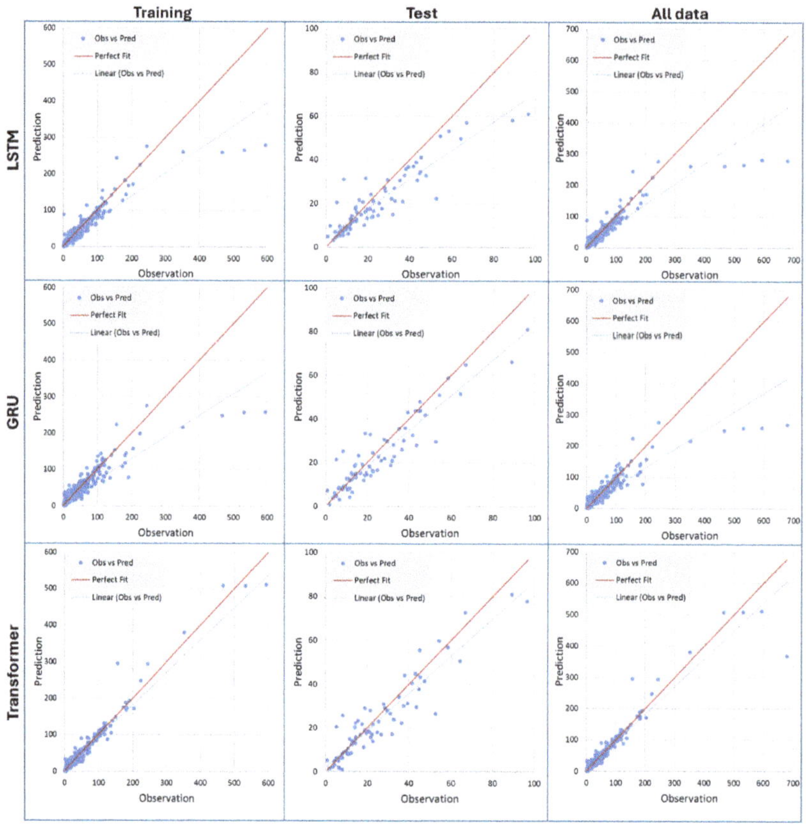

Figure 3. Scatter plots of observed chlorophyll-a concentration (x-axis) versus prediction Chl-a concentration (y-axis) for deep learning models (LSTM, GRU, and Transformer) using different data subsets (training, test, and all data).

To further evaluate the performance of the deep learning models, predictions for the peak HAB months of July, August, and September were analyzed separately as shown in Figure 4. These months are critical as they typically exhibit the highest HAB values, particularly in August [75]. In July, the GRU model demonstrated superior performance with the lowest errors and highest R-squared value. The Transformer model also showed strong results, despite having slightly higher errors compared to the GRU. The LSTM model had the highest errors, indicating lower predictive accuracy for this month.

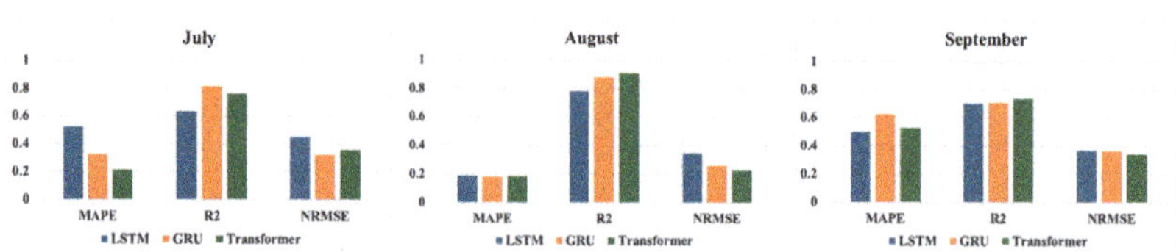

Figure 4. HAB prediction performance of tested models in peak months.

During August, when HAB values peaked, the Transformer model excelled, achieving the highest accuracy among the models. The GRU model maintained strong predictive capabilities but had somewhat higher errors in comparison to the Transformer. The LSTM model improved from its July performance but still did not match the accuracy of the Transformer and GRU. In September, the Transformer model once again led in performance, with the lowest errors and highest R-squared value. The GRU model continued to show robust predictive power, though its errors were marginally higher than those of the Transformer. The LSTM model, while better than its July performance, remained the least accurate among the deep learning models.

Overall, the monthly analysis highlights the consistent strength of the GRU and Transformer models in predicting HABs, with the Transformer model particularly excelling during the peak month of August and maintaining top performance in September. The LSTM model, although improved, generally exhibited higher errors, suggesting that the GRU and Transformer models are better suited for capturing the dynamics of HABs in Lake Erie during these critical months.

The use of SHAP analysis can enhance the interpretability of these models, providing a deeper understanding of the key water quality parameters influencing the predictions. This interpretability supports more informed ecological management decisions, highlighting the practical applicability of these advanced models in real-world environmental monitoring and decision-making. To better understand the underlying key factors that affect the models, we performed a feature importance analysis using the SHAP library.

The feature importance plot (Figure 5) for the Transformer model ranks the input variables based on their mean absolute SHAP values, highlighting the most critical factors influencing the model's predictions. Particulate Organic Nitrogen (PON), Particulate Organic Carbon (POC), and total phosphorus (TP) emerge as the top three features. The significant difference in the mean SHAP value for PON/POC compared to other factors indicates its dominant role in affecting chlorophyll-a concentrations. POC and PON are direct indicators of the availability of organic matter in aquatic ecosystems. Elevated levels of POC and PON often signal increased nutrient availability, which can fuel algal growth and lead to HABs. Biologically, PON is a critical source of nitrogen, an essential nutrient for algal metabolism [63,64]. Similarly, POC serves as a substrate for microbial activity, which can indirectly influence algal dynamics by altering the nutrient cycling within the water column [65]. These relationships align with the SHAP results, where higher POC and PON levels were associated with increased chlorophyll-a concentrations, a proxy for HAB intensity.

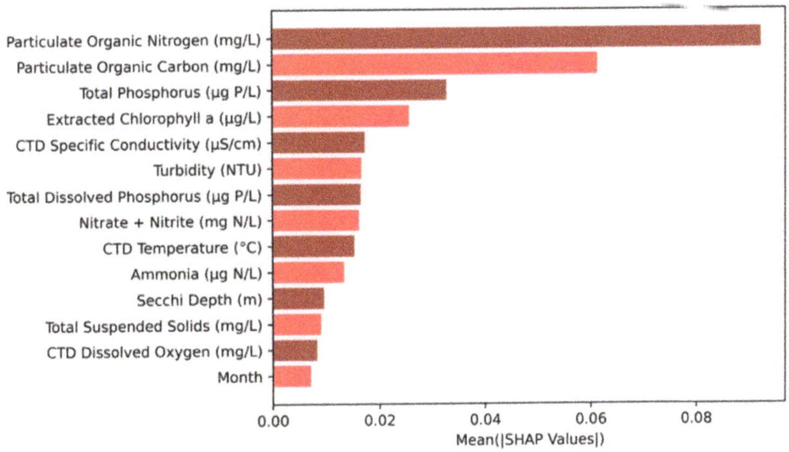

Figure 5. Feature importance analysis for Transformer model.

The SHAP value plot (Figure 6) for the Transformer model offers a more granular view, showing the individual SHAP values for each feature along with the corresponding feature values. Higher values of POC and PON, indicated by the red color, are associated with increased chlorophyll-a concentrations, demonstrating a positive correlation with harmful algal bloom occurrence. In contrast, lower values, shown in blue, are linked to decreased chlorophyll-a levels. Additionally, the SHAP analysis reveals that TP also plays a significant role in the model's predictions. High levels of TP contribute positively to chlorophyll-a concentrations, highlighting its importance as a nutrient that supports algal growth. Phosphorus is often the limiting nutrient in these ecosystems, and its presence in elevated concentrations can trigger rapid algal growth. This aligns with ecological understanding, as phosphorus is a key nutrient that often limits primary productivity in aquatic systems [101,102].

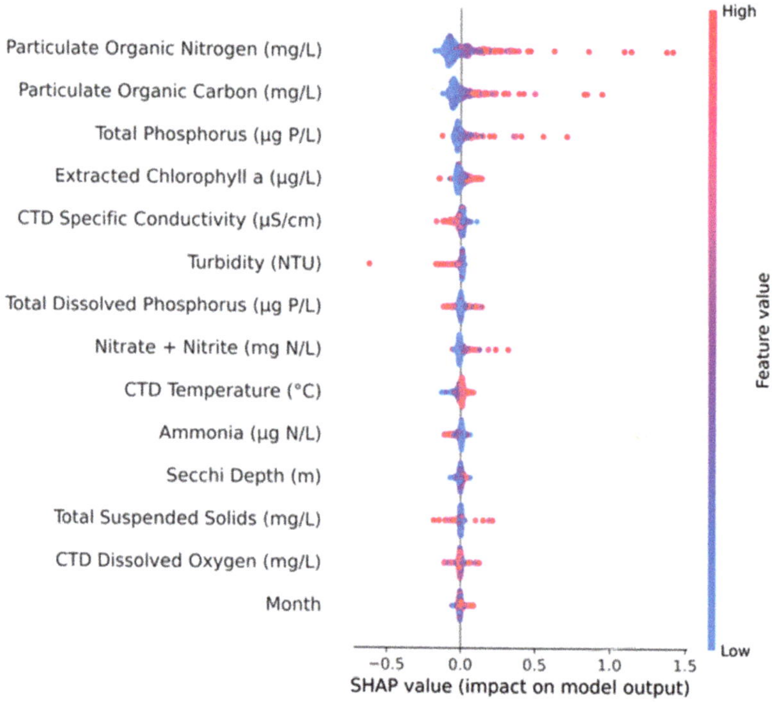

Figure 6. SHAP values of features and their impact on predictions for Transformer model.

The GRU model, similar to the Transformer model, highlighted the significance of several key water quality parameters in predicting chlorophyll-a concentrations. The SHAP analysis (Figures A1 and A2) for the GRU model indicates that POC, PON, and TP are the most influential features. These parameters consistently showed high mean absolute SHAP values, emphasizing their critical role in the model's predictions. Higher values of POC and PON were associated with increased chlorophyll-a concentrations, suggesting a positive correlation with HAB occurrence. Additionally, other features such as historical chlorophyll-a and Nitrate + Nitrite also contributed significantly to the model's output, reinforcing the importance of nutrient levels in driving algal blooms [61,103]. The GRU model's SHAP value plot demonstrated a similar pattern to the Transformer model, with higher nutrient levels pushing the predictions towards higher chlorophyll-a concentrations.

The LSTM model's SHAP analysis (Figures A3 and A4) further confirms the importance of certain water quality parameters in predicting chlorophyll-a concentrations. POC, PON, and historical chlorophyll-a emerged as the top contributing features, with consistently high mean absolute SHAP values. This finding underscores the role of particulate organic matter and phosphorus in influencing algal growth and bloom dynamics. Higher values of these features were positively correlated with increased chlorophyll-a concentrations, indicating their significant impact on HAB events. Additionally, the LSTM model highlighted the importance of other features such as TP and Nitrate + Nitrite, which also played a crucial role in the predictions. The SHAP value plot for the LSTM model revealed that higher levels of these nutrients consistently led to higher predicted chlorophyll-a concentrations, mirroring the patterns observed in both the Transformer and GRU models. This consistency across models reinforces the importance of these key parameters in the context of HAB forecasting.

The Transformer model distinctly outperforms its counterparts—LSTM, GRU, Ridge, and Lasso—in predicting HABs in Lake Erie. The comprehensive analyses, encapsulated in tables and figures provided throughout this section, consistently highlight the superior performance of the Transformer model. Notably, its ability to achieve higher R^2 values, along with lower MAE, MAPE, and NRMSE, underscores its robustness and precision in forecasting chlorophyll-a concentrations.

SHAP values play a crucial role in interpreting the predictions of deep learning models, making the decision-making process more transparent and actionable. By quantifying the contribution of each input variable to the model's output, SHAP values help identify the most influential factors driving HAB occurrences. In our study, SHAP analysis revealed that POC, PON, and TP are the primary drivers of chlorophyll-a concentrations, reinforcing their significance in HAB formation. This insight allows environmental managers to prioritize monitoring and mitigation efforts, focusing on controlling these key parameters to reduce bloom risks. Furthermore, SHAP values enhance trust in model predictions by explaining how different water quality parameters influence HAB forecasts. By providing both global feature importance and individual prediction explanations, SHAP analysis ensures that stakeholders can understand and validate model outputs. This interpretability is essential for developing targeted management strategies, optimizing water quality interventions, and informing policy decisions.

Our findings have significant ecological and policy implications, particularly for the management and mitigation of HABs in freshwater systems. The identification of key drivers reinforces the necessity of targeted nutrient management strategies to control bloom formation. These insights can inform watershed management policies by guiding nutrient reduction initiatives and improving regulatory frameworks aimed at limiting agricultural runoff and wastewater discharge. From a broader ecological perspective, the ability to accurately predict HAB occurrences enhances proactive ecosystem management, reducing the impact of toxic blooms on aquatic biodiversity and fisheries. By integrating deep learning models into early warning systems, water resource managers can implement timely mitigation strategies, thereby safeguarding drinking water supplies and recreational water bodies. These findings support data-driven decision-making and emphasize the need for interdisciplinary collaboration between researchers, policymakers, and environmental agencies to address the growing challenges posed by HABs in the context of climate change and anthropogenic influences.

5. Conclusions

In this study, we proposed a comprehensive analysis of deep learning methods for predicting chlorophyll-a concentrations, representing a significant shift from traditional methods by developing a model capable of providing integrated predictions across multiple monitoring stations rather than focusing on single-location predictions or satellite imagery. Our approach utilized comprehensive datasets encompassing physical, chemical, and biological water quality parameters from multiple monitoring stations in the western basin of Lake Erie for eight years (2013–2020). We have explored the effectiveness of advanced deep learning models, particularly the Transformer model, for predicting chlorophyll-a concentrations within the western basin of Lake Erie. The analysis emphasized the importance of an explainable AI approach to enhance model transparency and decision-making for ecological management.

Our comprehensive analysis, utilizing four key metrics—R^2, MAE, MAPE, and NRMSE—demonstrates the superior capability of the Transformer model in HAB prediction. The Transformer model outperformed the LSTM, GRU, Ridge, and Lasso models, achieving higher R^2 and lower MAE, MAPE, and NRMSE scores, thereby underscoring its precision and reliability in forecasting chlorophyll-a concentrations. Moreover, the SHAP analysis provided a reasonable interpretation of how each variable affects the HAB prediction. By using SHAP values, the effect of each variable on chlorophyll-a prediction could be interpreted more specifically. Our analysis revealed key water quality parameters significantly influencing the HAB formation, such as Particulate Organic Carbon (POC), Particulate Organic Nitrogen (PON), and total phosphorus (TP).

Our analysis also included a correlation between increases in these factors and rising chlorophyll-a concentration, which served as an indicator for HABs, highlighting a potential link that warrants further investigation This consistency underscores the importance of these parameters in the context of HAB prediction and supports more informed decision-making in ecological management and public safety. This study not only highlights the efficacy of the Transformer model in HAB prediction but also contributes significantly to the field of environmental modeling by showing the potential of advanced deep learning techniques to enhance prediction accuracy and interpretability. The integration of explainable AI methodologies ensures that the model's predictions are transparent and actionable, facilitating better management strategies for mitigating the impacts of HABs.

The early detection of harmful algal blooms (HABs) using deep learning models, particularly the Transformer model, has significant practical implications for water resource management and ecological sustainability. Accurate forecasting enables proactive decision-making, allowing water quality managers and environmental agencies to implement timely intervention strategies such as adjusting water treatment processes, issuing public health advisories, and reducing nutrient loads. This can mitigate the adverse effects of HABs on drinking water supplies, fisheries, and recreational activities. Furthermore, precise predictions improve resource allocation by identifying high-risk areas, allowing targeted monitoring and mitigation efforts, thereby reducing costs and inefficiencies. The insights from SHAP analysis highlight key drivers such as POC, PON, and TP, which can inform policy decisions on watershed management, agricultural runoff control, and regulatory frameworks for nutrient load reduction. By integrating these predictive capabilities into existing water quality management frameworks, this study shifts HAB management from reactive to proactive, enhancing public health protection and ecosystem sustainability.

While the overall prediction performance provided valuable insights, it is important to acknowledge certain limitations of the DL model in fully capturing the complexities of these relationships. For example, while SHAP analysis identifies the relative importance of these variables for model predictions, it does not explicitly account for the interactions between POC, PON, TP, and other environmental factors, such as light availability or hydrodynamic conditions. These interactions can play significant roles in regulating HAB formation and growth, highlighting the need for further exploration. The model was constructed using only a limited number of stations owing to the missing data values and the training period being relatively short. While this study focuses on validating the model using a temporally and spatially stratified dataset from the western basin of Lake Erie, future research will apply the DL models, especially the Transformer model, to independent datasets from additional locations and time periods to further strengthen the generalizability of our approach in different environmental conditions. Further studies should develop benchmark datasets that include cyanobacteria-specific indicators, such as phycocyanin concentrations or direct cyanobacteria cell counts, and extend the scope of the studies by preserving the data collected daily through continuous data management. Benchmark datasets are crucial for ensuring the reproducibility of research and facilitating the improvement of modeling studies by providing a standardized basis for comparison and validation [104,105].

Furthermore, parameters including meteorological data such as air temperature, wind speed, precipitation, and solar irradiance and hydrodynamic data such as water level and flow rate can be added to properly perform training to improve prediction performance. In addition, further development and refinement of the model's architecture and training process could also enhance its predictive capabilities. Such advancements will be crucial for improving environmental monitoring and management strategies, ultimately contributing to the sustainability of aquatic ecosystems in the face of evolving environmental challenges.

Author Contributions: B.Z.D.: conceptualization, methodology, software, data curation, visualization, and writing—original draft preparation; O.M.: conceptualization, methodology, validation, formal analysis, investigation, writing—review and editing, and visualization; Ö.B.: conceptualization, data curation, and writing—review and editing; I.D.: writing—review and editing, validation, supervision, project administration, and funding acquisition. All authors have read and agreed to the published version of the manuscript.

Funding: This research received no external funding.

Data Availability Statement: The data used in this study are available in the NOAA National Centers for Environmental Information (NCEI) data repository at https://doi.org/10.25921/11da-3x54 (accessed on 15 September 2023). Detailed descriptions of these data were also presented in Boegehold et al. [77].

Acknowledgments: O.M. would like to give special thanks to the Next Generation Internet Transatlantic Fellowship Program for their generous support and funding, which has been instrumental in the completion of this work. This research was supported in part by an appointment to the NRC Research Associateship Program at the National Oceanic and Atmospheric Administration—Global System Laboratory (NOAA-GSL), administered by the Fellowships Office of the National Academies of Sciences, Engineering, and Medicine. The statements, findings, conclusions, and recommendations expressed in this work are solely those of the author and do not necessarily reflect the views of NOAA or the U.S. Department of Commerce.

Conflicts of Interest: The authors declare no conflicts of interest.

Appendix A

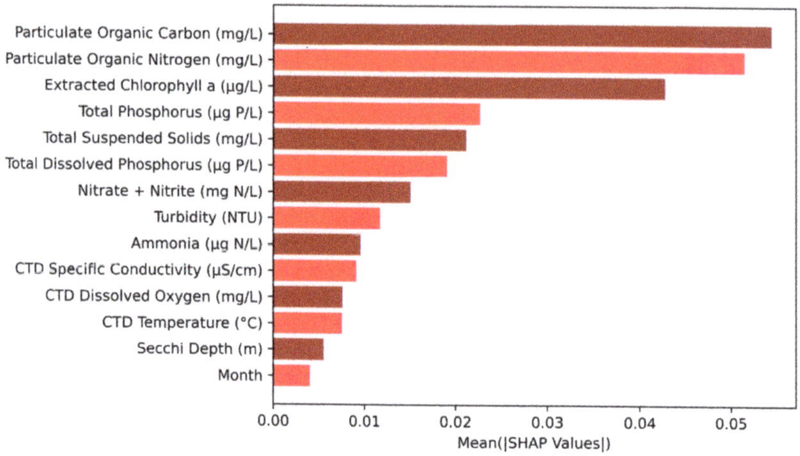

Figure A1. Feature importance analysis for GRU model.

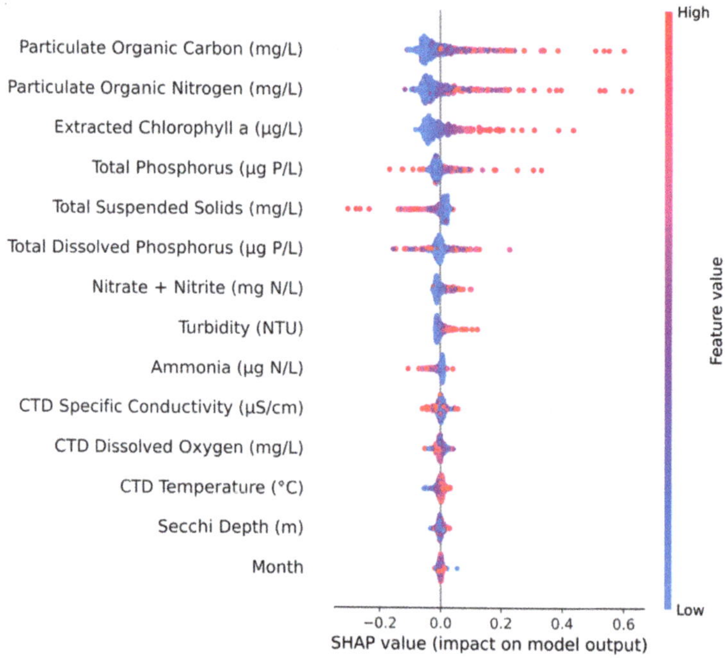

Figure A2. SHAP values of features and their impact on predictions for GRU model.

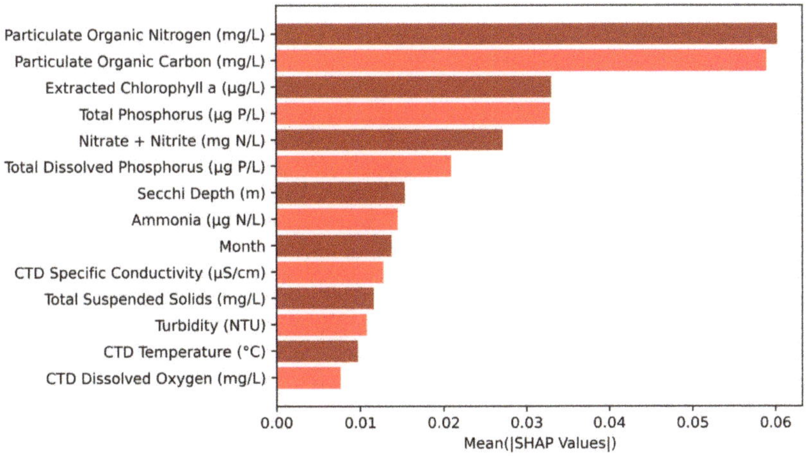

Figure A3. Feature importance analysis for LSTM model.

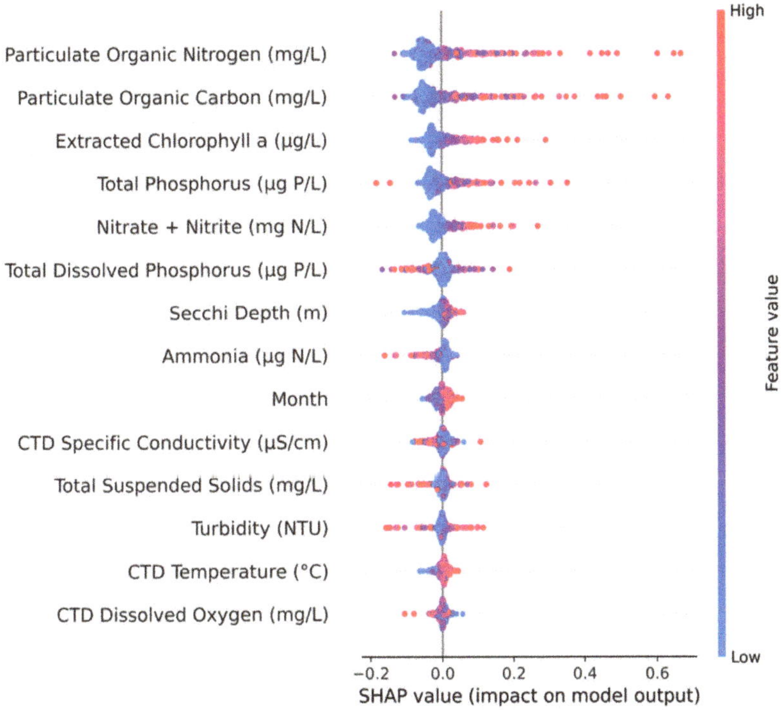

Figure A4. SHAP values of features and their impact on predictions for LSTM model.

References

1. Demir, I.; Beck, M.B. GWIS: A prototype information system for Georgia watersheds. In *Proceedings of the Georgia Water Resources Conference: Regional Water Management Opportunities*; UGA: Athens, GA, USA, 2009.
2. Paerl, H.W.; Paul, V.J. Climate change: Links to global expansion of harmful cyanobacteria. *Water Res.* **2012**, *46*, 1349–1363. [CrossRef] [PubMed]
3. Graham, J.L.; Dubrovsky, N.M.; Eberts, S.M. *Cyanobacterial Harmful Algal Blooms and US Geological Survey Science Capabilities*; US Department of the Interior, US Geological Survey: Reston, VA, USA, 2016; pp. 1–42.

4. Weirich, C.A.; Miller, T.R. Freshwater harmful algal blooms: Toxins and children's health. *Curr. Probl. Pediatr. Adolesc. Health Care* **2014**, *44*, 2–24. [CrossRef]
5. Yeşilköy, S.; Demir, I. Crop yield prediction based on reanalysis and crop phenology data in the agroclimatic zones. *Theor. Appl. Climatol.* **2024**, *155*, 7035–7048. [CrossRef]
6. Tanir, T.; Yildirim, E.; Ferreira, C.M.; Demir, I. Social vulnerability and climate risk assessment for agricultural communities in the United States. *Sci. Total Environ.* **2024**, *908*, 168346. [CrossRef]
7. Greene, S.B.D.; LeFevre, G.H.; Markfort, C.D. Improving the spatial and temporal monitoring of cyanotoxins in Iowa lakes using a multiscale and multi-modal monitoring approach. *Sci. Total Environ.* **2021**, *760*, 143327. [CrossRef] [PubMed]
8. Paerl, H.W.; Gardner, W.S.; Havens, K.E.; Joyner, A.R.; McCarthy, M.J.; Newell, S.E.; Scott, J.T. Mitigating cyanobacterial harmful algal blooms in aquatic ecosystems impacted by climate change and anthropogenic nutrients. *Harmful Algae* **2016**, *54*, 213–222. [CrossRef]
9. Ratté-Fortin, C.; Plante, J.F.; Rousseau, A.N.; Chokmani, K. Parametric versus nonparametric machine learning modelling for conditional density estimation of natural events: Application to harmful algal blooms. *Ecol. Model.* **2023**, *482*, 110415. [CrossRef]
10. Yan, Z.; Kamanmalek, S.; Alamdari, N. Predicting coastal harmful algal blooms using integrated data-driven analysis of environmental factors. *Sci. Total Environ.* **2024**, *912*, 169253. [CrossRef]
11. Boyer, J.N.; Kelble, C.R.; Ortner, P.B.; Rudnick, D.T. Phytoplankton bloom status: Chlorophyll a biomass as an indicator of water quality condition in the southern estuaries of Florida, USA. *Ecol. Indic.* **2009**, *9*, S56–S67.
12. Mellios, N.K.; Moe, S.J.; Laspidou, C. Using Bayesian hierarchical modelling to capture cyanobacteria dynamics in Northern European lakes. *Water Res.* **2020**, *186*, 116356. [CrossRef]
13. Wells, M.L.; Trainer, V.L.; Smayda, T.J.; Karlson, B.S.; Trick, C.G.; Kudela, R.M.; Cochlan, W.P. Harmful algal blooms and climate change: Learning from the past and present to forecast the future. *Harmful Algae* **2015**, *49*, 68–93. [CrossRef] [PubMed]
14. Glibert, P.M. Harmful algae at the complex nexus of eutrophication and climate change. *Harmful Algae* **2020**, *91*, 101583. [CrossRef] [PubMed]
15. Zhou, Z.X.; Yu, R.C.; Zhou, M.J. Evolution of harmful algal blooms in the East China Sea under eutrophication and warming scenarios. *Water Res.* **2022**, *221*, 118807. [CrossRef]
16. Maze, G.; Olascoaga, M.J.; Brand, L. Historical analysis of environmental conditions during Florida Red Tide. *Harmful Algae* **2015**, *50*, 1–7. [CrossRef]
17. Paerl, H.W.; Hall, N.S.; Calandrino, E.S. Controlling harmful cyanobacterial blooms in a world experiencing anthropogenic and climatic-induced change. *Sci. Total Environ.* **2011**, *409*, 1739–1745. [CrossRef]
18. Wells, M.L.; Karlson, B.; Wulff, A.; Kudela, R.; Trick, C.; Asnaghi, V.; Trainer, V.L. Future HAB science: Directions and challenges in a changing climate. *Harmful Algae* **2020**, *91*, 101632. [CrossRef]
19. Katin, A.; Del Giudice, D.; Hall, N.S.; Paerl, H.W.; Obenour, D.R. Simulating algal dynamics within a Bayesian framework to evaluate controls on estuary productivity. *Ecol. Model.* **2021**, *447*, 109497. [CrossRef]
20. Giere, J.; Riley, D.; Nowling, R.J.; McComack, J.; Sander, H. An investigation on machine-learning models for the prediction of cyanobacteria growth. *Fundam. Appl. Limnol.* **2020**, *194*, 85–94. [CrossRef]
21. Greer, B.; McNamee, S.E.; Boots, B.; Cimarelli, L.; Guillebault, D.; Helmi, K.; Campbell, K. A validated UPLC–MS/MS method for the surveillance of ten aquatic biotoxins in European brackish and freshwater systems. *Harmful Algae* **2016**, *55*, 31–40. [CrossRef] [PubMed]
22. Lombard, F.; Boss, E.; Waite, A.M.; Vogt, M.; Uitz, J.; Stemmann, L.; Appeltans, W. Globally consistent quantitative observations of planktonic ecosystems. *Front. Mar. Sci.* **2019**, *6*, 196. [CrossRef]
23. Cheng, K.H.; Chan, S.N.; Lee, J.H. Remote sensing of coastal algal blooms using unmanned aerial vehicles (UAVs). *Mar. Pollut. Bull.* **2020**, *152*, 110889. [CrossRef] [PubMed]
24. Kislik, C.; Dronova, I.; Grantham, T.E.; Kelly, M. Mapping algal bloom dynamics in small reservoirs using Sentinel-2 imagery in Google Earth Engine. *Ecol. Indic.* **2022**, *140*, 109041. [CrossRef]
25. Qiu, Y.; Liu, H.; Liu, F.; Li, D.; Liu, C.; Liu, W.; Duan, H. Development of a collaborative framework for quantitative monitoring and accumulation prediction of harmful algal blooms in nearshore areas of lakes. *Ecol. Indic.* **2023**, *156*, 111154. [CrossRef]
26. Rolim, S.B.A.; Veettil, B.K.; Vieiro, A.P.; Kessler, A.B.; Gonzatti, C. Remote sensing for mapping algal blooms in freshwater lakes: A review. *Environ. Sci. Pollut. Res.* **2023**, *30*, 19602–19616. [CrossRef]
27. Beck, M.B.; Jiang, F.; Shi, F.; Walker, R.V.; Osidele, O.O.; Lin, Z.; Demir, I.; Hall, J.W. Re-engineering cities as forces for good in the environment. In *Proceedings of the Institution of Civil Engineers—Engineering Sustainability*; Thomas Telford Ltd.: London, UK, 2010; Volume 163, pp. 31–46.
28. Shin, C.M.; Kim, D.; Song, Y. Analysis of hydraulic characteristics of Yeongsan River and estuary using EFDC model. *J. Korean Soc. Water Environ.* **2019**, *35*, 580–588.

29. Verhamme, E.M.; Redder, T.M.; Schlea, D.A.; Grush, J.; Bratton, J.F.; DePinto, J.V. Development of the Western Lake Erie Ecosystem Model (WLEEM): Application to connect phosphorus loads to cyanobacteria biomass. *J. Great Lakes Res.* **2016**, *42*, 1193–1205. [CrossRef]
30. Wynne, T.T.; Stumpf, R.P.; Tomlinson, M.C.; Fahnenstiel, G.L.; Dyble, J.; Schwab, D.J.; Joshi, S.J. Evolution of a cyanobacterial bloom forecast system in western Lake Erie: Development and initial evaluation. *J. Great Lakes Res.* **2013**, *39*, 90–99. [CrossRef]
31. Baek, S.S.; Kwon, Y.S.; Pyo, J.; Choi, J.; Kim, Y.O.; Cho, K.H. Identification of influencing factors of A. catenella bloom using machine learning and numerical simulation. *Harmful Algae* **2021**, *103*, 102007. [CrossRef]
32. Bui, H.H.; Ha, N.H.; Nguyen, T.N.D.; Nguyen, A.T.; Pham, T.T.H.; Kandasamy, J.; Nguyen, T.V. Integration of SWAT and QUAL2K for water quality modeling in a data scarce basin of Cau River basin in Vietnam. *Ecohydrol. Hydrobiol.* **2019**, *19*, 210–223. [CrossRef]
33. Zheng, L.; Wang, H.; Liu, C.; Zhang, S.; Ding, A.; Xie, E.; Wang, S. Prediction of harmful algal blooms in large water bodies using the combined EFDC and LSTM models. *J. Environ. Manag.* **2021**, *295*, 113060. [CrossRef]
34. Wool, T.; Ambrose Jr, R.B.; Martin, J.L.; Comer, A. WASP 8: The next generation in the 50-year evolution of USEPA's water quality model. *Water* **2020**, *12*, 1398. [CrossRef]
35. Liu, S.T.; Zhang, L. Surface Chaotic Theory and the Growth of Harmful Algal Bloom. In *Surface Chaos and Its Applications*; Springer: Singapore, 2022; pp. 299–320.
36. Baydaroğlu, Ö. Harmful algal bloom prediction using empirical dynamic modeling. *Sci. Total Environ.* **2025**, *959*, 178185. [CrossRef] [PubMed]
37. Franks, P.J. Recent advances in modelling of harmful algal blooms. In *Global Ecology and Oceanography of Harmful Algal Blooms*; Springer: Dordrecht, The Netherlands, 2018; pp. 359–377.
38. Janssen, A.B.; Janse, J.H.; Beusen, A.H.; Chang, M.; Harrison, J.A.; Huttunen, I.; Mooij, W.M. How to model algal blooms in any lake on earth. *Curr. Opin. Environ. Sustain.* **2019**, *36*, 1–10. [CrossRef]
39. Sit, M.; Seo, B.C.; Demiray, B.; Demir, I. EfficientRainNet: Leveraging EfficientNetV2 for memory-efficient rainfall nowcasting. *Environ. Model. Softw.* **2024**, *176*, 106001. [CrossRef]
40. Baydaroğlu, Ö.; Demir, I. Temporal and spatial satellite data augmentation for deep learning-based rainfall nowcasting. *J. Hydroinform.* **2024**, *26*, 589–607. [CrossRef]
41. Demiray, B.Z.; Sit, M.; Demir, I. EfficientTempNet: Temporal Super-Resolution of Radar Rainfall. *arXiv* **2023**, arXiv:2303.05552.
42. Tounsi, A.; Abdelkader, M.; Temimi, M. Assessing the simulation of streamflow with the LSTM model across the continental United States using the MOPEX dataset. *Neural Comput. Appl.* **2023**, *35*, 22469–22486. [CrossRef]
43. Wang, P.; Yao, J.; Wang, G.; Hao, F.; Shrestha, S.; Xue, B.; Peng, Y. Exploring the application of artificial intelligence technology for identification of water pollution characteristics and tracing the source of water quality pollutants. *Sci. Total Environ.* **2019**, *693*, 133440. [CrossRef] [PubMed]
44. Bayar, S.; Demir, I.; Engin, G.O. Modeling leaching behavior of solidified wastes using back-propagation neural networks. *Ecotoxicol. Environ. Saf.* **2009**, *72*, 843–850. [CrossRef]
45. Tao, Y.; Ren, J.; Zhu, H.; Li, J.; Cui, H. Exploring Spatiotemporal Patterns of Algal Cell Density in Lake Dianchi with Explainable Machine Learning. *Environ. Pollut.* **2024**, *356*, 124395 [CrossRef] [PubMed]
46. Yan, Z.; Kamanmalek, S.; Alamdari, N.; Nikoo, M.R. Comprehensive Insights into Harmful Algal Blooms: A Review of Chemical, Physical, Biological, and Climatological Influencers with Predictive Modeling Approaches. *J. Environ. Eng.* **2024**, *150*, 03124002. [CrossRef]
47. Rostam, N.A.P.; Malim, N.H.A.H.; Azmee, N.A.; Figueiredo, R.J.; Osman, M.A.; Abdullah, R. Harmful algal blooms (HAB) open issues: A review of ecological data challenges, factor analysis, and prediction approaches using data-driven methods. *Comput. Artif. Intell.* **2023**, *1*, 100. [CrossRef]
48. Busari, I.; Sahoo, D.; Harmel, R.D.; Haggard, B.E. Prediction of Chlorophyll-a as an Index of Harmful Algal Blooms Using Machine Learning Models. *J. Nat. Resour. Agric. Ecosyst.* **2024**, *2*, 53–61. [CrossRef]
49. Ai, H.; Zhang, K.; Sun, J.; Zhang, H. Short-term Lake Erie algal bloom prediction by classification and regression models. *Water Res.* **2023**, *232*, 119710. [CrossRef]
50. Lin, S.; Pierson, D.C.; Mesman, J.P. Prediction of algal blooms via data-driven machine learning models: An evaluation using data from a well-monitored mesotrophic lake. *Geosci. Model Dev.* **2023**, *16*, 35–46. [CrossRef]
51. Jeong, B.; Chapeta, M.R.; Kim, M.; Kim, J.; Shin, J.; Cha, Y. Machine learning-based prediction of harmful algal blooms in water supply reservoirs. *Water Qual. Res. J.* **2022**, *57*, 304–318. [CrossRef]
52. Izadi, M.; Sultan, M.; Kadiri, R.E.; Ghannadi, A.; Abdelmohsen, K. A remote sensing and machine learning-based approach to forecast the onset of harmful algal bloom. *Remote Sens.* **2021**, *13*, 3863. [CrossRef]
53. Han, J.W.; Kim, T.; Lee, S.; Kang, T.; Im, J.K. Machine learning and explainable AI for chlorophyll-a prediction in Namhan River Watershed, South Korea. *Ecol. Indic.* **2024**, *166*, 112361. [CrossRef]
54. Yu, P.; Gao, R.; Zhang, D.; Liu, Z.P. Predicting coastal algal blooms with environmental factors by machine learning methods. *Ecol. Indic.* **2021**, *123*, 107334. [CrossRef]

55. Busari, I.; Sahoo, D.; Harmel, R.D.; Haggard, B.E. A Review of Machine Learning Models for Harmful Algal Bloom Monitoring in Freshwater Systems. *J. Nat. Resour. Agric. Ecosyst.* **2023**, *1*, 63–76. [CrossRef]
56. Demiray, B.Z.; Sit, M.; Demir, I. DEM super-resolution with EfficientNetV2. *arXiv* **2021**, arXiv:2109.09661.
57. Sit, M.; Demiray, B.; Demir, I. Short-term hourly streamflow prediction with graph convolutional GRU networks. *arXiv* **2021**, arXiv:2107.07039.
58. Lee, S.; Lee, D. Improved prediction of harmful algal blooms in four major South Korea's rivers using deep learning models. *Int. J. Environ. Res. Public Health* **2018**, *15*, 1322. [CrossRef] [PubMed]
59. Hu, Z.; Zhang, Y.; Zhao, Y.; Xie, M.; Zhong, J.; Tu, Z.; Liu, J. A water quality prediction method based on the deep LSTM network considering correlation in smart mariculture. *Sensors* **2019**, *19*, 1420. [CrossRef]
60. Marndi, A.; Patra, G.K. Chlorophyll prediction using ensemble deep learning technique. In *Progress in Computing, Analytics and Networking: Proceedings of ICCAN 2019*; Springer: Singapore, 2020; pp. 341–349.
61. Lee, D.; Kim, M.; Lee, B.; Chae, S.; Kwon, S.; Kang, S. Integrated explainable deep learning prediction of harmful algal blooms. *Technol. Forecast. Soc. Change* **2022**, *185*, 122046. [CrossRef]
62. Qian, J.; Pu, N.; Qian, L.; Xue, X.; Bi, Y.; Norra, S. Identification of driving factors of algal growth in the South-to-North Water Diversion Project by Transformer-based deep learning. *Water Biol. Secur.* **2023**, *2*, 100184. [CrossRef]
63. Du, Y.; An, S.; He, H.; Wen, S.; Xing, P.; Duan, H. Production and transformation of organic matter driven by algal blooms in a shallow lake: Role of sediments. *Water Res.* **2022**, *219*, 118560. [CrossRef] [PubMed]
64. Zhou, Z.X.; Yu, R.C.; Zhou, M.J. Resolving the complex relationship between harmful algal blooms and environmental factors in the coastal waters adjacent to the Changjiang River estuary. *Harmful Algae* **2017**, *62*, 60–72. [CrossRef]
65. Wang, G.; Zhou, W.; Cao, W.; Yin, J.; Yang, Y.; Sun, Z.; Zhao, J. Variation of particulate organic carbon and its relationship with bio-optical properties during a phytoplankton bloom in the Pearl River estuary. *Mar. Pollut. Bull.* **2011**, *62*, 1939–1947. [CrossRef] [PubMed]
66. Humbert, J.F.; Fastner, J. Ecology of cyanobacteria. In *Handbook of Cyanobacterial Monitoring and Cyanotoxin Analysis*; CRC Press: Boca Raton, FL, USA, 2016; pp. 9–18.
67. Hartshorn, N.; Marimon, Z.; Xuan, Z.; Cormier, J.; Chang, N.B.; Wanielista, M. Complex interactions among nutrients, chlorophyll-a, and microcystins in three stormwater wet detention basins with floating treatment wetlands. *Chemosphere* **2016**, *144*, 408–419. [CrossRef] [PubMed]
68. Hollister, J.W.; Kreakie, B.J. Associations between chlorophyll-a and various microcystin health advisory concentrations. *F1000Research* **2016**, *5*, 151. [PubMed]
69. Castelvecchi, D. Can we open the black box of AI? *Nature* **2016**, *538*, 20. [CrossRef]
70. Saeed, W.; Omlin, C. Explainable AI (XAI): A systematic meta-survey of current challenges and future opportunities. *Knowl.-Based Syst.* **2023**, *263*, 110273. [CrossRef]
71. Goodman, B.; Flaxman, S. European Union regulations on algorithmic decision-making and a "right to explanation". *AI Mag.* **2017**, *38*, 50–57. [CrossRef]
72. Arrieta, A.B.; Díaz-Rodríguez, N.; Del Ser, J.; Bennetot, A.; Tabik, S.; Barbado, A.; Herrera, F. Explainable Artificial Intelligence (XAI): Concepts, taxonomies, opportunities and challenges toward responsible AI. *Inf. Fusion* **2020**, *58*, 82–115. [CrossRef]
73. Langer, M.; Oster, D.; Speith, T.; Hermanns, H.; Kästner, L.; Schmidt, E.; Baum, K. What do we want from Explainable Artificial Intelligence (XAI)? A stakeholder perspective on XAI and a conceptual model guiding interdisciplinary XAI research. *Artif. Intell.* **2021**, *296*, 103473. [CrossRef]
74. Magnuson, J.J.; Webster, K.E.; Assel, R.A.; Bowser, C.J.; Dillon, P.J.; Eaton, J.G.; Quinn, F.H. Potential effects of climate changes on aquatic systems: Laurentian Great Lakes and Precambrian Shield Region. *Hydrol. Process.* **1997**, *11*, 825–871. [CrossRef]
75. Tewari, M.; Kishtawal, C.M.; Moriarty, V.W.; Ray, P.; Singh, T.; Zhang, L.; Tewari, K. Improved seasonal prediction of harmful algal blooms in Lake Erie using large-scale climate indices. *Commun. Earth Environ.* **2022**, *3*, 195. [CrossRef]
76. Sterner, R.W.; Keeler, B.; Polasky, S.; Poudel, R.; Rhude, K.; Rogers, M. Ecosystem services of Earth's largest freshwater lakes. *Ecosyst. Serv.* **2020**, *41*, 101046. [CrossRef]
77. Boegehold, A.G.; Burtner, A.M.; Camilleri, A.C.; Carter, G.; DenUyl, P.; Fanslow, D.; Fyffe Semenyuk, D.; Godwin, C.M.; Gossiaux, D.; Johengen, T.H.; et al. Routine monitoring of western Lake Erie to track water quality changes associated with cyanobacterial harmful algal blooms. *Earth Syst. Sci. Data Discuss.* **2023**, *15*, 3853–3868. [CrossRef]
78. Stumpf, R.P.; Johnson, L.T.; Wynne, T.T.; Baker, D.B. Forecasting annual cyanobacterial bloom biomass to inform management decisions in Lake Erie. *J. Great Lakes Res.* **2016**, *42*, 1174–1183. [CrossRef]
79. Carmichael, W.W.; Boyer, G.L. Health impacts from cyanobacteria harmful algae blooms: Implications for the North American Great Lakes. *Harmful Algae* **2016**, *54*, 194–212. [CrossRef] [PubMed]
80. Buratti, F.M.; Manganelli, M.; Vichi, S.; Stefanelli, M.; Scardala, S.; Testai, E.; Funari, E. Cyanotoxins: Producing organisms, occurrence, toxicity, mechanism of action and human health toxicological risk evaluation. *Arch. Toxicol.* **2017**, *91*, 1049–1130. [CrossRef]

81. Kouakou, C.R.; Poder, T.G. Economic impact of harmful algal blooms on human health: A systematic review. *J. Water Health* **2019**, *17*, 499–516. [CrossRef] [PubMed]
82. Demiray, B.Z.; Sit, M.; Mermer, O.; Demir, I. Enhancing hydrological modeling with transformers: A case study for 24-h streamflow prediction. *Water Sci. Technol.* **2024**, *89*, 2326–2341. [CrossRef] [PubMed]
83. Tibshirani, R. Regression shrinkage and selection via the lasso. *J. R. Stat. Soc. Ser. B (Methodol.)* **1996**, *58*, 267–288. [CrossRef]
84. Hoerl, A.E.; Kennard, R.W. Ridge regression: Biased estimation for nonorthogonal problems. *Technometrics* **1970**, *12*, 55–67. [CrossRef]
85. Hochreiter, S.; Schmidhuber, J. Long short-term memory. *Neural Comput.* **1997**, *9*, 1735–1780. [CrossRef]
86. Cho, K.; Van Merriënboer, B.; Gulcehre, C.; Bahdanau, D.; Bougares, F.; Schwenk, H.; Bengio, Y. Learning phrase representations using RNN encoder-decoder for statistical machine translation. *arXiv* **2014**, arXiv:1406.1078.
87. Sit, M.; Demiray, B.Z.; Demir, I. A systematic review of deep learning applications in streamflow data augmentation and forecasting. *EarthArXiv* **2022**. [CrossRef]
88. Kratzert, F.; Klotz, D.; Brenner, C.; Schulz, K.; Herrnegger, M. Rainfall–runoff modelling using long short-term memory (LSTM) networks. *Hydrol. Earth Syst. Sci.* **2018**, *22*, 6005–6022. [CrossRef]
89. Zhang, Z.; Zhang, S.; Chen, C.; Yuan, J. A systematic survey of air quality prediction based on deep learning. *Alexandria Eng. J.* **2024**, *93*, 128–141. [CrossRef]
90. Liu, P.; Wang, J.; Sangaiah, A.K.; Xie, Y.; Yin, X. Analysis and prediction of water quality using LSTM deep neural networks in IoT environment. *Sustainability* **2019**, *11*, 2058. [CrossRef]
91. Vaswani, A.; Shazeer, N.; Parmar, N.; Uszkoreit, J.; Jones, L.; Gomez, A.N.; Polosukhin, I. Attention is all you need. In *Advances in Neural Information Processing Systems*; NIPS Foundation: La Jolla, CA, USA, 2017; Volume 30.
92. Hittawe, M.M.; Harrou, F.; Togou, M.A.; Sun, Y.; Knio, O. Time-series weather prediction in the Red sea using ensemble transformers. *Appl. Soft Comput.* **2024**, *164*, 111926. [CrossRef]
93. Wu, H.; Xu, J.; Wang, J.; Long, M. Autoformer: Decomposition transformers with auto-correlation for long-term series forecasting. *Adv. Neural Inf. Process. Syst.* **2021**, *34*, 22419–22430.
94. Zhou, T.; Ma, Z.; Wen, Q.; Wang, X.; Sun, L.; Jin, R. Fedformer: Frequency Enhanced Decomposed Transformer for Long-Term Series Forecasting. In Proceedings of the International Conference on Machine Learning, Baltimore, MD, USA; 2022; Volume 162, pp. 27268–27286.
95. Demiray, B.Z.; Demir, I. Towards Generalized Hydrological Forecasting using Transformer Models for 120-Hour Streamflow Prediction. *arXiv* **2024**, arXiv:2406.07484.
96. Lin, T.; Wang, Y.; Liu, X.; Qiu, X. A survey of transformers. *AI Open* **2022**, *3*, 111–132. [CrossRef]
97. Lundberg, S.M.; Erion, G.; Chen, H.; DeGrave, A.; Prutkin, J.M.; Nair, B.; Lee, S.I. From local explanations to global understanding with explainable AI for trees. *Nat. Mach. Intell.* **2020**, *2*, 56–67. [CrossRef]
98. Stubblefield, J.; Hervert, M.; Causey, J.L.; Qualls, J.A.; Dong, W.; Cai, L.; Huang, X. Transfer learning with chest X-rays for ER patient classification. *Sci. Rep.* **2020**, *10*, 20900. [CrossRef]
99. Burkart, N.; Huber, M.F. A survey on the explainability of supervised machine learning. *J. Artif. Intell. Res.* **2021**, *70*, 245–317. [CrossRef]
100. Lundberg, S.M.; Lee, S.I. A unified approach to interpreting model predictions. In *Advances in Neural Information Processing Systems*; NIPS Foundation: La Jolla, CA, USA, 2017; Volume 30.
101. Havens, K.E. Phosphorus–algal bloom relationships in large lakes of south Florida: Implications for establishing nutrient criteria. *Lake Reserv. Manag.* **2003**, *19*, 222–228. [CrossRef]
102. Zhang, X.; Li, Y.; Zhao, J.; Wang, Y.; Liu, H.; Liu, Q. Temporal dynamics of the Chlorophyll a–Total phosphorus relationship and algal production efficiency: Drivers and management implications. *Ecol. Indic.* **2024**, *158*, 111339. [CrossRef]
103. Rosales, D.; Ellett, A.; Jacobs, J.; Ozbay, G.; Parveen, S.; Pitula, J. Investigating the relationship between nitrate, total dissolved nitrogen, and phosphate with abundance of pathogenic Vibrios and harmful algal blooms in Rehoboth Bay, Delaware. *Appl. Environ. Microbiol.* **2022**, *88*, e00356-22. [CrossRef] [PubMed]
104. Demir, I.; Xiang, Z.; Demiray, B.; Sit, M. Waterbench: A large-scale benchmark dataset for data-driven streamflow forecasting. *Earth Syst. Sci. Data Discuss.* **2022**, *14*, 5605–5616. [CrossRef]
105. Sit, M.; Seo, B.C.; Demir, I. Iowarain: A statewide rain event dataset based on weather radars and quantitative precipitation estimation. *arXiv* **2021**, arXiv:2107.03432.

water

MDPI

Article

The Effects of Data Quality on Deep Learning Performance for Aquatic Insect Identification: Advances for Biomonitoring Studies

Predrag Simović [1], Aleksandar Milosavljević [2], Katarina Stojanović [3], Dimitrija Savić-Zdravković [4], Ana Petrović [1], Bratislav Predić [2] and Djuradj Milošević [4,*]

1 Department of Biology and Ecology, Faculty of Science, University of Kragujevac, Radoja Domanovića 12, 34000 Kragujevac, Serbia; predrag.simovic@pmf.kg.ac.rs (P.S.); ana.petrovic@pmf.kg.ac.rs (A.P.)
2 Faculty of Electronic Engineering, University of Niš, Aleksandra Medvedeva 14, 18000 Niš, Serbia; aleksandar.milosavljevic@elfak.ni.ac.rs (A.M.); bratislav.predic@elfak.ni.ac.rs (B.P.)
3 Department of Zoology, Faculty of Biology, University of Belgrade, Studentski trg 16, 11000 Belgrade, Serbia; k.bjelanovic@bio.bg.ac.rs
4 Department of Biology and Ecology, Faculty of Sciences and Mathematics, University of Niš, Višegradska 33, 18000 Niš, Serbia; dimitrija.savic@pmf.edu.rs
* Correspondence: djuradj.milosevic@pmf.edu.rs

Abstract: Deep learning models, known as convolutional neural networks (CNNs), have paved the way for reliable automated image recognition. These models are increasingly being applied in research on freshwater biodiversity, aiming to enhance efficiency and taxonomic resolution in biomonitoring. However, insufficient or imbalanced datasets remain a significant bottleneck for creating high-precision classifiers. The highly imbalanced data, where some species are rare and others are common, are typical of the composition of most benthic communities. In this study, a series of CNN models was built using 33 species of aquatic insects, with datasets ranging from 10 to 80 individuals, to determine the optimal number of individuals each class should have to build a high-precision classifier. We also consider the effect of class imbalance in the training dataset and the use of oversampling technique. The results showed that a robust model with acceptable accuracy (99.45%) was achieved with at least 30 individuals per class. A strongly imbalanced dataset caused an approximately 2% decrease in classification accuracy, while a moderately imbalanced dataset had no significant effect. The application of the oversampling technique enhanced in 1.88% the accuracy of strongly imbalanced models. These findings can help effectively tailor future aquatic macroinvertebrate training datasets.

Keywords: convolutional neural network; imbalance dataset; data size; oversampling technique; Ephemeroptera; Plecoptera; Trichoptera; EPT taxa

1. Introduction

Biomonitoring with high taxonomic resolution is fundamental for understanding ecosystem dynamics, especially in an era where freshwater ecosystems are facing increasing pressures from climate change and anthropogenic stressors [1]. While conventional methods based on physical and chemical parameters provide information on current environmental conditions and pollutant levels, biological monitoring offers a profound understanding of ecosystem health and functioning over time [2]. The biomonitoring approach uses biological indicators to assess the quality of habitats and ecosystems, offering valuable insights into the impacts of human activities and natural changes [3]. In freshwater ecosystems research, the larvae of insects from the orders Ephemeroptera, Plecoptera, and Trichoptera (collectively referred to as EPT) are considered the best bioindicators of water quality because they typically dominate benthic fauna in terms of both abundance and species richness [4,5].

Ecologists have most often relied on traditional identification approaches based on morphological features, which often result in data with low or unverifiable taxonomic precision [6]. Furthermore, this approach is quite time and labor intensive and requires specialized knowledge and experience [6]. Additionally, challenges in identification are also attributed to phenotypic plasticity, the existence of cryptic species, and variations in size and sexual dimorphism [7,8]. Therefore, the challenges associated with the traditional identification of these larvae highlight the need for more reliable and efficient ways of processing benthic samples [1].

In recent years, there has been an exponential increase in the number of scientific publications combining the terms "biodiversity" and "deep learning" [9,10]. Special attention has been paid to the type of deep learning models known as convolutional neural networks (CNNs), which have proven effective in domains such as image classification and recognition [11,12]. A Convolutional Neural Network (CNN) analyzes images in segments, learning by identifying approximate feature correspondences at corresponding locations between paired images [13]. The main benefit of deep learning CNN architecture is its ability to self-learn and self-organize without the need for manual supervision [14].

Many authors have demonstrated the application of deep learning for the automatic identification of aquatic macroinvertebrates, both in laboratory settings [15–17] and under field conditions [18], which is vital for monitoring species populations. If widely adopted, automatic classification techniques could drastically reduce the costs of water quality analysis in aquatic ecosystems, while also decreasing the time and effort required by experts [2]. However, several technical challenges currently hinder the widespread application of computer vision in biomonitoring programs, biodiversity, and conservation studies. To achieve reliable predictions, CNN models often require many samples and the creation of balanced datasets for training [19,20]. The task of acquiring an extensive array of datasets covering relevant biomonitoring taxa is often hindered by the localized and uneven distribution of some species. This can lead to the 'imbalanced dataset problem', where some classes, the minority groups, contain significantly fewer examples than the other classes, the majority groups [21,22]. Therefore, typically rare and endangered species tend to perform poorly in automatic species identification due to limited training data [17]. Addressing this issue by excluding or misclassifying rare species can directly impact the calculation of multimetric indexes used to assess ecosystem status, potentially leading to inaccurate evaluations [23]. At the same time, CNNs tend to exhibit bias towards the majority class [19], i.e., common species that have a disproportionately large number of training data, typically those that are cosmopolitan and conspicuous [24]. More closely related species are harder to separate in classification models [17,25], and a severe problem may arise if one of these species is represented in lower numbers during the training of CNN.

Despite the increasing application of deep learning and computer vision in biomonitoring [16,17,26], very little empirical work in this field examines the effects of sample size and dataset imbalance on the performance of species classification models [25,27]. The objectives of this study are as follows: (1) to examine how classification performance changes with increasing training sample sizes, using a balanced dataset; (2) to test the influence of imbalanced datasets on the accuracy of CNN models, using a varying number of image samples for certain target species; and (3) to evaluate the effectiveness of oversampling techniques applied to minority class samples during training in improving class balance. This study is significant as it proposes a new design of standards for reference image datasets, enabling cost-effective training of more accurate algorithms for the automatic identification of indicator taxa.

2. Materials and Methods

2.1. Dataset Acquisition

Figure 1 provides an overview of the steps taken to construct the training datasets for the CNN models. The larvae of EPT species were collected alongside other macroinvertebrates as part of hydrobiological studies in aquatic ecosystems within Serbia, during the development of a freshwater biomonitoring program for Serbian surface waters. The sampling sites included various lotic and lentic habitats from the Middle Danube Basins [28]. The larvae were collected from different substrate types using a Surber sampler and preserved in 96% ethanol. They were transported and stored at the Department of Biology and Ecology, Faculty of Science, University of Kragujevac, Republic of Serbia. After separating the collected samples from extraneous materials, taxonomic identification was performed. The material was identified at the species level using relevant identification literature [29–32] by experts for this group of insects. Morphological identification of bioindicators is especially challenging for immature and early life stages [17]. To address this, our study focused on late instar larvae, which have well-developed and clearly visible morphological traits that enable accurate identification. In total, we gathered 100 individuals for each of the 33 target species (Table 1; Figure 2). Each individual was positioned in a dorsal view and photographed using either a Nikon SMZ800 stereomicroscope (Nikon Corporation, Tokyo, Japan) equipped with a Leica Flexacam C3 microscope camera (Leica Microsystems, Wetzlar, Germany) or a Nikon D900 camera (Nikon Corporation, Tokyo, Japan). According to the previous study [17] the dorsal view of these larvae provides morphological characteristics that are sufficiently informative for reliable identification. We completed our training dataset with a total of 3330 photos (one image per individual).

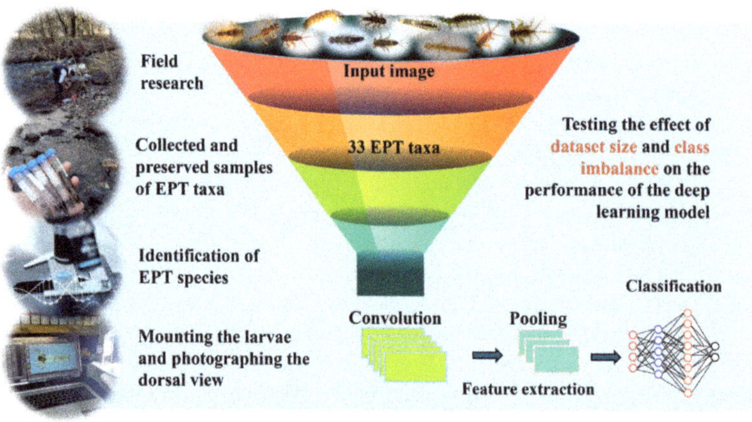

Figure 1. Schematic overview of research methodology.

Table 1. The EPT species included in the study.

Code	Order	Family	Species
1	Ephemeroptera	Baetidae	*Baetis alpinus* (Pictet, 1843)
2	Ephemeroptera	Baetidae	*Baetis rhodani* (Pictet, 1843)
3	Ephemeroptera	Caenidae	*Caenis macrura* (Stephens, 1835)
4	Ephemeroptera	Ephemeridae	*Ephemera danica* (Müller, 1764)
5	Ephemeroptera	Ephemerellidae	*Ephemerella mucronata* (Bengtsson, 1909)
6	Ephemeroptera	Ephemerellidae	*Ephemerella (Serratella) ignita* (Poda, 1761)
7	Ephemeroptera	Ephemerellidae	*Torleya major* (Klapálek, 1905)

Table 1. *Cont.*

Code	Order	Family	Species
8	Ephemeroptera	Heptagenidae	*Rhithrogena semicolorata* (Curtis, 1834)
9	Ephemeroptera	Heptagenidae	*Epeorus assimilis* (Eaton, 1865)
10	Ephemeroptera	Leptophlebidae	*Habrophlebia fusca* (Curtis, 1834)
11	Ephemeroptera	Oligoneuriidae	*Oligoneuriella rhenana* (Imhoff, 1852)
12	Ephemeroptera	Potamantidae	*Potamanthus luteus* (Linneaus, 1767)
13	Ephemeroptera	Siphlonuridae	*Siphlonurus aestivalis* (Eaton, 1903)
14	Plecoptera	Leuctridae	*Leuctra hippopus* Kempny, 1899
15	Plecoptera	Nemouridae	*Protonemura hrabei* Raušer, 1956
16	Plecoptera	Perlidae	*Perla palida* Guérin-Méneville, 1838
17	Plecoptera	Perlodidae	*Isoperla tripartita* Illies, 1954
18	Trichoptera	Brachycentridae	*Oligoplectrum maculatum* (Fourcroy, 1785)
19	Trichoptera	Brachycentridae	*Lepidostoma basale* (Kolenati, 1848)
20	Trichoptera	Goeridae	*Lithax obscurus* (Hagen, 1859)
21	Trichoptera	Goeridae	*Silo pallipes* Fabricius, 1781
22	Trichoptera	Hydropsychidae	*Hydropsyche incognita* Pitshc, 1993
23	Trichoptera	Hydropsychidae	*Cheumatopsyche lepida* Pictet, 1834
24	Trichoptera	Limnephilidae	*Limnephilus rhombicus* Linnaeus, 1758
25	Trichoptera	Limnephilidae	*Potamophylax luctuosus* (Piller & Mitterpacher, 1783)
26	Trichoptera	Limnephilidae	*Halesus digitatus* (von Paula Schrank, 1781)
27	Trichoptera	Psychomiidae	*Tinodes unicolor* (Pictet, 1834)
28	Trichoptera	Psychomiidae	*Psychomyia pusilla* (Fabricius, 1781)
29	Trichoptera	Philopotamidae	*Philopotamus montanus* (Donovan, 1813)
30	Trichoptera	Rhyacophilidae	*Rhyacophila fasciata* Hagen, 1859
31	Trichoptera	Rhyacophilidae	*Rhyacophila tristis* Pictet, 1834
32	Trichoptera	Sericostomatidae	*Sericostoma flavicorne* Schneider, 1845
33	Trichoptera	Ueonidae	*Thremma anomalum* McLachlan, 1876

Figure 2. Dorsal view of larvae from the 33 EPT species included in the training dataset (species codes are detailed in Table 1).

2.2. Model Development

We trained a series of CNN models based on EfficientNetB2 architecture using different training image datasets [33]. A comprehensive description of the model training is provided in the study by Simović et al. [17]. For model validation and testing, 20% (10% for validation and 10% for testing) from each species were consistently used, while the training percentages varied across models.

To assess how the number of images (individuals) affects classification accuracy, we trained a series of eight CNN models with three different configurations. In the first configuration we varied the number of individuals in the training set from 10 to 80. In the second configuration, eight subsets were randomly created to form a series of smaller training sets, allowing us to investigate the effects of dataset imbalance on the performance of species classification models. For all eight models, one species group contained 80 individuals, while the other species had progressively smaller sample sizes of 60, 40, 20, or 10 individuals, following the pattern outlined in Figure 3. In the third configuration, we used the same unbalanced dataset and applied the oversampling technique, which duplicates random examples from the minority class to balance the dataset.

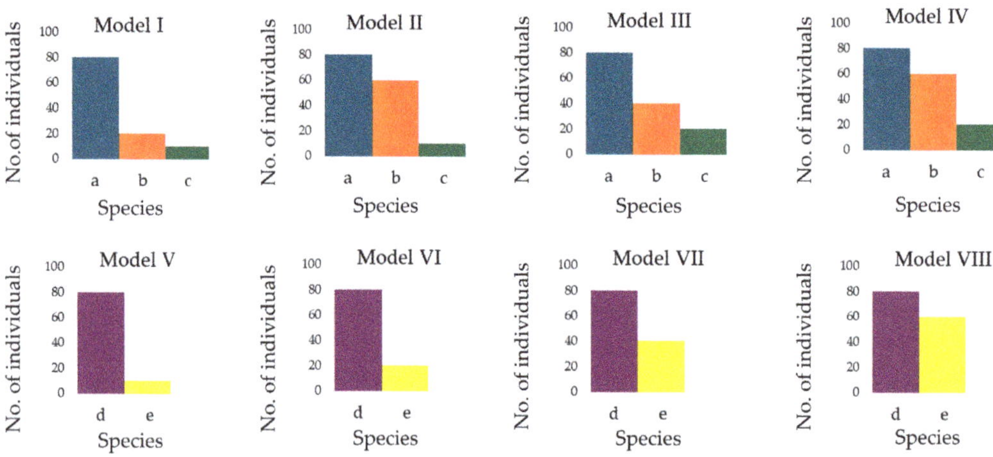

Species codes according to Table 1.
a - 1, 2, 4, 5, 11, 12, 15, 17, 22, 24, 26, 28, 33
b - 3, 9, 10, 13, 14, 18, 19, 21, 23, 31
c - 6, 7, 8, 16, 20, 25, 27, 29, 30, 32
d - 2, 3, 4, 5, 8, 9, 10, 11, 12, 13, 14, 15, 16, 17, 18, 19, 21, 22, 24, 26, 31, 33
e - 1, 6, 7, 20, 23, 25, 27, 28, 29, 30, 32

Figure 3. Sample sizes and composition of training and validation data in eight imbalanced datasets.

The minimum number of training individuals used for certain species was set at 10, as preliminary analysis indicated that sample sizes below this threshold led to challenges in cross-validation parameter tuning due to the limited representation within the class. Since model training starts with random weights, to statistically test our hypothesis, we trained five models for each experimental setup to obtain mean accuracy and standard deviation. The differences in classification accuracy of the CNN models were evaluated using a one-way Analysis of Variance (ANOVA) to assess the effects of sample size and degree of imbalance. Post hoc pairwise comparisons were conducted using Tukey's Honest Significant Difference (HSD) test to identify significant differences between groups. To evaluate the effect of the oversampling technique on classification accuracy, an independent samples *t*-test was conducted for each of the eight models. Prior to the ANOVA and *t*-test, the data were tested for normality and homogeneity of variances to ensure the validity of the analysis. All statistical analyses were performed using R software (accessed on 10 December 2024) [34].

A confusion matrix (CM) was generated for each dataset size and class imbalance in the deep learning model to reveal the true and incorrect categories of predictions. A CM is a graphical representation that visualizes the class-wise distribution of a classification model's predictive performance, showing how the errors are distributed [35]. Python scripts used to train the model and produce visualizations are open-sourced on the project's GitHub page (https://github.com/a-milosavljevic/aiaquami-ept33, accessed on 12 December 2024). The training results used in this research are available for download in the Releases section of the GitHub project (https://github.com/a-milosavljevic/aiaquami-ept33/releases, accessed on 12 December 2024).

3. Results

3.1. The Effect of Varied Sample Size

The classification accuracy of the CNN model across five training repetitions with varying sample sizes per class (ranging from 10 to 80) significantly varied (ANOVA, $F = 49{,}149$, $p < 0.001$) (Figure 4). The model achieved a remarkable mean classification accuracy of 97.09% (range: 96.36–97.58%) when trained on only 10 individuals per each of the 33 EPT species. This accuracy was significantly improved to 99.45% with larger training dataset sizes of 30 individuals per species (Tukey's HSD, $p < 0.01$). The highest accuracy was 99.70% (range: 99.39–100%) when 50 individuals per species were used for training. The same highest average accuracy of 99.70% was also confirmed with 70 individuals per species used for training, but it decreased to 99.58% when the number of individuals increased to 80. However, the accuracy percentage remained high, so these oscillations can be considered negligible (Figure 4). In addition, low standard deviation values (Std.) indicate that the results are consistent across different tests, which means that the model is stable and reliable.

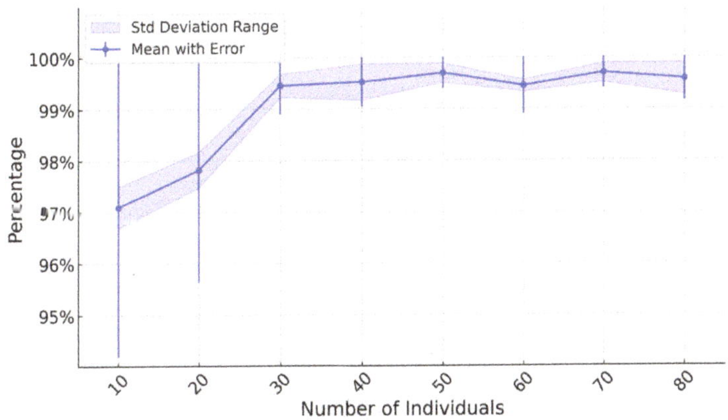

Figure 4. The variability in CNN model accuracy with increasing training set sizes (ranging from 10 to 80 individuals per species) for classifying 33 EPT species.

The deep CNN model produced confusion matrices displaying the accuracy percentage for each species, with classification errors for models with balanced datasets (Appendix A). When the CNN model was trained on 10 individuals per species, classification errors ranged from 10% to 30% across 11 species (Appendix A). However, as the training dataset size increased, error rates for individual species decreased substantially. With a test dataset of 40 individuals per species, classification errors for individual species did not exceed 10%. Only the species *Sericostoma flavicorne* and *Torleya major* showed classification errors of 10% when tested with 50 individuals per species. Misclassified individuals were identified as *Limnephilus rhombicus* and *Caenis macrura*, respectively. Similar misclassification patterns were observed in models tested with 60, 70, and 80 individuals of each species. Additionally, in some cases, *Ephemerella mucronata* showed a classification error, with 10% of individuals being misclassified as *Ephemerella ignita* (Appendix A).

3.2. The Effect of Imbalanced Databases

The performance of the CNN model on the test set was influenced by the class imbalance levels in the training sets, resulting in significant changes in classification accuracy (ANOVA, F = 86.816, $p < 0.001$). The model achieved the best performance when the imbalance ratio in the tested EPT species was the smallest (99.82 ± 0.41% Model VIII and 99.88 ± 0.15 Model VI) (Figure 5). In those cases, we noted low classification errors for only two species, *Thremma anomalum* and *Torleya major* (Appendix B). When the degree of imbalance among individuals used for the tested EPT species was highest (Models I and II), the model's performance was significantly the lowest (Tukey's HSD, $p < 0.01$), with an accuracy of 97.09 ± 0.41% and 97.94 ± 0.59, respectively (Figure 5). When considering individual species, the CNN models generally struggled to accurately identify species with the fewest available individuals (only 10), such as *Lithax obscurus*, *T. major*, *S. flavicorne*, and *Philopotamus montanus* (Appendix B). Species *L. obscurus* and *T. major* exhibited the highest error rates, ranging between 10% and 40%. The misclassified individuals of *L. obscurus* were assigned to the image class of the species *Silo pallipes*, while *T. major* individuals were misclassified as belonging to the species *E. mucronata* (Appendix B).

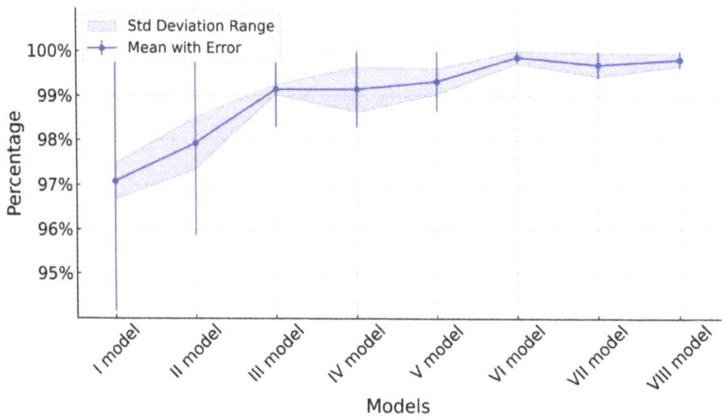

Figure 5. The variability in CNN model accuracy with increasing imbalance levels for the classification of 33 EPT species.

3.3. The Effect of Oversampling on Imbalanced Databases

The application of the oversampling technique significantly improved classification accuracy in Models I, II, and III (*t*-test, $p < 0.05$; Figure 6). Accuracy increased from 97.09% ± 0.41% to 98.97% ± 0.31% in Model I, from 97.94% ± 0.59% to 98.85% ± 0.23% in Model II, and from 99.15% ± 0.12% to 99.76% ± 0.12% in Model III (Figure 6). In contrast, Model IV showed a slight but significant decrease in accuracy, dropping from 99.88% ± 0.15% to 99.39% ± 0.27% (*t*-test, $p < 0.05$).

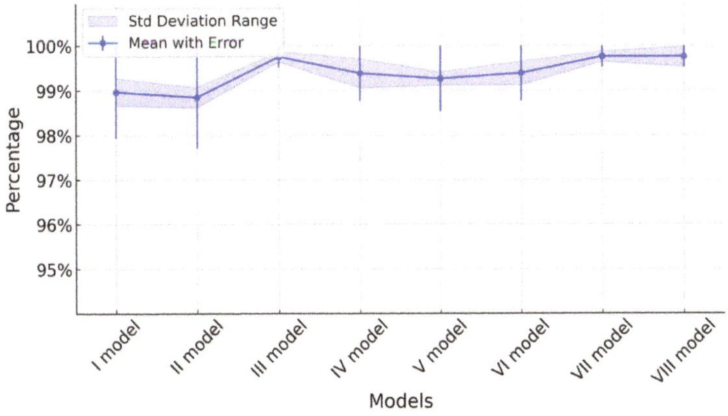

Figure 6. The variability in CNN model accuracy with increasing imbalance levels for the classification of 33 EPT with oversampling techniques applied to the minority class.

4. Discussion

Deep learning has the potential to enhance investigation processes in various ecological fields, including biomonitoring; however, a significant bottleneck in this process is the need for accurate and balanced training data [36]. Obtaining a dataset of a size sufficient for automatic classification of high accuracy is challenging, particularly when sample collection is conducted during field campaigns [37]. Establishing a sufficiently large and balanced dataset for aquatic insects requires extensive sampling along various spatial and temporal gradients. This process can be time-consuming and particularly challenging when it comes to locating and collecting species with limited geographical distribution ranges. So far, only a few authors have explored the sufficient number of images per individual required to achieve a highly accurate classifier for the automatic identification of aquatic insects [25,38]. As expected, the classification accuracy of our model gradually increased with an increasing number of individuals used for training. The accuracy of our models increased from 97.09 $\pm$ 0.41% with the smallest training dataset to 99.70 $\pm$ 0.19% with the largest training dataset, which is similar to the results reported in other studies. Høye et al. [25], demonstrated a 97% classification accuracy when the CNN model was trained on a balanced set of 10 individuals from 16 aquatic macroinvertebrate taxa and a striking 99.2% classification accuracy when trained on 50 individuals from the same 16 taxa. In this study, the model was trained on species from seven morphologically and phylogenetically distinct taxonomic orders, including mollusks and crustaceans, which makes the classification task less challenging than one in our study. Although it is expected that the similarity among taxa could make classification more challenging for CNN algorithms, our analysis of performance on a balanced dataset found no evidence that morphologically similar and closely related species within the same genus (e.g., *Baetis rhodani* and *B. alpinus*) or family (e.g., *Silo pallipes* and *Lithax obscurus* from the family Glossosomatidae) were more difficult to classify than species with greater morphological differences. CNNs typically show improved performance with increased data volume, but beyond a certain saturation point, the rate of improvement can slow and eventually diminish [39]. However, the number of data considered 'sufficient' for optimal CNN performance depends on factors such as model complexity, task requirements, and data quality [25,38]. Accordingly, our findings suggest that a sample size of 30 individuals per species was optimal for achieving high classification precision. Training with a larger sample size introduced fluctuations and even a slight decrease in model accuracy, indicating diminishing returns beyond this threshold in our case.

The frequency of species within freshwater macroinvertebrate communities is naturally unbalanced, with fewer dominant taxa and a higher number of rare ones [40]. At the same time, the minority class is usually more important from a data mining perspective, as despite its rarity, it may contain significant and useful information [41]. This imbalance is consequently reflected in the construction of training data for deep learning models. The overall performance of the CNN models after the decrease of the imbalance ratio increased from 97.09% to 99.88%. Various studies have observed that CNN models struggle to accurately identify classes with small and imbalanced training samples [22,42]. The findings of some authors indicate that sensitivity to class imbalance increases as problem complexity increases (i.e., the number of training variables and the degree of class imbalance) [22,43,44]. When we analyzed the performance on an imbalanced dataset, we found that the species with the poorest performance were those with the lowest number of images in the dataset (Models I and II, in which the smallest number of individuals for some species was 10), which is consistent with other observations [45]. We acknowledge the potential impact of class imbalance on classification accuracy within CNN models; while our findings showed a significant decrease in accuracy due to imbalanced samples, the loss was deemed acceptable and unlikely to compromise our primary objective of accurate taxa identification. In our study, we observed a reduction in accuracy, approximately 2%, due to imbalanced samples, a trend also noted by Thabtah et al. [44] across various domains, including medical diagnosis and credit card fraud detection, where minor classes contributed to a 2% accuracy drop. Similarly, in Lee et al.'s [46] study on plankton image classification, balancing the class distribution led to a 5% improvement in accuracy for minority classes. Despite this increase, the study's overall classification accuracy remained lower than ours at around 88%, reinforcing that a well-tuned model can achieve strong performance even with some degree of class imbalance. Given these findings, we believe that the current level of accuracy in our model is sufficiently high, and the minor error introduced by sample imbalance—estimated to be under 2%—is unlikely to detract from the reliability of our taxa classification outcomes. Krawczyk [41] suggests that the imbalance ratio itself may not be the primary source of learning challenges; rather, satisfactory results can be achieved if each class has sufficient representation. In experiments with artificially generated data of varying complexity and degrees of imbalance, some authors [47] demonstrate that when each class has at least 50 examples, even at high levels of concept complexity, the classification error remains below 1%, indicating that robust performance can be maintained with adequate representation across classes. Therefore, while we recognize the potential for imbalance-induced errors in CNN models, the 2% accuracy loss observed in our study indicates that taxa identification remains robust and effective, ensuring reliable classification outcomes even in the presence of a slight class imbalance.

When observing individual species, the common classification errors included misclassifying *L. obscurus* as *S. pallipes* and *T. major* as *E. mucronata*. Both cases resulted in confusion among species within the families Glossosomatidae and Ephemerellidae, respectively. This is not surprising because species that are morphologically similar and closely related (within the same genus or family) are considered more difficult to separate, as has been previously shown, not only for humans but also for deep learning models [17,25,45,48]. Although the misclassified species are closely related, we cannot conclude that morphological or phylogenetic similarity is the cause of misclassification. In fact, for some morphologically similar species within the same genera (*Baetis rhodani* and *B. alpinus*, *Rhyacophila tristis* and *R. fasciata*, *Ephemerella mucronata* and *E. ignita*), our models achieved a perfect classification rate of 100%.

Misclassification of individuals during routine species monitoring can lead to inaccurate calculations of multimetric index scores, which, according to the Water Framework Directive (WFD) regulations, are used across Europe to assess the health of aquatic ecosystems [49]. Closely related species within the same genus may exhibit different preferences for various water qualities [49]. For example, *B. rhodani* is a cosmopolitan species that inhabits waters of varying quality, while the occurrence of *B. alpinus* in some ecosystems is more indicative of high-quality waters. The presence of *S. pallipes* indicates oligotrophic waters, while *L. obscurus* is a bioindicator of oligotrophic to beta-mesosaprobic waters [50]. This emphasizes that reliable species identification is a crucial step in assessing the health of water ecosystems and recognizing areas of high conservation priority.

Synthetic data generation like data augmentation, including transformations such as rotation, flipping, scaling, and cropping, can significantly increase the number of images in a dataset for CNN training, enhancing model generalization without additional data collection [51,52]. Our study demonstrates that, with the exception of Models I and II, species in all models were represented by a sufficient number of data. Class imbalance had a significant impact on the performance of Models I and II, leading to an approximate 2% reduction in classification accuracy. By applying the oversampling technique, we successfully mitigated the errors caused by imbalanced samples, almost completely restoring the species classification accuracy to 98.97 $\pm$ 0.31% and 98.85 $\pm$ 0.23%, respectively. This technique, which duplicates random examples from the minority class to balance the dataset, is widely supported in the literature as an effective approach to improving model performance [53,54]. Therefore, the oversampling technique offers a viable solution to address the challenges associated with detecting rare taxa in automated species monitoring.

In conclusion, building a robust model with acceptable accuracy for automated species identification largely depends on having a sufficient sample size in the training dataset, with at least 30 images per class. The impact of sample imbalance on the classification accuracy of CNN models in our study was observed as a 2% reduction; however, this level remains within an acceptable range for achieving the primary goal of sufficient taxonomic resolution in routine biomonitoring of surface waters, with the adverse effects further mitigated through the application of oversampling techniques. Our findings demonstrate the feasibility of automated classification for less common species, but further research is needed to understand the effects of training convolutional neural network (CNN) models on a broader range of species, particularly with high-class imbalance and increased diversity among closely related species.

The findings of this research provide valuable insights into determining the optimal amount of information needed to develop an effective deep-learning model capable of automatically identifying species essential for biomonitoring, biodiversity, and conservation studies at high taxonomic resolution. This research supports the organization of reference datasets to improve existing models, enhance their quality, and streamline the application of deep learning techniques for accurate species identification. These approaches have the potential to tackle critical biomonitoring challenges by significantly reducing both time and effort.

Author Contributions: Conceptualization, P.S., D.M. and A.M.; Methodology, P.S., D.M. and A.M.; Software, A.M.; Validation, P.S., A.M. and D.M.; Formal Analysis, P.S.; Investigation, P.S.; Resources, P.S., K.S., A.M., D.S.-Z., B.P., A.P. and D.M.; Data Curation, P.S. and K.S.; Writing—Original Draft Preparation, P.S.; Writing—Review & Editing, P.S. and D.M.; Visualization, P.S.; Supervision, D.M.; Project Administration, D.S.-Z. All authors have read and agreed to the published version of the manuscript.

Funding: This research was funded by the Science Fund of the Republic of Serbia (Grant #7751676) for the project 'Application of Deep Learning in Bioassessment of Aquatic Ecosystems: Toward the

Construction of an Automatic Identifier for Aquatic Macroinvertebrates—AIAQUAMI'. Additional support was provided by the Ministry of Science, Technological Development and Innovation of the Republic of Serbia under Grant No. 451-03-66/2024-03/200122.

Data Availability Statement: The training results, including the images used in this research, are available for download in the Releases section of the GitHub project: https://github.com/a-milosavljevic/aiaquami-ept33/releases (accessed on 12 December 2024).

Conflicts of Interest: The authors declare no conflicts of interest. The funders were not involved in the study design, data collection, analysis, interpretation, manuscript writing, or the decision to publish the results.

Appendix A

Classification errors and their distribution among species for each model using balanced datasets of 10 to 80 individuals per EPT taxon, across five replicates.

Appendix B

Classification errors and their distribution among species for each model using imbalanced datasets, both with and without applying the oversampling technique for the minority class in the dataset.

References

1. Besson, M.; Alison, J.; Bjerge, K.; Gorochowski, T.E.; Høye, T.T.; Jucker, T.; Mann, H.M.R.; Clements, C.F. Towards the fully automated monitoring of ecological communities. *Ecol. Lett.* **2022**, *25*, 2753–2775. [CrossRef] [PubMed]
2. Riabchenko, E.; Meissner, K.; Ahmad, I.; Iosifidis, A.; Tirronen, V.; Gabbouj, M.; Kiranyaz, S. Learned vs. engineered features for fine-grained classification of aquatic macroinvertebrates. In Proceedings of the 2016 23rd International Conference on Pattern Recognition (ICPR), Cancun, Mexico, 4–8 December 2016; pp. 2276–2281. [CrossRef]
3. Batzias, F.A.; Siontorou, C.G.A. Knowledge-Based Approach to Environmental Biomonitoring. *Environ. Monit. Assess.* **2006**, *123*, 167–197. [CrossRef] [PubMed]
4. Amaral, P.H.M.d.; Silveira, L.S.d.; Rosa, B.F.J.V.; Oliveira, V.C.d.; Alves, R.d.G. Influence of Habitat and Land Use on the Assemblages of Ephemeroptera, Plecoptera, and Trichoptera in Neotropical Streams. *J. Insect Sci.* **2015**, *15*, 60. [CrossRef] [PubMed]
5. Simović, P.; Milošević, D.; Simić, V.; Stojanović, K.; Atanacković, A.; Jakovljević, A.; Petrović, A. Benthic macroinvertebrates in a tufa-depositing environment: A case study of highly vulnerable karst lotic habitats in Southeast Europe. *Hydrobiologia* **2024**, *851*, 4761–4779. [CrossRef]
6. Haase, P.S.; Pauls, K.; Schindehu, T.T.E.; Sundermann, A. First audit of macroinvertebrate samples from an EU Water Framework Directive monitoring program: Human error greatly lowers precision of assessment results. *J. N. Am. Benthol. Soc.* **2010**, *29*, 1279–1291. [CrossRef]
7. Zhou, X.; Jacobus, L.M.; DeWalt, R.E.; Adamowicz, S.J.; Hebert, P.D.N. Ephemeroptera, Plecoptera, and Trichoptera fauna of Churchill (Manitoba, Canada): Insights into biodiversity patterns from DNA barcoding. *J. N. Am. Benthol.* **2010**, *29*, 814–837. [CrossRef]
8. Suh, K.I.; Hwang, J.M.; Bae, Y.J.; Kang, J.H. Comprehensive DNA barcodes for species identification and discovery of cryptic diversity in mayfly larvae from South Korea: Implications for freshwater ecosystem biomonitoring. *Entomol. Res.* **2019**, *49*, 46–54. [CrossRef]
9. Villon, S.; Iovan, C.; Mangeas, M.; Vigliola, L. Confronting Deep-Learning and Biodiversity Challenges for Automatic Video-Monitoring. *Sensors* **2022**, *10*, 497. [CrossRef]
10. Chang, G.J. Biodiversity estimation by environment drivers using machine/deep learning for ecological management. *Ecol. Inform.* **2023**, *78*, 102319. [CrossRef]
11. LeCun, Y.; Bengio, Y.; Hinton, G. Deep learning. *Nature* **2015**, *521*, 436–444. [CrossRef]
12. Chandrakar, R.; Raja, R.; Miri, R.; Tandan, S.R.; Ramya Laxmi, K. Detection and Identification of Animals in Wildlife Sanctuaries using Convolutional Neural Network. *Int. J. Recent Technol. Eng.* **2020**, *8*, 2277–3878. [CrossRef]
13. Badre, P.; Chandanshive, P.; Bandiwadekar, S.; Chaudhari, A.; Jadhav, S. Automatically Identifying Animals Using Deep Learning. *Int. J. Recent Innov. Trends Comput. Commun.* **2018**, *6*, 194–197.
14. Rauf, H.T.; Lali, M.I.U.; Zahoor, S.; Shah, S.Z.H.; Rehman, A.U.; Bukhari, S.A.C. Visual features based automated identification of fish species using deep convolutional neural networks. *Comput. Electron. Agric.* **2019**, *167*, 105075. [CrossRef]

15. Larios, N.; Lin, J.; Zhang, M.; Moldenke, A.; Shapiro, L.; Dietterich, T. Stacked spatial-pyramid kernel: An object-class recognition method to combine scores from random trees. In Proceedings of the 2011 IEEE Workshop on Applications of Computer Vision (WACV), Kona, HI, USA, 5–7 January 2011; pp. 329–335. [CrossRef]
16. Raitoharju, J.; Riabchenko, E.; Ahmad, I.; Iosifidis, A.; Gabbouj, M.; Kiranyaz, S.; Tirronen, V.; Ärje, J.; Kärkkäinen, S.; Meissner, K. Benchmark database for fine-grained image classification of benthic macroinvertebrates. *Image Vis. Comput.* **2018**, *78*, 73–83. [CrossRef]
17. Simović, P.; Milosavljević, A.; Stojanović, K.; Radenković, M.; Predić, B.; Božanić, M.; Petrović, A.; Milošević, D. Automated identification of aquatic insects: A case study using deep learning and computer vision techniques. *Sci. Total Environ.* **2024**, *935*, 172877. [CrossRef]
18. Jaballah, S.; Fernandez Garcia, S.; Martignac, F.; Parisey, N.; Jumel, S.; Roussel, J.-M.; Dézerald, O. A deep learning approach to detect and identify live freshwater macroinvertebrates. *Aquat. Ecol.* **2023**, *57*, 933–949. [CrossRef]
19. Johnson, J.M.; Khoshgoftaar, T.M. Survey on deep learning with class imbalance. *J. Big Data* **2019**, *6*, 29. [CrossRef]
20. Ghosh, K.; Bellinger, C.; Corizzo, R.; Branco, P.; Krawczyk, B.; Japkowicz, N. The class imbalance problem in deep learning. *Mach. Learn.* **2024**, *113*, 4845–4901. [CrossRef]
21. Li, J.; Fong, S.; Mohammed, S.; Fiaidhi, J. Improving the classification performance of biological imbalanced datasets by swarm optimization algorithms. *J. Supercomput.* **2016**, *72*, 3708–3728. [CrossRef]
22. Sotiropoulos, D.N.; Tsihrintzis, G.A. The Class Imbalance Problem. In *Machine Learning Paradigms;* Springer: Cham, Switzerland, 2016; Volume 118, pp. 51–78. [CrossRef]
23. Wach, M.; Guéguen, J.; Christian, C.; Delmas, F.; Dagens, N.; Feret, T.; Loriot, S.; Tison-Rosebery, J. Probability of misclassifying river ecological status: A large-scale approach to assign uncertainty in macrophyte and diatom-based biomonitoring. *Ecol. Indic.* **2019**, *101*, 285–295. [CrossRef]
24. Høye, T.T.; Ärje, J.; Bjerge, K.; Hansen, O.L.P.; Iosifidis, A.; Leese, F.; Mann, H.M.R.; Meissner, K.; Melvad, C.; Raitoharju, J. Deep learning and computer vision will transform entomology. *Proc. Natl. Acad. Sci. USA* **2021**, *118*, e2002545117. [CrossRef] [PubMed]
25. Høye, T.T.; Dyrmann, M.; Kjær, C.; Nielsen, J.; Bruus, M.; Mielec, C.L.; Vesterdal, M.S.; Bjerge, K.; Madsen, S.A.; Jeppesen, M.R.; et al. Accurate image-based identification of macroinvertebrate specimens using deep learning—How much training data is needed? *PeerJ* **2022**, *10*, e13837. [CrossRef] [PubMed]
26. Kiranyaz, S.; Ince, T.; Pulkkinen, J.; Gabbouj, M.; Ärje, J.; Kärkkäinen, S.; Tirronen, V.; Juhola, M.; Turpeinen, T.; Meissner, K. Classification and retrieval on macroinvertebrate image databases. *Comput. Biol. Med.* **2011**, *41*, 463–472. [CrossRef] [PubMed]
27. Durden, J.M.; Hosking, B.; Bett, B.J.; Cline, D.; Ruhl, H.A. Automated classification of fauna in seabed photographs: The impact of training and validation dataset size, with considerations for the class imbalance. *Progr. Oceanogr.* **2021**, *196*, 102612. [CrossRef]
28. Simić, V.; Bănăduc, D.; Curtean-Bănăduc, A.; Petrović, A.; Veličković, T.; Stojković-Piperac, M.; Simić, S. Assessment of the ecological sustainability of river basins based on the modified theESHIPPOfish model on the example of the Velika Morava basin (Serbia, Central Balkans). *Front. Environ. Sci.* **2022**, *10*, 952692. [CrossRef]
29. Aubert, J. Plecoptera. In *Insecta Helvetica, Fauna;* Imprimerie La Concorde: Lausanne, Switzerland, 1959; Volume 1, pp. 91–136.
30. Eiseler, B. Identification key to the mayfly larvae of the German Highlands und Lowlands. *Lauterbornia* **2005**, *53*, 1–112.
31. Rozkošný, R. *Klíč Vodnich Larev Hmyzu;* Academia Nakladatelstvi Československé Akademie Véd: Praha, Czech Republic, 1959.
32. Waringer, J.; Graf, W. *Atlas der Mitteleuropäischen Köcherfiegenlarven: Atlas of Central-European Trichoptera Larvae;* Erik Mauch Verlag: Dinkelscher, Germany, 2011.
33. Tan, M.; Le, Q. EfficientNet: Rethinking Model Scaling for Convolutional Neural Networks. In Proceedings of the 36th International Conference on Machine Learning, PMLR, Long Beach, CA, USA, 9–15 June 2019; Volume 97, pp. 6105–6114.
34. R Core Team. *R: A Language and Environment for Statistical Computing;* R Foundation for Statistical Computing: Vienna, Austria, 2023. Available online: https://www.R-project.org/ (accessed on 10 December 2024).
35. Fernandes, J.A.; Irigoien, X.; Boyra, G.; Lozano, J.A.; Inza, I. Optimizing the number of classes in automated zooplankton classification. *J. Plankton Res.* **2009**, *31*, 19–29. [CrossRef]
36. Benkendorf, D.J.; Hawkins, C.P. Effects of sample size and network depth on a deep learning approach to species distribution modeling. *Ecol. Inform.* **2020**, *60*, 101137. [CrossRef]
37. Ramezan, C.A.; Warner, T.A.; Maxwell, A.E.; Price, B.S. Effects of Training Set Size on Supervised Machine-Learning Land-Cover Classification of Large-Area High-Resolution Remotely Sensed Data. *Remote Sens.* **2021**, *13*, 368. [CrossRef]
38. Ärje, J.; Melvad, C.; Jeppesen, M.R.; Madsen, S.A.; Raitoharju, J.; Meissner, K.; Rasmussen, M.S.; Iosifidis, A.; Tirronen, V.; Gabbouj, M.; et al. Automatic image-based identification and biomass estimation of invertebrates. *Methods Ecol. Evol.* **2020**, *11*, 922–931. [CrossRef]
39. Davidian, M.; Lahav, A.; Joshua, B.-Z.; Wand, O.; Lurie, Y.; Mark, S. Exploring the Interplay of Dataset Size and Imbalance on CNN Performance in Healthcare: Using X-rays to Identify COVID-19 Patients. *Diagnostics* **2024**, *14*, 1727. [CrossRef] [PubMed]
40. Magurran, A.E. Measuring biological diversity. *Curr. Biol.* **2021**, *31*, R1174–R1177. [CrossRef] [PubMed]

41. Krawczyk, B. Learning from imbalanced data: Open challenges and future directions. *Prog. Artif. Intell.* **2016**, *5*, 221–232. [CrossRef]
42. Buda, M.; Maki, A.; Mazurowski, M.A. A systematic study of the class imbalance problem in convolutional neural networks. *Neural Netw.* **2018**, *106*, 249–259. [CrossRef]
43. Japkowicz, N. The class imbalance problem: Significance and strategies. In Proceedings of the 2000 International Conference on Artificial Intelligence (ICAI), Vancouver, BC, Canada, 13–15 November 2000; pp. 111–117.
44. Thabtah, F.; Hammoud, S.; Kamalov, F.; Gonsalves, A. Data imbalance in classification: Experimental evaluation. *Inf. Sci.* **2020**, *513*, 429–441. [CrossRef]
45. Valan, M.; Makonyi, K.; Maki, A.; Vondráček, D.; Ronquist, F. Automated Taxonomic Identification of Insects with Expert-Level Accuracy Using Effective Feature Transfer from Convolutional Networks. *Syst. Biol.* **2019**, *68*, 876–895. [CrossRef]
46. Lee, H.; Park, M.; Kim, J. Plankton classification on imbalanced large-scale database via convolutional neural networks with transfer learning. In Proceedings of the 2016 IEEE International Conference on Image Processing (ICIP), Phoenix, AZ, USA, 25–28 September 2016; pp. 3713–3717. [CrossRef]
47. An, A.; Cercone, N.; Huang, X. Case Study for Learning from Imbalanced Data. In *Advances in Artificial Intelligence: 14th Biennial Conference of the Canadian Society for Computational Studies of Intelligence, AI 2001 Ottawa, Canada, June 7–9, 2001 Proceedings*; Stroulia, E., Matwin, S., Eds.; Springer: Berlin/Heidelberg, Germany, 2001. [CrossRef]
48. Larios, N.; Deng, H.; Zhang, W.; Sarpola, M.; Yuen, J.; Paasch, R.; Moldenke, A.; Lytle, D.A.; Ruiz Correa, S.; Mortensen, E.N.; et al. Automated insect identification through concatenated histograms of local appearance features: Feature vector generation and region detection for deformable objects. *Mach. Vis. Appl.* **2008**, *19*, 105–123. [CrossRef]
49. *Directive 2000/60/EC of the European Parliament and of the Council of 23 October 2000 Establishing a Framework for Community Action in the Field of Water Policy*; European Union: Brussels, Belgium, 2000; pp. 1–73.
50. Sládeček, V. System of water quality from the biological point of view. *Arch. Hydrobiol.* **1973**, *7*, 1–218.
51. Han, D.; Liu, Q.; Fan, W. A new image classification method using CNN transfer learning and web data augmentation. *Expert Syst. Appl.* **2018**, *95*, 43–56. [CrossRef]
52. Talukdar, J.; Biswas, A.; Gupta, S. Data Augmentation on Synthetic Images for Transfer Learning using Deep CNNs. In Proceedings of the 5th International Conference on Signal Processing and Integrated Networks (SPIN), Noida, India, 22–23 February 2018; pp. 215–219. [CrossRef]
53. Yang, Z.; Zhu, Y.; Liu, T.; Zhao, S.; Wang, Y.; Tao, D. Output Layer Multiplication for Class Imbalance Problem in Convolutional Neural Networks. *Neural Process. Lett.* **2020**, *52*, 2637–2653. [CrossRef]
54. Dablain, D.; Jacobson, K.N.; Bellinger, C.; Roberts, M.; Chawla, N. Understanding CNN fragility when learning with imbalanced data. *Mach. Learn.* **2024**, *113*, 4785–4810. [CrossRef]

water

MDPI

Article

Correlations Between Spatiotemporal Variations in Phytoplankton Community Structure and Physicochemical Parameters in the Seungchon and Juksan Weirs

Hyeonsu Chung, Misun Son, Taesung Kim *, Jonghwan Park and Won-Seok Lee

Yeongsan River Environment Research Center, National Institute of Environmental Research, Gwangju 61011, Republic of Korea; jys7246@korea.kr (H.C.); miza03@korea.kr (M.S.); thanks@korea.kr (J.P.); boystone@korea.kr (W.-S.L.)

* Correspondence: kimts3@korea.kr

Abstract: The Yeongsan River is one of the four major rivers in South Korea. Since the construction of two weirs as part of the Four Major Rivers Project to secure water resources in 2011, issues with algal blooms have frequently arisen, prompting the Ministry of Environment of Korea to conduct continuous monitoring of water quality and algal outbreaks. This study, conducted between 2019 and 2023, examined the relationship between the phytoplankton community structure and physicochemical factors at the Seungchon and Juksan weirs. Phytoplankton were categorized into four groups (Bacillariophyceae, Chlorophyceae, *Cyanophyceae*, and other phytoplankton), and 20 dominant genera were selected for analysis. As microalgal species vary depending on environmental conditions, understanding the specific relationships among the microalgae observed in the study area can help explain their occurrence mechanisms and contribute to the development of effective management strategies. Therefore, we used principal component analysis (PCA) to analyze the seasonal variation patterns of the four microalgal groups and visualize key data features through dimensionality reduction. Additionally, PCA was employed to identify and visualize environmental factors related to seasonal variations in phytoplankton communities. PCA helped elucidate how different environmental factors influence phytoplankton fluctuations across seasons. We used canonical correspondence analysis (CCA) to investigate the relationships among the 20 dominant genera in each group and environmental factors. Additionally, CCA was used to analyze the relationship between the distribution of the top five dominant phytoplankton taxa in each group and various environmental factors. CCA allowed for a detailed examination of how these dominant taxa interact with environmental conditions. PCA revealed significant correlations between other phytoplankton and Chl-*a* in spring and *Cyanophyceae* and water temperature in summer. Bacillariophyceae was positively correlated with nitrogen-based nutrients but negatively with phosphate phosphorus (PO_4-P). CCA revealed significant correlations between dominant genera and environmental factors. *Stephanodiscus* sp. was associated with nitrogen-based nutrients, whereas *Microcystis* sp. and *Dolichospermum* sp. were associated with water temperature and PO_4-P. *Stephanodiscus* sp. affected water treatment through filtration and sedimentation issues, whereas *Microcystis* sp. and *Dolichospermum* sp. produced the toxin microcystin. These findings offer valuable insights for water quality management.

Keywords: phytoplankton; Yeongsan River; Seungchon Weir; Juksan Weir; water quality

1. Introduction

Freshwater ecosystems provide water resources, purify pollutants, and serve as habitats for diverse biological communities [1]. However, because of agricultural activities and urbanization, several natural rivers have been transformed into artificial environments designed to serve various purposes such as preventing floods, ensuring adequate water supply, and providing a venue for aquatic recreational activities [2]. The construction of weirs, which obstruct natural river flow, extend water retention times, and transform lotic ecosystems into lentic ones, is an anthropogenic factor that significantly affects river ecosystems [3,4] because these artificial structures hinder pollutant dispersion, leading to the accumulation of soluble materials and a reduction in the self-purification capacity of rivers. Consequently, eutrophication and excessive phytoplankton growth ensue, severely affecting the aquatic environment and biota [5].

The Yeongsan River, the study area of this research, flows through southwestern South Korea and is among the four major rivers in the country, along with the Han, Geum, and Nakdong Rivers. With an area of 3371 km^2 and an annual average water availability of 3 billion m^3, which constitutes only 3.9% of the national total, the Yeongsan River basin is the smallest among South Korea's four major rivers. This relatively small watershed size renders the river vulnerable to water quality fluctuations, even from minor pollutant inputs, highlighting its sensitivity to pollution [6]. The Yeongsan River has a higher level of water pollution than the other watersheds, with frequent occurrences of water quality issues such as eutrophication and low dissolved oxygen levels [7]. Furthermore, the watershed area and runoff are small, and the channel coefficient is high because of the topographical characteristics. Moreover, the Yeongsan River is the only major river that relies solely on agricultural reservoirs for maintaining river flow, making it vulnerable to flow rate instability [8]. In the context of the Four Major Rivers Restoration Project, which was implemented in 2010–2011 to prevent disasters, ensure adequate water supply, and improve water quality, two weirs—Seungchon and Juksan—were constructed on the main channel of the Yeongsan River [9]. Consequently, water retention times have increased and flow rates have decreased because the river's dynamics have shifted to those of a lentic water body [10]. Additionally, the upstream region of the Yeongsan River, which includes the densely populated Gwangju Metropolitan City (with approximately 1.4 million residents) and its industrial complexes, has encountered challenges such as eutrophication, excessive phytoplankton proliferation, foul odor, and fish kills due to pollution discharge [11,12].

As primary producers, phytoplankton are crucial food web components in aquatic ecosystems [13]. Phytoplankton primary production serves as an indicator for evaluating ecosystem trophic status as it responds sensitively to changes in species composition and cell counts due to environmental shifts [14]. Moreover, phytoplankton communities are influenced by a range of complex physicochemical factors, including watershed characteristics, water temperature, and nutrient levels [15]. Long-term analysis of phytoplankton community structures and corresponding physicochemical factors provides essential data for assessing and monitoring the impacts of external influences on aquatic ecosystems [16,17]. Consequently, understanding the spatiotemporal distribution of phytoplankton and its relationships with environmental factors is crucial for evaluating the current state of aquatic ecosystems [18]. Despite this critical role of phytoplankton communities, most related research has been concentrated in the Yeongsan River estuary [19–21], leaving a gap in the knowledge on the relationships between the spatiotemporal dynamics of the phytoplankton community structure and environmental factors in the river's weir sections, which are heavily influenced by urbanization. In particular, studies that correlate dominant species—key shapers of community structure—with environmental factors are notably scarce. We expected the findings of this study to contribute to identifying the key environmental factors influencing phytoplankton blooms at weirs in temperate regions, including South Korea, providing baseline data for effective aquatic ecosystem management.

2. Materials and Methods

2.1. Study Area

The Yeongsan River originates from Yongchubong Peak (elevation: 560 m) in Yong-myeon, Damyang-gun, Jeollanam-do, South Korea, and stretches 136 km to its estuary. Its basin hosts several major tributaries, including the Hwangryong River and the Jiseok, Gomakwon, Hampyeong, and Gwangju Streams [22]. Between 2010 and 2011, as part of

South Korea's Four Major Rivers Restoration Project, two weirs were constructed in the middle and lower reaches of the river (Figure 1). The Seungchon Weir (SC) is situated in Seungchon-dong, Nam-gu, Gwangju Metropolitan City; it has a drainage area of 1327 km^2 and a reservoir capacity of 9,000,000 m^3 [23]. Downstream, the Juksan Weir (JS) is located in Juksan-ri, Dasi-myeon, Naju-si, Jeollanam-do; it has a drainage area of 2359 km^2 and a reservoir capacity of 25,700,000 m^3 (Table 1) [24]. Both weirs supply agricultural water.

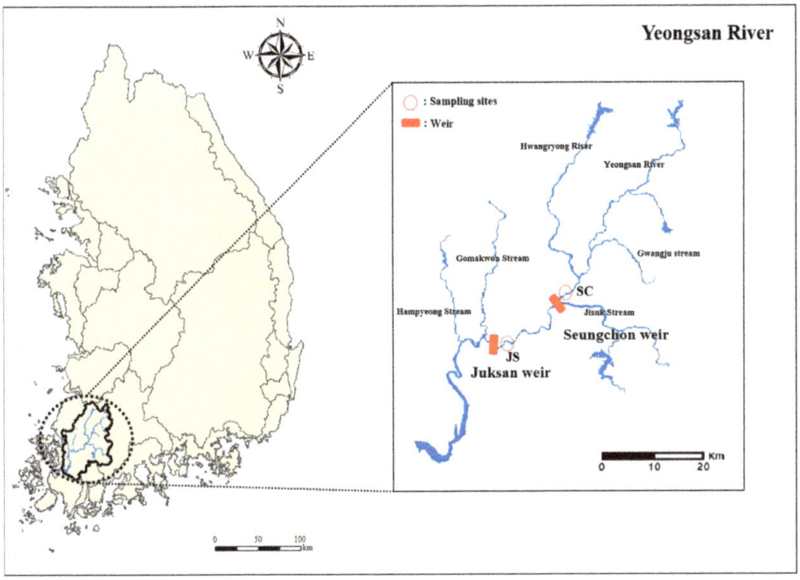

Figure 1. Sampling sites at the Seungchon Weir (SC) and Juksan Weir (JS) in the Yeongsan River.

Table 1. Physical parameters of the study sites.

Site	Elevation (m)	Capacity (10^6 m^3)	Drainage Area (km^2)	Average Depth (m)	Geographic Coordinates (Latitude/Longitude)
SC	7.5	9.0	1327.0	7.5	35°04′13.90″ N/126°46′34.09″ E
JS	3.5	25.7	2359.0	7.1	34.976675° N/126.632778° E

Note(s): SC, Seungchon Weir; JS, Juksan Weir.

2.2. Physicochemical Parameter Analysis

Surface water samples were collected by boat weekly between 2019 and 2023 at SC (n = 249) and JS (n = 250) to analyze the phytoplankton community structure and assess the water quality of the part of the Yeongsan River between the two weirs. Samples were collected from the surface to a depth of 0.5 m. In situ measurements of water temperature, pH, dissolved oxygen (DO), and electrical conductivity (EC) were conducted using a calibrated multiparameter instrument (EXO2 Sonde; YSI, Yellow Springs, OH, USA). The water samples were immediately cooled and transported to the laboratory in an ice box. The following analytical tests were conducted in accordance with standard water quality assessment methods [25]: biochemical oxygen demand (BOD), chemical oxygen demand (COD), suspended solids (SS), total nitrogen (TN), nitrate nitrogen (NO_3-N), ammonium nitrogen (NH_3-N), total phosphorus (TP), phosphate phosphorus (PO_4-P), chlorophyll-*a* (Chl-*a*), and total organic carbon (TOC). BOD, COD, and TOC were determined using the diaphragm electrode method, the potassium permanganate method, and high-temperature combustion, respectively. TN, TP, and PO_4-P were analyzed using a microflow chemistry analyzer (QuAAtro 39 Autoanalyzer; BLTEC, Osaka, Japan). Ion chromatography was used for NO_3-N (Integrion; Dionex, Sunnyvale, CA, USA) and NH_3-N (AA3; BLTEC, Osaka, Japan) measurements. Chl-*a* levels were measured using a UV–Vis spectrophotometer (LAMBDA™ 25; Perkin-Elmer, Shelton, CT, USA).

2.3. Phytoplankton Analysis

For phytoplankton identification and enumeration, the collected samples were preserved by adding Lugol's solution to achieve a final concentration of 2–3%. The samples were examined under an optical microscope at magnifications ranging from 100× to 1000×. Cell counts per milliliter were determined using a Sedgwick-Rafter chamber (Wildco, MI, USA), following standard operating procedures [26]. Based on previously established methodologies [27,28] and the classification system developed by Simonsen [29], phytoplankton were taxonomically identified and classified into four groups: Bacillariophyceae (diatoms), Chlorophyceae (green algae), *Cyanophyceae* (cyanobacteria), and other phytoplankton. Other phytoplankton were not classified into groups such as Dinophyceae, Chrysophyceae, and Cryptophyceae because over the 5-year study period, only 11 genera were observed at the sampling sites. Total cell counts for the remaining genera, excluding the top 5 with the highest cell counts, were less than 700 cells/mL, which was considered too low for detailed classification.

2.4. Statistical Analysis

PCA was performed using the R software (version 4.3.2) to identify the key environmental factors influencing seasonal phytoplankton dynamics. The number of principal components was determined based on eigenvalues greater than 1.0, which indicate the variance explained by each principal component [30]. Before conducting PCA, the Kaiser–Meyer–Olkin (KMO) test was used to assess the sampling adequacy of the variables. The KMO test measures the degree of interrelationships among variables, with values closer to 1 indicating greater adequacy for factor analysis; a minimum value of 0.5 is required to proceed with the analysis. The KMO values for SC and JS were 0.74 and 0.77, respectively, indicating that the sampled data were suitable for factor analysis in both cases. To determine correlations between the dominant phytoplankton groups and environmental factors, CCA was conducted using the PAST software (version 4.03). In the CCA biplot, arrows indicate the relative correlations of environmental variables as a measure of the relationship between the axes and environmental variables, and arrow length represents the relative explanatory power of the variable [31]. Additionally, five key genera were selected from each of the above four taxonomic groups and subjected to further analysis.

3. Results

3.1. Environmental Characteristics

Table 2 shows the values of the environmental factors analyzed by season at SC and JS during the study period. Among the most critical variables influencing the growth and succession of phytoplankton communities, the mean water temperature at SC was 17.7 °C (range: 5.2–30.9 °C), whereas that at JS was 17.1 °C (3.3–31.4 °C). The mean DO concentrations at SC and JS were 9.4 mg/L (range: 1.5–16.5 mg/L) and 10.0 mg/L (4.1–18.0 mg/L), respectively, suggesting that DO concentrations are inversely correlated with water temperature [32]. The pH was neutral, ranging from 7.3 to 7.7.

Nutrient fluctuations generally significantly affect phytoplankton communities [33]. Nitrogen-based nutrient (TN, NO_3-N, and NH_3-N) concentrations exhibited a consistent U-shaped pattern at both weirs; their values increased during winter and decreased during summer. In summer, the mean TN, NO_3-N, and NH_3-N concentrations in SC and JS decreased to 3.66, 1.44, and 1.42 mg/L, and 2.68, 1.36, and 0.57 mg/L, respectively. This pattern is presumably the result of nutrient dilution due to high rainfall in summer, which accounts for more than 40% of the annual rainfall [34]. The annual average rainfall in the study area is 1460.5 mm, and more than 61.6% of this total, amounting to 899.0 mm, is concentrated during the summer months (June to August). The annual mean TN, NO_3-N, and NH_3-N concentrations at SC and JS were 5.65, 2.12, and 2.59 mg/L, and 4.33, 2.16, and 1.28 mg/L, respectively, which were slightly higher than previously reported values [35,36].

Table 2. Physicochemical parameters at SC and JS during different seasons.

Site	Season	WT (°C)	pH	DO (mg/L)	TN (mg/L)	NO_3-N (mg/L)	NH_3-N (mg/L)	TP (mg/L)	PO_4-P (mg/L)
SC	Spring	16.5 (10.2–26.8)	7.4 (6.6–9.0)	9.5 (1.6–16.3)	6.54 (2.74–12.95)	2.03 (0.75–3.18)	3.43 (0.74–9.23)	0.19 (0.09–0.58)	0.04 (0.01–0.27)
	Summer	26.4 (20.6–30.9)	7.2 (6.2–8.7)	6.7 (1.5–12.4)	3.66 (2.01–6.42)	1.44 (0.76–2.41)	1.42 (0.21–4.24)	0.17 (0.08–0.38)	0.06 (0.00–0.23)
	Autumn	19.7 (10.7–28.3)	7.2 (6.2–8.9)	9.1 (4.0–16.5)	4.41 (2.22–6.99)	2.11 (0.76–3.35)	1.47 (0.17–3.49)	0.13 (0.05–0.28)	0.04 (0.00–0.20)
	Winter	7.8 (5.2–12.7)	7.3 (6.7–8.2)	12.4 (8.9–15.6)	8.14 (3.63–11.81)	2.96 (2.04–3.96)	4.12 (0.70–7.13)	0.16 (0.06–0.31)	0.03 (0.00–0.16)
	Total	17.7 (5.2–30.9)	7.3 (6.2–9.0)	9.4 (1.5–16.5)	5.65 (2.01–12.95)	2.12 (0.75–3.96)	2.59 (0.17–9.23)	0.16 (0.05–0.58)	0.05 (0.00–0.27)
JS	Spring	15.8 (8.0–23.9)	7.7 (6.9–9.0)	9.1 (4.4–15.5)	5.08 (2.77–9.22)	2.08 (0.83–3.22)	1.99 (0.08–5.23)	0.13 (0.06–0.24)	0.03 (0.00–0.13)
	Summer	26.3 (20.9–31.4)	7.4 (6.6–8.9)	7.1 (4.1–11.4)	2.68 (1.51–4.03)	1.36 (0.83–2.03)	0.57 (0.04–1.73)	0.13 (0.07–0.23)	0.05 (0.00–0.13)
	Autumn	19.5 (9.6–28.2)	7.7 (6.5–9.2)	9.8 (6.2–14.2)	3.31 (1.64–5.57)	2.02 (1.04–2.01)	0.48 (0.02–2.01)	0.11 (0.04–0.26)	0.04 (0.00–0.14)
	Winter	6.3 (3.3–9.9)	8.1 (7.0–9.2)	14.2 (9.4–18.0)	6.41 (3.21–9.19)	3.26 (2.14–4.75)	2.15 (0.49–4.29)	0.12 (0.04–0.24)	0.02 (0.00–0.19)
	Total	17.1 (3.3–31.4)	7.7 (6.5–9.2)	10.0 (4.1–18.0)	4.33 (1.51–9.22)	2.16 (0.83–4.75)	1.28 (0.02–5.23)	0.12 (0.04–0.26)	0.03 (0.00–0.19)

Note(s): Data are means (range) over 5 years. WT, water temperature; DO, dissolved oxygen; TN, total nitrogen; NO_3-N, nitrate nitrogen; NH_3-N, ammonium nitrogen; TP, total phosphorus; PO_4-P, phosphate phosphorus.

Phosphorus-based nutrients (TP and PO_4-P) did not exhibit significant seasonal patterns; the annual mean TP and PO_4-P concentrations at SC and JS were 0.16 and 0.05 mg/L, and 0.12 and 0.03 mg/L, respectively.

3.2. Phytoplankton Community

The relative distribution of each taxonomic group indicates the succession patterns of phytoplankton throughout the study period (Figure 2). During winter, when the water temperature is low, Bacillariophyceae were predominant at both SC (58.8%) and JS (48.6%); however, their abundance gradually decreased from March to April, whereas that of Chlorophyceae and other phytoplankton increased (SC: Chlorophyceae 25.8%, other phytoplankton 16.6%; JS: Chlorophyceae 26.5%, other phytoplankton 37.8%). From May to August, when the water temperature rises, Chlorophyceae became dominant at both weirs (SC: 52.8%; JS: 42.7%). From August to September, *Cyanophyceae* reached their annual peak (SC: 11.3%; JS: 24.5%), followed by the gradual dominance of Bacillariophyceae as the water temperature gradually decreased, showing a typical succession pattern of phytoplankton in temperate aquatic systems [37,38]. Throughout the study period, Bacillariophyceae and Chlorophyceae were the most frequently occurring groups at both weirs (SC 88.6%; JS 75.9%), whereas *Cyanophyceae* and other phytoplankton appeared less frequently (SC: *Cyanophyceae* 3.3%, other phytoplankton 8.1%; JS: *Cyanophyceae* 9.6%, other phytoplankton 14.6%). Despite the relatively low annual mean abundance of *Cyanophyceae* (6.5%) at SC and JS in the Yeongsan River compared with that in other domestic water bodies [39,40], their presence is critically important in water quality management because of their potential toxicity, making the management of *Cyanophyceae* blooms a top environmental priority.

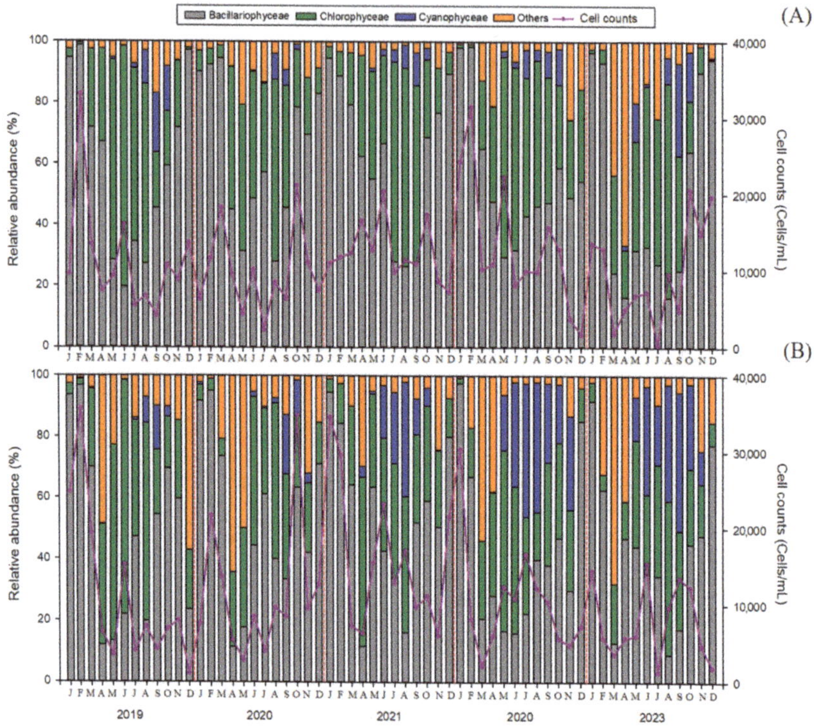

Figure 2. Relative abundance and cell count for the four taxonomic groups at (**A**) SC and (**B**) JS in the Yeongsan River.

Cell counts for the dominant genera within each taxonomic group are presented in Figures 3 and 4. To simplify the analysis, species-level data were aggregated at the genus level, and five key genera selected from each of the four phytoplankton groups (Bacillariophyceae, Chlorophyceae, *Cyanophyceae*, and other phytoplankton) were analyzed. These dominant genera collectively comprised a substantial portion of the phytoplankton community, accounting for an average of 93.5% of the total cell count at SC (Bacillariophyceae: 99.1%; Chlorophyceae: 81.2%; *Cyanophyceae*: 95.4%; other phytoplankton: 98.4%) and 98.8% at JS (Bacillariophyceae: 98.5%; Chlorophyceae: 98.7%; *Cyanophyceae*: 99.7%; other phytoplankton: 98.5%). Their high representation underscores the importance of these genera as reference taxa in characterizing the phytoplankton community structure in the Yeongsan River. The key genera in *Bacillariophyceae* included *Stephanodiscus, Cyclotella, Aulacoseira, Skeletonema,* and *Synedra.* The key genera in Chlorophyceae were *Micratinium, Scenedesmus,* and *Actinastrum,* whereas those in *Cyanophyceae* were *Microcystis, Pseudanabaena, Merismopedia, Aphanizomenon,* and *Dolichospermum.* Among other phytoplankton, *Cryptomonas, Euglena, Mallomonas,* and *Peridinium* were the dominant genera. The key genera in each taxonomic group were similar at both weirs.

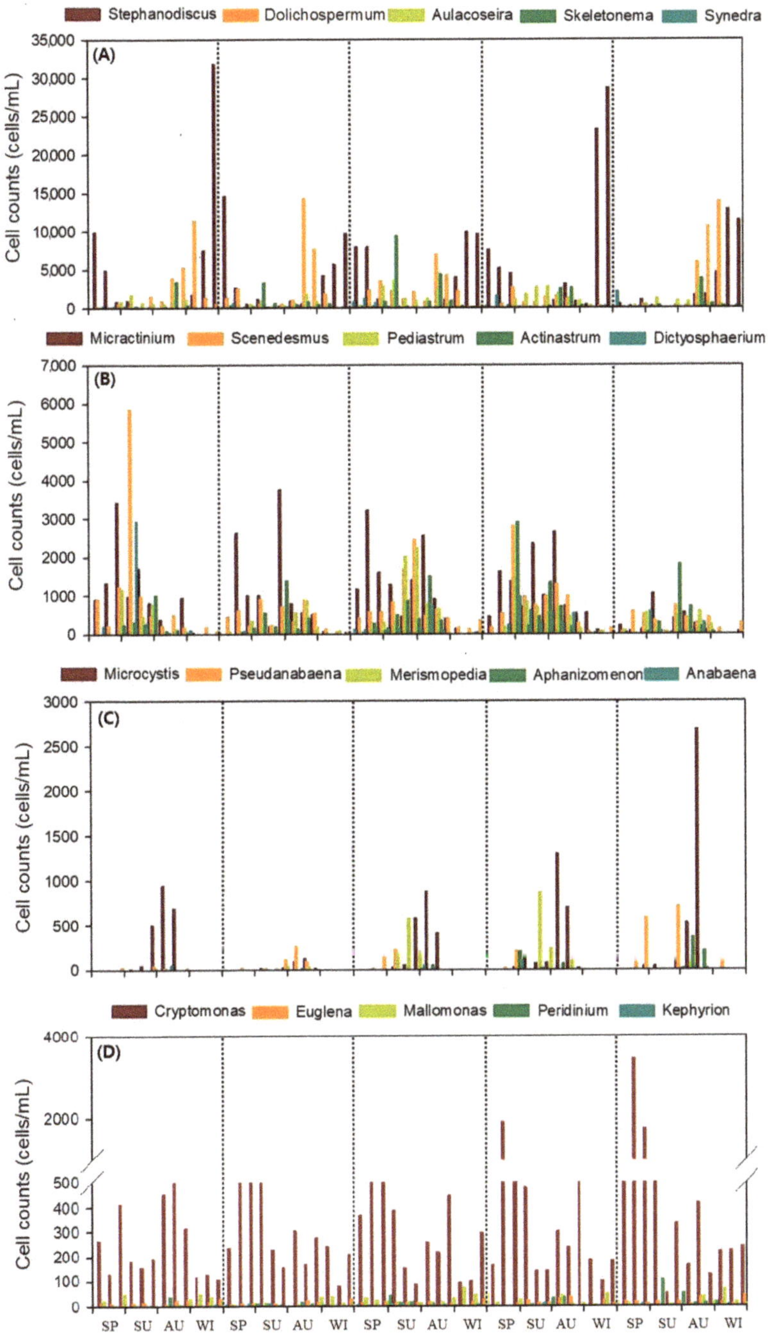

Figure 3. Cell counts of dominant genera within (**A**) Bacillariophyceae, (**B**) Chlorophyceae, (**C**) *Cyanophyceae*, and (**D**) other phytoplankton at SC. SP, spring; SU, summer; AU, autumn; WI, winter.

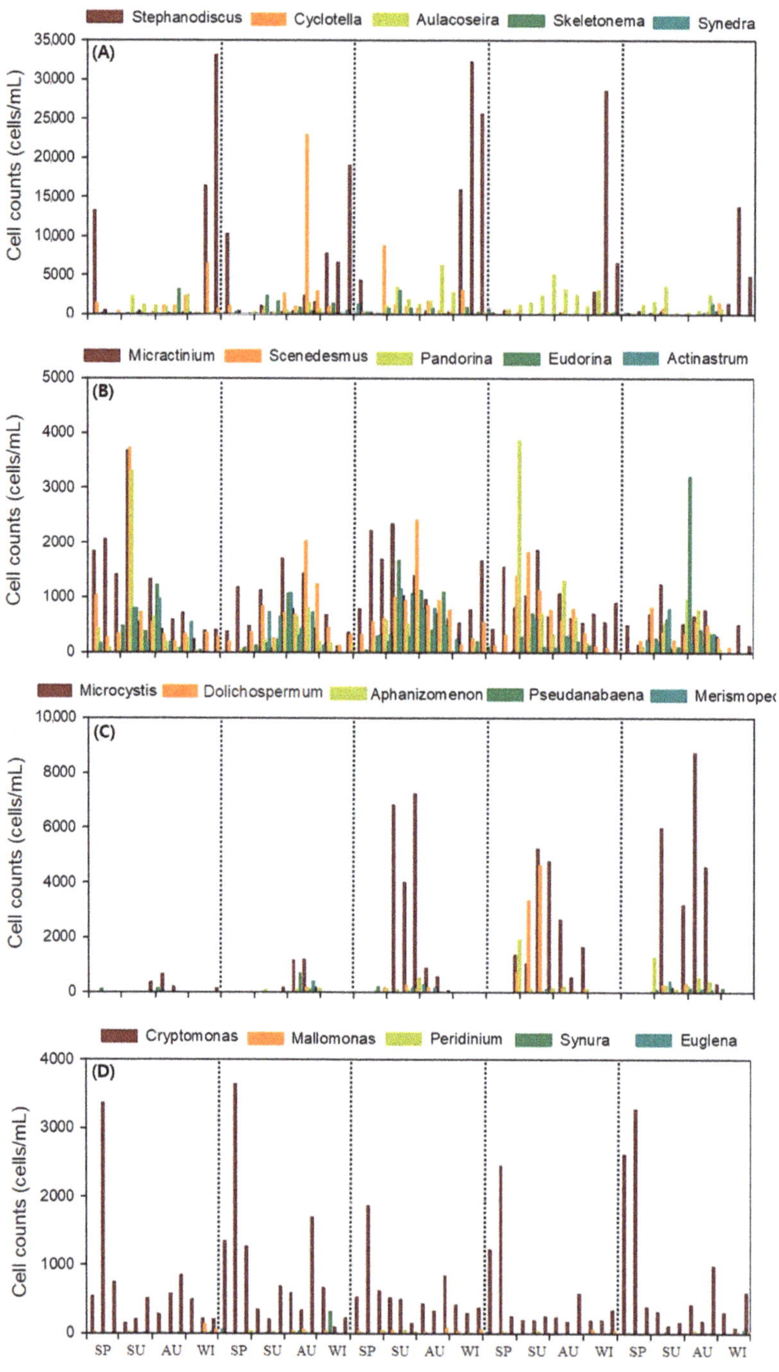

Figure 4. Cell counts of dominant genera within (**A**) Bacillariophyceae, (**B**) Chlorophyceae, (**C**) *Cyanophyceae*, and (**D**) other phytoplankton at JS. SP, spring; SU, summer; AU, autumn; WI, winter.

Upon examining the dominance rate of the most frequently observed dominant genus among the top five dominant genera for each taxon, a comparison of the total cell counts between the predominant genus and the four subdominant genera in each taxonomic group yielded the following results at SC: Bacillariophyceae (*Stephanodiscus* sp.: 248,334 cells/mL vs. four subdominant genera: 222,860 cells/mL), Chlorophyceae (*Micratinium* sp.: 48,967 cells/mL vs. four subdominant genera: 84,860 cells/mL), *Cyanophyceae* (*Microcystis* sp.: 10,109 cells/mL vs. four subdominant genera: 6414 cells/mL), and other phytoplankton (*Cryptomonas* sp.: 24,574 cells/mL vs. four subdominant genera: 2651 cells/mL). The results for JS were as follows: Bacillariophyceae (*Stephanodiscus* sp.: 253,516 cells/mL vs. four subdominant genera: 160,813 cells/mL), Chlorophyceae (*Micractinium* sp.: 53,232 cells/mL vs. four subdominant genera: 89,108 cells/mL), *Cyanophyceae* (*Microcystis* sp.: 63,687 cells/mL vs. four subdominant genera: 20,750 cells/mL), and other phytoplankton (*Cryptomonas* sp.: 41,657 cells/mL vs. four subdominant genera: 2146 cells/mL). Except for Chlorophyceae, the predominant genus at both weirs consistently exhibited higher cell counts than the combined counts of the four subdominant genera. Notably, in the other phytoplankton group, *Cryptomonas* sp. was particularly dominant, with a prevalence of 92.5% (SC: 90%; JS: 95%). This high prevalence is attributed to the fact that *Cryptomonas* sp. frequently appears in spring when water temperatures are low, resulting in peak counts during this season at SC (9802 cells/mL of 27,225 cells/mL) and JS (22,948 cells/mL of 43,802 cells/mL) [41].

3.3. Multifunctional Analysis

To elucidate the relationships between the phytoplankton structure and environmental factors at SC and JS, PCA was conducted using the cell counts of each phytoplankton taxonomic group and the values of water temperature, pH, DO, BOD, COD, TOC, EC, Chl-*a*, and nutrients (TN, NO_3-N, NH_3-N, TP, and PO_4-P) (Figure 5A,B). Four principal components with eigenvalues exceeding 1.0 were identified for both weirs. The first component accounted for 31.1% and 36.9% of the total variance at SC and JS, respectively, whereas the second to fourth components explained 17.9% and 18.8%, 13.7% and 11.7%, and 8.4% and 9.0% of the total variance at SC and JS, respectively. In spring at SC, the "other phytoplankton" group was positively correlated with the Chl-*a* concentration, suggesting that *Cryptomonas* sp. was predominant within this group (Figure 5A). In summer, *Cyanophyceae* abundance was positively correlated with the water temperature. During winter, similar environmental patterns influenced Bacillariophyceae at both weirs. Specifically, at SC, Bacillariophyceae abundance was positively correlated with DO, EC, and nitrogen-based nutrients (TN, NH_3-N, and NO_3-N) but negatively correlated with PO_4-P. At JS, Bacillariophyceae abundance was positively correlated with nitrogen-based nutrients, pH, BOD, and Chl-*a* but negatively correlated with PO_4-P (Figure 5B).

CCA was employed to identify the environmental factors affecting the distribution of the dominant genera within the study area, using the 14 environmental variables derived from PCA to evaluate total cell counts and relationships among the dominant genera, including five key genera from each taxonomic group, totaling twenty genera. Figure 6 presents the CCA biplot for the phytoplankton community. At SC, axes 1 and 2 explained 83.7% of the variance among the phytoplankton groups (Figure 6A). Axis 1, which accounted for 65.3% of the total variance, was positively correlated with water temperature for *Merismopedia* sp., *Dolichospermum* sp., *Aphanizomenon* sp., *Pseudanabaena* sp., *Microcystis* sp., *Actinastrum* sp., *Pediastrum* sp., *Peridinium* sp., and *Aulacoseira* sp. and with nitrogen-based nutrients, DO, and EC for *Stephanodiscus* sp., but it was negatively correlated with water temperature and PO_4-P. Axis 2, which explained 18.3% of the variance, had a weak positive correlation with TP, BOD, COD, and TOC for *Synedra* sp. At JS, axes 1 and 2 accounted for 74.8% of the total variance (Figure 6B). Axis 1, which explained 58.8% of the total variance, revealed that *Microcystis* sp. and *Dolichospermum* sp. were positively correlated with water temperature, whereas *Stephanodiscus* sp. was positively correlated with nitrogen-based nutrients and EC. Axis 2, which explained 16.0% of the total variance, indicated that *Cyclotella* sp. was positively correlated with SS.

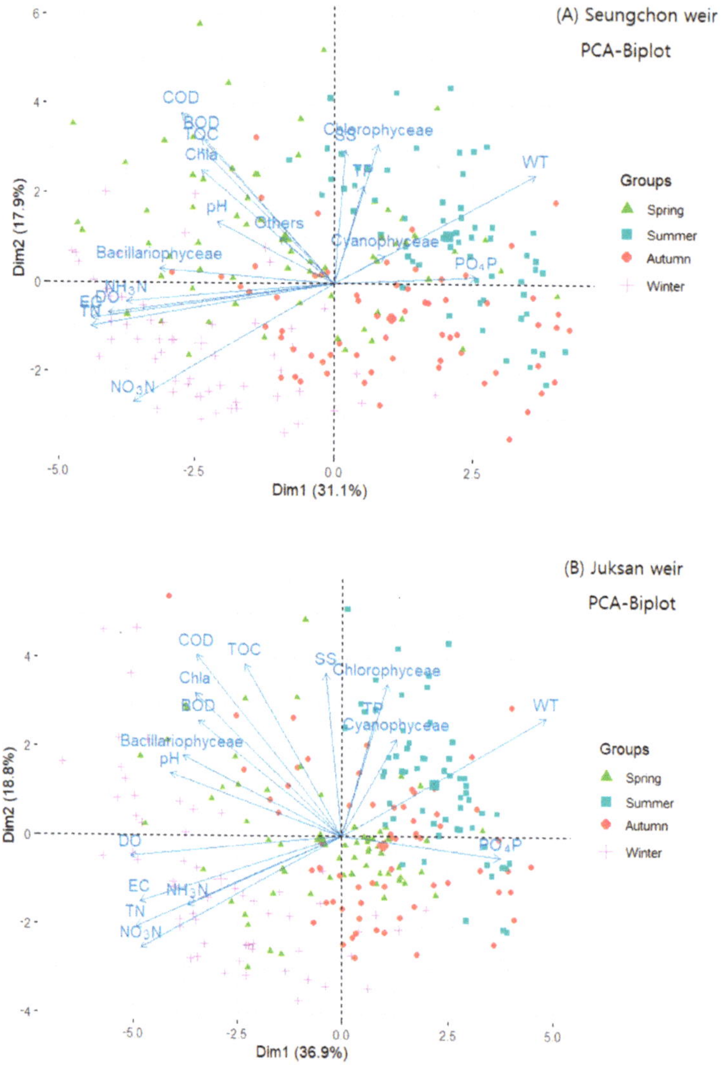

Figure 5. Biplots of PCA results grouped by season at (**A**) SC and (**B**) JS in the Yeongsan River. PCA, principal components analysis.

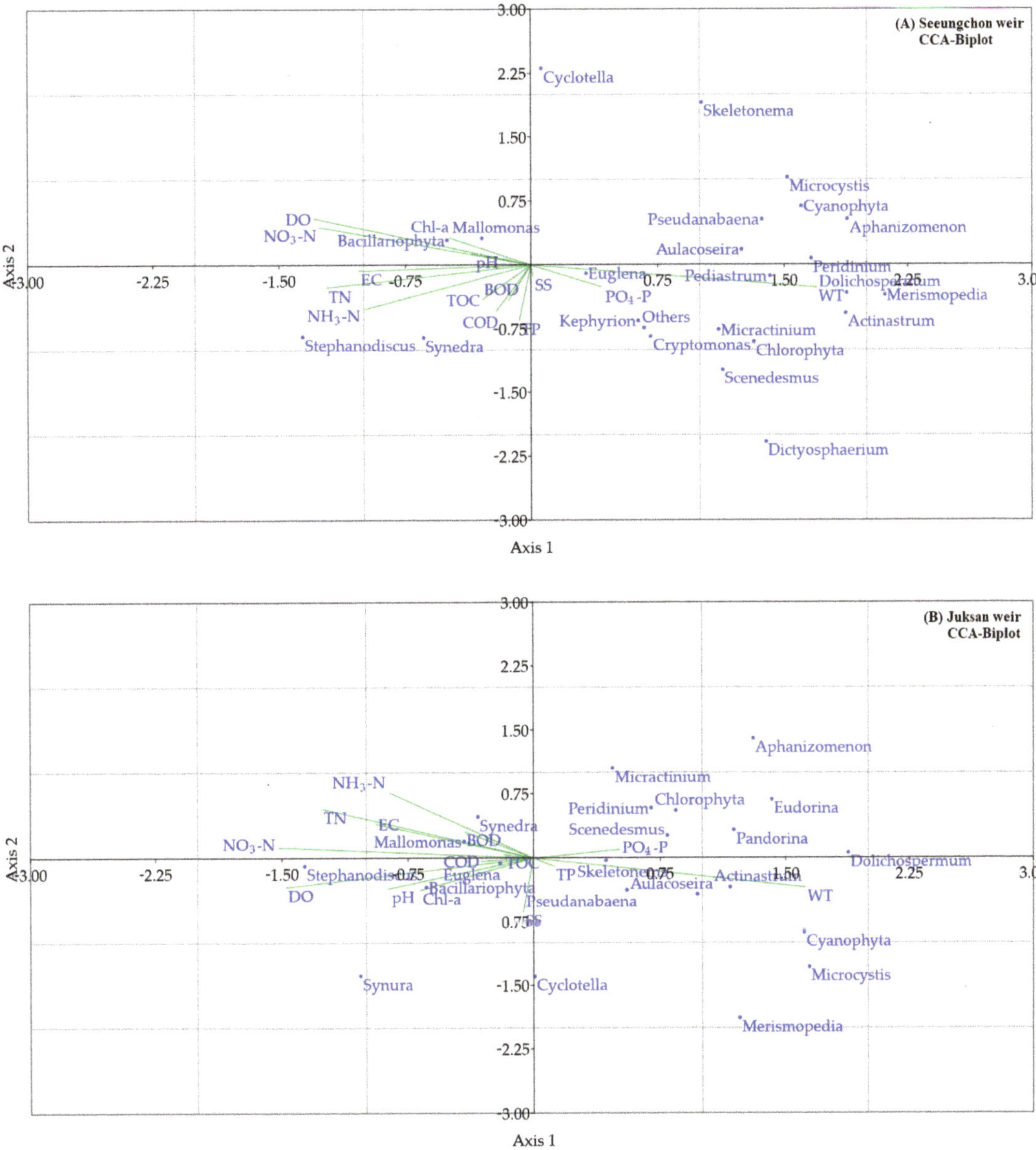

Figure 6. Biplots of CCA results at (**A**) SC and (**B**) JS in the Yeongsan River. CCA, canonical correspondence analysis.

4. Discussion

4.1. Environmental Trends

The Seungchon and Juksan Weirs were established to secure agricultural water resources. However, the water quality of the Yeongsan River is poor because of pollutant discharge from agricultural lands and industrial complexes [42]. At SC, the water temperature in 2023 increased by 0.2 °C compared to the average water temperature over the previous four years, whereas the TN concentration rose by 0.100 mg/L, and the TP concentration decreased by 0.017 mg/L. At JS, the water temperature increased by 0.1 °C compared to the average over the previous four years, and the TN concentration increased by 0.129 mg/L, whereas the TP concentration decreased by 0.008 mg/L. Thus, at both weirs on the Yeongsan River, the water temperature and TN concentration showed an increasing trend, whereas the TP concentration showed a decreasing trend. Because water temperatures are rising because of global warming and climate issues, and directly controlling them is challenging, managing nutrient levels is becoming increasingly important [43]. To this end, it is essential to promote the establishment and expansion of wastewater treatment facilities and public livestock manure treatment centers. Additionally, continuous efforts in water quality improvement, such as ecological river restoration, non-point-source pollution reduction, and river mouth cleanups, are necessary.

4.2. Algal Bloom Dynamics

The Yeongsan River is experiencing high levels of non-point-source pollution from agricultural lands (79.4% based on BOD, 52 t/day), resulting in conditions favorable for algal growth due to eutrophication. Consequently, blue-green algae blooms occur every summer [44]. Additionally, the two weirs reduce water flow, further increasing algal blooms. Although *Cyanophyceae* were not dominant in the study area, they can produce toxins that affect humans and higher organisms and are therefore a key monitoring target. Toxin-producing *Cyanophyceae* growths are referred to as "harmful algal blooms", and four genera (*Microcystis* sp., *Aphanizomenon* sp., *Anabaena* sp., and *Oscillatoria* sp.) have been designated as management targets. The occurrence of harmful algae has been increasing at both weirs each year, which is exacerbated by global warming, and has led to prolonged late heat, allowing blooms to appear until November, after which water temperatures drop rapidly [44]. To manage algal blooms, research on the characteristics and mechanisms of phytoplankton growth is necessary [45]. Previous studies have mainly focused on the correlations between nitrogen and phosphorus and algal blooms [46,47]; however, given the reduced diversity of phytoplankton in artificial environments, future research should prioritize identifying the primary dominant genera within these environments and assessing their succession patterns and distribution characteristics. This approach is vital for predicting algal blooms and effectively managing water resources. The focus of this study on selecting five dominant genera from each taxonomic group for environmental factor and time-series analyses represents a significant methodological contribution, particularly in understanding the impact of human-induced changes in hydrology on phytoplankton communities.

4.3. Statistical Insights

The occurrence of algal blooms involves a complex mechanism that is influenced by various environmental factors and the characteristics of phytoplankton species. Therefore, using statistical techniques to preemptively assess correlations would be beneficial. PCA was used to identify the key environmental factors affecting the composition and distribution of phytoplankton communities and to analyze the correlations between them. This analysis allows for a systematic understanding of how seasonal environmental changes impact phytoplankton and provides essential information for evaluating and managing ecosystem health. PCA revealed factors that significant influence phytoplankton. We found a positive correlation between other phytoplankton and the Chl-*a* concentration, suggesting a predominance of *Cryptomonas* sp. in this group as *Cryptomonas* sp. reportedly has higher per-cell Chl-*a* concentrations than other algal genera [48]. In summer, *Cyanophyceae* abundance was positively correlated with the water temperature, aligning with findings that rising temperatures contribute to the proliferation of *Cyanophyceae* [49,50]. Cyanobacteria can adjust their buoyancy using intracellular gas vesicles, allowing them to remain in the surface layer where they receive adequate light and optimal temperatures for growth [51]. During winter, the environmental factors affecting Bacillariophyceae exhibited similar patterns at both weirs. At SC, Bacillariophyceae abundance was positively correlated with DO, EC, and nitrogen-based nutrients (TN, NH_3-N, and NO_3-N) but negatively correlated with PO_4-P. At JS, Bacillariophyceae abundance showed a positive correlation with nitrogen-based nutrients, pH, BOD, and Chl-*a* but a negative correlation with PO_4-P. These findings emphasize the importance of nitrogen-based nutrients in structuring Bacillariophyceae communities. The sensitivity of phytoplankton community composition to environmental changes underscores the potential for using these organisms as indicators of ecosystem health, especially for identifying the causes of algal blooms.

CCA was used to identify the relationships between the distribution of phytoplankton and environmental variables to assess the influences of specific environmental factors on phytoplankton community structure. As we selected the dominant genera at the two weirs, this analysis allowed for a nuanced understanding of how these genera respond to environmental changes, providing essential information for water resource management and ecosystem health evaluation. Furthermore, this analysis contributes to understanding the interactions between water quality changes and the ecosystem and can aid in predicting ecosystem variability due to climate change. The results revealed significant environmental influences on the dominant genera in Bacillariophyceae and *Cyanophyceae*. Among Bacillariophyceae, *Cyclotella* sp. exhibited relatively low correlations with environmental variables, whereas *Aulacoseira* sp. was positively correlated with the water temperature. This result is consistent with findings by Parakkandi et al. [52] and Roshith et al. [53], who found that *Aulacoseira* sp., which is capable of cold adaptation, can also thrive during summer, when *Cyanophyceae* are prevalent. *Stephanodiscus* sp. was strongly positively correlated with nitrogen-based nutrients but negatively correlated with the water temperature and PO_4-P concentration, consistent with previous findings indicating its dominance in cooler, nutrient-rich conditions [54,55]. Therefore, nitrogen-based nutrients and low water temperatures promote the growth of *Stephanodiscus* sp. Among *Cyanophyceae*, *Microcystis* sp. and *Dolichospermum* sp. were positively correlated with the water temperature and PO_4-P concentration. These genera, along with Cylindrospermopsis, are known for contributing to harmful cyanobacterial blooms globally [56]. Several studies have linked blooms of *Microcystis* sp. and *Dolichospermum* sp. to fluctuations in water temperature and nutrient levels [57,58]. The massive proliferation of *Stephanodiscus* sp., *Microcystis* sp., and *Dolichospermum* sp. negatively affects water resource management; an excessive growth of *Stephanodiscus* sp. can lead to decreased DO concentrations and problems with filtration and sedimentation in water treatment and supply infrastructure [59]. Meanwhile, *Microcystis* sp. and *Dolichospermum* sp. produce the toxin microcystin, which can induce mitochondrial dysfunction, hepatotoxicity, and nephrotoxicity [60]. Our study focused on the relationship between phytoplankton and environmental factors in areas that have transitioned from a flowing river to a still water environment. The results of the canonical correspondence analysis (CCA) reveal a strong link between harmful cyanobacteria, water temperature, and residence time. These findings align with existing research highlighting the impact of agricultural activities on phytoplankton composition in river ecosystems. For instance, studies conducted in Lake Erie, USA, indicate that agricultural non-point-source pollution significantly affects harmful algal blooms and phytoplankton communities [61]. Similarly, research in China emphasizes that agricultural activities contribute to nutrient influx in rivers, further altering phytoplankton dynamics [62]. Collectively, these studies underscore the extensive influence of agricultural practices on riverine ecosystems and highlight the need for ongoing management and water quality improvement efforts. By continuing to monitor these harmful algal blooms, we aim to develop strategies to mitigate their occurrence and maintain the health of aquatic ecosystems.

5. Conclusions

We investigated the relationships between the phytoplankton community structure and environmental factors at two weirs (SC and JS) in the Yeongsan River basin over 5 years (2019–2023). We classified phytoplankton into four taxonomic groups (Bacillariophyceae,

Chlorophyceae, *Cyanophyceae*, and other phytoplankton), enabling a systematic analysis of their community structure and succession patterns. PCA was conducted to identify seasonal influences of environmental factors on each taxonomic group. CCA was performed on twenty dominant genera (five from each group) to determine the environmental factors affecting the genera commonly found at both SC and JS. The results of this study are summarized as follows.

The key genera in Bacillariophyceae included *Stephanodiscus*, *Cyclotella*, *Aulacoseira*, *Skeletonema*, and *Synedra*. The key genera in Chlorophyceae were *Micratinium*, *Scenedesmus*, and *Actinastrum*, and those in *Cyanophyceae* were *Microcystis*, *Pseudanabaena*, *Merismopedia*, *Aphanizomenon*, and *Dolichospermum*. Among other phytoplankton, *Cryptomonas*, *Euglena*, *Mallomonas*, and *Peridinium* were the dominant genera. The key genera of each taxonomic group were similar at both weirs.

During the study period, similar phytoplankton succession patterns were observed at both weirs. Bacillariophyceae dominated during winter. The abundance of Chlorophyceae and other phytoplankton gradually increased from March to April. Chlorophyceae became dominant from May to August, as water temperatures increased, followed by an increase in the abundance of *Cyanophyceae* from August to September, exhibiting a typical succession pattern of phytoplankton in temperate aquatic systems.

PCA revealed three significant correlations. (1) In spring, "other phytoplankton" were positively correlated with the Chl-*a* concentration, probably because of the predominance of *Cryptomonas* sp., which have higher per-cell Chl-*a* concentrations than other phytoplankton genera. (2) In summer, *Cyanophyceae* were strongly positively correlated with the water temperature. This finding was attributed to the ability of *Cyanophyceae* to buoy to the water surface layer using gas vesicles to capitalize on sufficient sunlight and warmer temperatures conducive to their growth. (3) In winter, Bacillariophyceae were positively correlated with nitrogen-based nutrients (TN, NH_3-N, and NO_3-N) but negatively correlated with PO_4-P, suggesting that nitrogen-based nutrients are more impactful than PO_4-P on shaping the community structure of Bacillariophycea.

CCA revealed that *Stephanodiscus* sp., which was the dominant genus in Bacillariophyceae, was positively correlated with nitrogen-based nutrients, indicating that these nutrients serve as a nutrient source for *Stephanodiscus* sp. *Microcystis* sp. and *Dolichospermum* sp., which were the dominant genera in *Cyanophyceae*, were positively correlated with the water temperature and PO_4-P concentration, indicating that increased water temperatures and PO_4-P levels were the primary drivers of *Microcystis* sp. and *Dolichospermum* sp. growth during the study period and corroborating the high sensitivity of these species to these environmental factors. Given that *Stephanodiscus* sp. can disrupt water treatment processes by causing filtration and sedimentation problems and that *Microcystis* sp. and *Dolichospermum* sp. produce microcystin, identifying the environmental factors influencing these dominant genera is crucial for effective algal bloom control and water resource management.

Author Contributions: Conceptualization, H.C.; methodology, H.C. and M.S.; validation, H.C.; formal analysis, H.C. and M.S.; investigation, H.C., M.S., T.K., J.P. and W.-S.L.; data curation, H.C.; writing—original draft, H.C. and T.K.; writing—review and editing, H.C., T.K., J.P. and W.-S.L.; visualization, H.C. and T.K.; supervision, W.-S.L.; project administration, T.K. and W.-S.L. funding acquisition, W.-S.L. All authors have read and agreed to the published version of the manuscript.

Funding: This work was supported by a grant from the National Institute of Environmental Research, funded by the Ministry of Environment (ME) of the Republic of Korea (NIER-2024-01-01-075).

Data Availability Statement: The data that support the findings of this study are available from the corresponding author upon reasonable request.

Conflicts of Interest: The authors declare no conflicts of interest.

References

1. Kakore, B.G.; Atique, U.; An, K.G. Serial installation of massive weirs impacts the nutrients, solids, chlorophyll, and water chemistry along the river gradient. *Int. J. Environ. Sci. Technol.* **2023**, *20*, 8261–8276. [CrossRef]
2. Royer, T.V. Human-dominated rivers and river management in the Anthropocene. In *Stream Ecosystems in a Changing Environment*; Jones, J.B., Stanley, E.H., Eds.; Academic Press: Cambridge, MA, USA, 2016; pp. 491–524. [CrossRef]
3. Lee, J.; Hong, S.; An, S.A.; Khim, J.S. Methodological advances and future directions of microalgal bioassays for evaluation of potential toxicity in environmental samples: A review. *Environ. Int.* **2023**, *173*, 107869. [CrossRef] [PubMed]
4. Ralston, D.K.; Brosnahan, M.L.; Fox, S.E.; Lee, K.D.; Anderson, D.M. Temperature and residence time controls on an estuarine harmful algal bloom: Modeling hydrodynamics and *Alexandrium fundyense* in Nauset estuary. *Estuaries Coasts* **2015**, *38*, 2240–2258. [CrossRef]
5. Janse van Vuuren, S.; Swanepoel, A. Role of the Lethabo Weir in altering the phytoplankton community structure of the Vaal River, South Africa. *Afr. J. Aquat. Sci.* **2023**, *48*, 300–314. [CrossRef]
6. Yoon, S.J.; Lee, J.; Kim, H.G.; Kwon, B.O.; Kim, J.; Hong, S.; Khim, J.S. Phytoplankton assemblage responses to massive freshwater inputs and anthropogenic toxic substances contamination in the Geum River Estuary, South Korea. *Mar. Pollut. Bull.* **2024**, *199*, 116020. [CrossRef] [PubMed]
7. Shin, J.K.; Kang, B.G.; Hwang, S.J. Limnological study on spring-bloom of a green algae, *Eudorina elegans* and weirwater pulsed-flows in the Midstream (Seungchon Weir Pool) of the Yeongsan River, Korea. *Korean J. Ecol. Environ.* **2016**, *49*, 320–333. [CrossRef]
8. Chong, S.A.; Yi, H.S.; Hwang, H.S.; Kim, H.J. Modeling the flushing effect of multi-purpose weir operation on algae removal in Yeongsan River. *J. Korean Soc. Environ. Eng.* **2015**, *37*, 563–572. [CrossRef]
9. Kaown, D.; Koh, D.C.; Mayer, B.; Mahlknecht, J.; Ju, Y.J.; Rhee, S.K.; Kim, J.H.; Park, D.K.; Park, I.; Lee, H.L.; et al. Estimation of nutrient sources and fate in groundwater near a large weir-regulated river using multiple isotopes and microbial signatures. *J. Hazard. Mater.* **2023**, *446*, 130703. [CrossRef]
10. Plaas, H.E.; Paerl, R.W.; Baumann, K.; Karl, C.; Popendorf, K.J.; Barnard, M.A.; Chang, N.Y.; Curtis, N.P.; Huang, H.; Mathieson, O.L.; et al. Harmful cyanobacterial aerosolization dynamics in the airshed of a eutrophic estuary. *Sci. Total Environ.* **2022**, *852*, 158383. [CrossRef]
11. Abbas, M.; Dia, S.; Deutsch, E.S.; Alameddine, I. Analyzing eutrophication and harmful algal bloom dynamics in a deep Mediterranean hypereutrophic reservoir. *Environ. Sci. Pollut. Res. Int.* **2023**, *30*, 37607–37621. [CrossRef]
12. Cheng, Y.; Jiao, L.; Cheng, Q.; He, J.; Xhang, Y.; Ding, S. The evolution of a typical plateau lake from macrophyte to algae leads to the imbalance of nutrient retention. *Water Res.* **2023**, *236*, 119937. [CrossRef] [PubMed]
13. Tian, Y.; Jiang, Y.; Liu, Q.; Xu, D.; Liu, Y.; Song, J. The impacts of local and regional factors on the phytoplankton community dynamics in a temperate river, northern China. *Ecol. Indic.* **2021**, *123*, 107352. [CrossRef]
14. Liu, Y.; Li, C.Y.; Jian, S.L.; Miao, S.Y.; Li, K.M.; Guan, H.T.; Mao, Y.Q.; Wang, Z.Y.; Li, C.Z. Hydrodynamics regulate longitudinal plankton community structure in an alpine cascade reservoir system. *Front. Microbiol.* **2021**, *12*, 749888. [CrossRef]
15. Navas-Parejoa, J.C.C.; Corzoa, A.; Papaspyroua, S. Seasonal cycles of phytoplankton biomass and primary production in a tropical temporarily open-closed estuarine lagoon—The effect of an extreme climatic event. *Sci. Total Environ.* **2020**, *723*, 138014. [CrossRef] [PubMed]
16. Yoon, B.B.; Jung, Y.H.; Sin, Y.S. Assessing nutrient limitation in Yeongsan River estuary using bioassay experiments. *J. Mar. Sci. Eng.* **2023**, *11*, 1337. [CrossRef]
17. Wang, R.; Liu, X.; Wu, J.; Wai, T.C.; Shen, P.; Lam, P.K.S. Long-term variations of phytoplankton community in relations to environmental factors in Deep Bay, China, from 1994 to 2016. *Mar. Pollut. Bull.* **2020**, *153*, 111010. [CrossRef]
18. Kunpradid, T.; Chaimongkhon, P.; Tagun, R. Seasonal dynamics and environmental drivers of phytoplankton composition in a tropical dam over 5 years in Chiang Mai, Thailand. *Trends Sci.* **2024**, *21*, 7930. [CrossRef]
19. Kim, D.Y.; Sung, J.W.; Kim, T.H.; Cho, H.M.; Kim, J.H.; Park, H.J. Comparative seasonality of phytoplankton community in two contrasting temperate estuaries on the western coast of Korea. *Front. Marine Sci.* **2023**, *10*, 1257904. [CrossRef]
20. Kim, S.H.; Sin, Y.S. Response of size-fractionated phytoplankton to humic acids in the seawater of Yeongsan River Estuary. *Ocean Polar Res.* **2023**, *45*, 113–123. [CrossRef]
21. Sin, Y.S.; Kim, S.H. Effects of humic acids on size and species composition of phytoplankton in a eutrophic temperate estuary. *Appl. Sci.* **2022**, *12*, 10223. [CrossRef]
22. Kwak, S.D.; Choi, J.W.; An, K.G. Chemical water quality and fish component analyses in the periods of before and after the weir constructions in Yeongsan River. *Ecol. Environ.* **2016**, *39*, 99–110. [CrossRef]
23. Kim, D.; Shin, C. Algal Boom Characteristics of Yeongsan River Based on Weir and Estuary Dam Operating Conditions Using EFDC-NIER Model. *Water* **2021**, *13*, 2295. [CrossRef]
24. *K-Water Development of Methods for Optimal Algae Control Reflecting the Algae Growth Characteristics of Weirs in Four Major Rivers*; Korea Water Resources Corporation: Daejeon, Republic of Korea, 2012.
25. *ME Standard Method for Water and Wastewater*; Ministry of Environment: Sejong, Republic of Korea, 2017.
26. MOE. *Standard Method for the Examination of Water Pollution*; Ministry of Environment: Sejong, Republic of Korea, 2011.
27. Krammer, K.; Lange-Bertalot, H. *Süßwasserflora von Mitteleuropa, Band 2/1: Bacillariophyceae 1. Teil: Naviculaceae*; Ettl, H., Gerloff, J., Heying, H., Mollenhauer, D., Eds.; Elsevier Book Co.: Berlin, Germany, 2007.

28. Masaru, A.; Teru, I.; Kozo, I.; Hideo, K.; Shigeru, K.; Hiromu, K.; Eigi, T.; Kohei, T.; Minoru, H.; Takaaki, Y. *Illustration of the Japanese Fresh-water Algae*; Uchidarokakuho Publishing Company: Tokyo, Japan, 1997.
29. Simonsen, R. The diatom system: Ideas on phylogeny. *Bacillaria* **1979**, *2*, 9–71.
30. Soltani, N.; Khodaei, K.; Alnajar, N.; Shahsavari, A.; Ashja Ardalan, A. Cyanobacterial community patterns as water quality bioindicators. *Iran. J. Fish. Sci.* **2012**, *11*, 876–891.
31. Li, X.; Yu, H.; Wang, H.; Ma, C. Phytoplankton community structure in relation to environmental factors and ecological assessment of water quality in the upper reaches of the Genhe River in the Greater Hinggan Mountains. 2019. *Environ. Sci. Pollut. Res. Int.* **2019**, *26*, 17512–17519. [CrossRef]
32. Jeong, E.J.; Na, J.E.; Kim, G.M.; Shim, S.S.; Lee, H.Y. Water temperature and community of phytoplankton in Youngsan River, Korea. *Korean J. Environ. Biol.* **2010**, *28*, 56–63. (In Korean)
33. Zhang, Y.; Yu, H.; Liu, J.; Guo, Y. Analysis of water quality and the response of phytoplankton in the low-temperature environment of Majiagou Urban River, China. *Heliyon* **2024**, *10*, e25955. [CrossRef] [PubMed]
34. Chung, H.S.; Son, M.S.; Ryu, H.S.; Park, C.H.; Lee, R.R.; Cho, M.S.; Lim, C.H.; Park, J.H.; Kim, K.H. Variation of cyanobacteria occurrence pattern and environmental factors in Lake Juam. *Korean J. Environ. Biol.* **2019**, *37*, 640–651. (In Korean) [CrossRef]
35. Seo, K.A.; Na, J.E.; Ryu, H.S.; Kim, K.H. Characteristics of nitro-nutrients and phytoplankton dynamics in the Yeongsan River after weir construction. *J. Korean Soc. Water Environ.* **2018**, *34*, 423–430. (In Korean) [CrossRef]
36. Son, M.S.; Chung, H.S.; Park, C.H.; Park, J.H.; Lim, C.H.; Kim, K.H. The change of phytoplankton community structure and water quality in the Juksan Weir of the Yeongsan River watershed. *Korean J. Environ. Biol.* **2018**, *36*, 591–600. (In Korean) [CrossRef]
37. Grover, J.P.; Chrzanowski, T.H. Seasonal dynamics of phytoplankton in two warm temperate reservoirs: Association of taxonomic composition with temperature. *J. Plankton Res.* **2006**, *28*, 1–17. [CrossRef]
38. Ryu, H.S.; Park, H.K.; Lee, H.J.; Shin, R.Y.; Cheon, C.U. Occurrence and succession pattern of cyanobacteria in the upper region of the Nakdong River: Factors influencing *Aphanizomenon* bloom. *J. Korean Soc. Water Environ.* **2016**, *32*, 52–59. (In Korean) [CrossRef]
39. Kim, Y.H.; Cho, I.H.; Kim, H.K.; Hwang, E.A.; Han, B.H.; Kim, B.H. Assessing the impact of weirs on water quality and phytoplankton dynamics in the south Han River: A two-year study. *Water* **2024**, *16*, 833. [CrossRef]
40. Woo, C.Y.; Yun, S.L.; Kim, S.; Lee, W. Occurrence of harmful blue-green algae at the algae alert system and water quality forecast system sites in Daegu and Gyeongsangbuk-do between 2012 and 2019. *J. Korean Soc. Environ. Eng.* **2020**, *42*, 664–673. [CrossRef]
41. Huang, G.; Wang, X.; Chen, Y.; Deng, L.; Xu, D. Survival strategies of phytoplankton functional groups to environmental factors in a drinking water reservoir, central China. *Intl. J. Limnol.* **2021**, *57*, 18. [CrossRef]
42. Kim, J.; Kim, J.; Seo, D. Effect of major pollution sources on algal blooms in the seungchon weir in the Yeongsan River using EFDC. *J. Korea Water Resour. Assoc.* **2020**, *5*, 369–3814. [CrossRef]
43. Wang, Z.; Shao, D.; Westerhoff, P. Wastewater discharge impact on drinking water sources along the Yangtze River (China). *Sci. Total Environ.* **2017**, *599–600*, 1399–1407. [CrossRef]
44. Water Environment Information System (WEIS). 2024. Available online: http://water.nier.go.kr/ (accessed on 18 January 2024).
45. Wang, S.; Zhang, X.; Chen, N.; Tian, L.; Zhang, Y.; Nam, W.H. A systematic review and quantitative meta-analysis of the relationships between driving forces and cyanobacterial blooms at global scale. *Environ. Res.* **2023**, *216*, 114670. [CrossRef]
46. Zhao, B.; Peng, W.; Zhu, X.; Zhang, H.; Zhuang, X.; Wang, J.; Xi, S.; Lue, T. Research on the relationship between eukaryotic phytoplankton community structure and key physiochemical properties of water in the western half of the Chaohu Lake using high-throughput sequencing. *Water* **2024**, *16*, 2318. [CrossRef]
47. Perez-Gonzalez, R.; Soria-Perpinya, X.; Soria, J.; Sendra, M.D.; Vicente, E. Relationship between cyanobacterial abundance and physicochemical variables in the Ebro Basin Reservoirs (Spain). *Water* **2023**, *15*, 2538. [CrossRef]
48. Im, J.K.; Sim, Y.B.; Hwang, S.J.; Byeon, M.S.; Kang, T.G. Temporal and seasonal variations in a phytoplankton community structure in artificial Lake Uiam, South Korea. *Water* **2023**, *15*, 4118. [CrossRef]
49. Majhi, P.; Pranab, G.; Lianthuamluaia, L.; Chayna, J.; Tasso, T.; Suman, K.; Uttam, K.S.; Basanta, K.D. Assessing spatiotemporal dynamics of phytoplankton assemblage and ecohydrological interaction in a community managed reservoir using GIS for sustainable fisheries management. *Ecohydrol. Hydrobiol.* **2023**, *23*, 109–126. [CrossRef]
50. Pitois, F.; Thoraval, I.; Baures, E.; Thomas, O. Geographical patterns in cyanobacteria distribution: Climate influence at regional scale. *Toxins* **2014**, *6*, 509–522. [CrossRef]
51. Huber, S.T.; Jakobi, A.J. Structural biology of microbial gas vesicles: Historical milestones and current knowledge. *Biochem. Soc. Trans.* **2024**, *52*, 205–215. [CrossRef] [PubMed]
52. Parakkandi, J.; Saha, A.; Sarkar, U.K.; Das, B.K.; Puthiyottil, M.; Muhammadali, S.A.; Ramteke, M.; Johnson, C.; Kumari, S. Spatial and temporal dynamics of phytoplankton in association with habitat parameters in a tropical reservoir, India. *Arab. J. Geosci.* **2021**, *14*, 827. [CrossRef]
53. Roshith, C.M.; Meena, D.K.; Manna, R.K.; Sahoo, A.K.; Swain, H.S.; Raman, R.K.; Sengupta, A.; Das, B.K. Phytoplankton community structure of the Gangetic (Hooghly-Matla) estuary: Status and ecological implications in relation to eco-climatic variability. *Flora* **2018**, *240*, 133–143. [CrossRef]
54. Griffiths, K.; Duda, M.P.; Antoniades, D.; Smol, J.P.; Gregory-Eaves, I. Diatom species responses along gradients of dissolved inorganic carbon, total phosphorus, and lake depth from lakes across Canada. *J. Phycol.* **2024**, *60*, 843–852. [CrossRef]

55. Kong, X.; Seewald, M.; Dadi, T.; Friese, K.; Mi, C.; Boehrer, B.; Schultze, M.; Rinke, K.; Shatwell, T. Unravelling winter diatom blooms in temperate lakes using high frequency data and ecological modeling. *Water Res.* **2021**, *190*, 116681. [CrossRef]
56. Guedes, I.A.; Rachid, C.T.; Rangel, L.M.; Silva, L.H.; Bisch, P.M.; Azevedo, S.M.; Pacheco, A.B. Close link between harmful cyanobacterial dominance and associated bacterioplankton in a tropical eutrophic reservoir. *Front. Microbiol.* **2018**, *9*, 424. [CrossRef]
57. Wei, J.; Wang, M.; Chen, C.; Wu, H.; Lin, L.; Li, M. Seasonal succession of phytoplankton in two temperate artificial lakes with different water sources. *Environ. Sci. Pollut. Res. Int.* **2020**, *27*, 42324–42334. [CrossRef]
58. Lorenzi, A.S.; Chia, M.A.; Lopes, F.A.C.; Silva, G.G.Z.; Edwards, R.A.; do Carmo Bittencourt-Oliveira, M. Cyanobacterial biodiversity of semiarid public drinking water supply reservoirs assessed via next-generation DNA sequencing technology. *J. Microbiol.* **2019**, *57*, 450–460. [CrossRef] [PubMed]
59. Reavie, E.D.; Cai, M.; Twiss, M.R.; Carrick, H.J.; Davis, T.W.; Johengen, T.H.; Gossiaux, D.; Smith, D.E.; Palladino, D.; Burtner, A.; et al. Winter-spring diatom production in Lake Erie is an important driver of summer hypoxia. *J. Great Lakes Res.* **2016**, *42*, 608–618. [CrossRef]
60. Chia, M.A.; Ameh, I.; George, K.C.; Balogun, E.O.; Akinyemi, S.A.; Lorenzi, A.S. Genetic diversity of microcystin producers (cyanobacteria) and microcystin congeners in aquatic resources across Africa: A review paper. *Toxics* **2022**, *10*, 772. [CrossRef] [PubMed]
61. Fraker, M.E.; Aloysius, N.R.; Martin, J.F.; Keitzer, S.C.; Dippold, D.A.; Yen, H.; Arnold, J.; Daggupati, P.; Johnson, M.; Robertson, D.; et al. Agricultural conservation practices could help offset climate change impacts on cyanobacterial harmful algal blooms in Lake Erie. *J. Great Lakes Res.* **2023**, *49*, 209–219. [CrossRef]
62. Huang, Y.; Zhang, S.; Li, Y.; Sun, Z.; Feng, M.; Li, R.; Zhang, J.; Tian, X.; Zhang, W. Characteristics of Phytoplankton Community Structure and Indication to Water Quality in the Lake in Agricultural Areas. *Front. Environ. Sci.* **2022**, *10*, 833409. [CrossRef]

water

MDPI

Article

Chironomid Pupal Exuviae Technique in Ecological Research of Man-Made Water Bodies

Viktorija Ergović [1], Dubravka Čerba [2,*], Natalija Vučković [3] and Zlatko Mihaljević [3]

1 Department of Biology, University of Osijek, Cara Hadrijana 8/a, 31000 Osijek, Croatia; viktorija.ergovic@biologija.unios.hr
2 Department of Assessment and Aquatic Ecosystems Research, National Water Reference Laboratory of Slovakia, Water Research Institute, Nábr. arm. gen. L. Svobodu 5, 81249 Bratislava, Slovakia
3 Department of Biology, Faculty of Science, University of Zagreb, Horvatovac 102a, 10000 Zagreb, Croatia; natalija.vuckovic@biol.pmf.hr (N.V.); zlatko.mihaljevic@biol.pmf.unizg.hr (Z.M.)
* Correspondence: dubravka.cerba@vuvh.sk

Abstract: Reservoirs serve functional purposes such as irrigation and power generation. However, concerns are raised due to the alterations of the connected riverine ecosystems. Chironomidae (Diptera), a diverse aquatic macroinvertebrate group, are vital to the functioning of ecosystems and serve as water quality indicators. Their holometabolous development includes the pupal stage after four larval stages. The chironomid pupal skin (exuvia) is used in environmental assessments, where the Chironomid Pupal Exuvial Technique (CPET) is a recognized standard. The CPET method is adaptable to different freshwater environments and here was applied in the study of 28 man-made lakes in the Pannonian Lowlands and Dinaric Western Balkan Ecoregion in Croatia to obtain information on chironomid diversity and analyze the potential influence of environmental factors on the chironomid community. The lake surface was skimmed with an exuvial hand net (mesh size of 300 μm) along the lake edge with a transect length of 10 m in the area of accumulated debris of organic and inorganic matter. Individual exuviae were mounted in a Berlese mounting medium and identified by morphological characteristics to the lowest taxonomic level. During the study, 5698 chironomid pupal skins were collected, and 141 taxa (including 97 species) belonging to five subfamilies were identified. The tribe Tanytarsini comprised 40% of the identified taxa, with *Paratanytarsus* spp. being the most abundant. In the Dinaric ecoregion, *Paratanytarsus bituberculatus* dominated, while *Microchironomus tener* and the genus *Cricotopus* were the dominant taxa in the Pannonian ecoregion. Community structure in the Pannonian ecoregion was influenced by total organic carbon (TOC) and orthophosphates (PO_4^{3-}), indicating higher anthropogenic pressure compared to the Dinaric ecoregion, where water conductivity influenced Chironomidae assemblages. The research has provided valuable and useful information on the chironomid diversity in man-made and highly altered water bodies, as some of the most vulnerable aquatic habitats to anthropogenic influence. The CPET method could be a useful tool for the ecological studies and bioassessment of water quality in Croatia.

Keywords: artificial lakes; reservoirs; Croatia; CPET method; water quality

1. Introduction

Artificial lakes, such as reservoirs, are primarily created for the benefit of humans and have various benefits and uses, such as agricultural irrigation, water storage, power generation, or fish farming [1,2]. Today, these lakes are mainly used for relaxation and recreation purposes in/near urban areas. Artificial lakes for wider human use can also be created by excavating gravel or clay, dug as separated fishpond pools, fortified reservoirs for water storage, etc. However, reservoirs are generally built by constructing dams to store water in certain river sections. In Croatia, about 40 reservoirs were built in the last century [3]. Sometimes, the disadvantages of structures such as dams outweigh the benefits mentioned, and there is a need to closely monitor the impact of dams on the entire aquatic ecosystem. Construction alters the existing state of the river [4], leading to significant changes in the hydrological regime and water quality (e.g., increase in water temperature and increase in fluctuations and sedimentation) [5–7], and consequently, the biota. The newly formed lake with a higher water temperature may have an increased nutrient uptake, a higher degree of eutrophication, and oxygen deficiency in combination with a lower water transparency [8,9]. All this can lead to shifts in the existing biological communities and changes in their structures and interactions, which affects the functioning of the entire ecosystem [10,11]. According to some researchers, the dominance of tolerant taxa and the loss of sensitive taxa is evident [12,13]. However, Vilenica et al. [14–16] proposed that man-made lakes (in Croatia) can be biodiversity hotspots that harbor many species and could even contribute to species conservation.

The Chironomidae (Diptera) are one of the most diverse and abundant taxonomic groups of freshwater macroinvertebrates, occurring at all trophic levels and playing an essential role in food webs, energy flow, and aquatic ecosystem functioning [17–21]. Some chironomid species can even be found in marine (e.g., *Halocladius* (s.str.) *varians* (Staeger, 1839) and *Thalassosmittia thalassophila* (Bequaert and Goetghebuer, 1913)) and terrestrial (e.g., *Georthocladius luteicornis* (Goetghebuer, 1941)) environments [22]. Chironomidae inhabit heterogeneous microhabitats, including some very specific environments, e.g., glacial brooks [23], hot springs [24], or live as mayfly parasites [25,26]. Chironomidae larvae are known to feed in a variety of ways and are, therefore, often categorized as collectors, grazers, shredders, and predators [27]; some are even known to feed on the hemolymph/blood of aquatic invertebrates or juvenile vertebrates [25,28]. Due to their considerable diversity and high contribution to the biomass of the total macrozoobenthos (both in terms of number of individuals and number of species), they can also be used as water quality indicators in the bioassessment of aquatic environments [17,29,30].

Chironomids are a holometabolous insect group that develops a pupal stage after four larval stages and before the emergence of adults, which can last from a few hours to several days. After emergence, the skin, known as the pupal exuvia, remains on the water surface for up to seven days, depending on water currents, atmospheric conditions, bacterial activity, etc. [17]. The use of chironomid pupal exuviae for the environmental impact assessment of water bodies is increasingly recognized in aquatic environmental studies [31–34]. The Chironomid Pupal Exuvial Technique (CPET) [35,36] was introduced as the standard method for determining water quality in accordance with the European Union Water Framework Directive [29]. The CPET was initially developed for river biomonitoring [35,37] but is now used in all types of freshwater research [36,38]. CPET is an easy method to use [39–41], and its advantages are seen especially in environments with difficult sampling conditions [42]. In a reservoir or lake, the species that are able to maintain a population reflect the physical and chemical characteristics of the sediments and water [43].

This study was part of an extensive project: "Development of a classification system for assessing the ecological potential of artificial and heavily modified surface waters", with the aim of obtaining ecological data on invertebrate communities in man-made and modified lakes in Croatia. Although water bodies adapted for human utilization can have rather uniform conditions, studied lakes and reservoirs differed in their geographical position, type, degradation level, etc., which could affect biotic communities within in different ways. In Croatia, there was no knowledge of the level of changes in those water bodies from the "natural lake conditions", nor which are the main parameters influencing the macroinvertebrates. There were no data on the benthic biodiversity in such ecosystems, and studying diversity, especially in the profundal zone, can be quite challenging. By applying the CPET method, we mainly aimed to obtain information on chironomid diversity, detect the presence of non-littoral species, analyze the potential influence of environmental factors on the chironomid community, and also to compare communities between the two different ecoregions. The present study is the first to apply the chironomid pupal exuviae method in man-made lakes in Croatia.

2. Materials and Methods

2.1. Study Area

We analyzed 28 man-made or heavily modified lakes in Croatia (Figure 1). These lakes are mainly reservoirs for irrigation, power generation, and water supply. Some of the lakes are natural but modified water bodies (e.g., Sakadaš Lake, Prološko Blato), and some are gravel pits that are now mainly used for recreational purposes (e.g., Koprivnica Šoderica, Novo Čiče, Rakitje). Hydropower plant artificial lakes (e.g., Dubrava, Varaždin, Čakovec) are lakes with high seasonal water level fluctuations (Table 1).

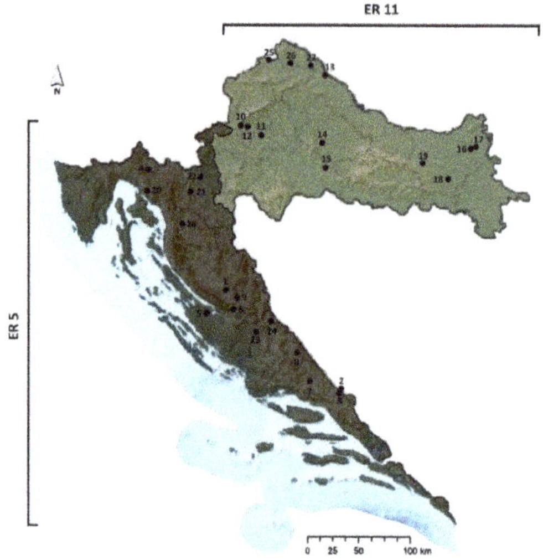

Figure 1. Positions of the study sites of the 28 man-made lakes (ER 5—Dinaric Western Balkan: 1—Opsenica, 2—Ričice-Zeleno Lake, 3—Prološko Blato, 4—Lokve, 5—Vlačine, 6—Razovac, 7—Pranjčević, 8—Peruča, 9—Štikada, 20—Tribalj, 21—Sabljaci, 22—Lešće, 23—Brljan, 24—Golubić, 28—Gusić Polje; ER 11—Pannonian Lowlands: 10—Rakitje, 11—Novo Čiče, 12—Jarun, 13—Koprivnica Šoderica, 14—Popovac, 15—Pakra, 16—Biljsko Lake, 17—Sakadaš Lake, 18—Jošava, 19—Lapovac, 25—Varaždin, 26—Čakovec, 27—Dubrava).

Table 1. Study sites with geographical coordinates, altitude, surface area, and depth of man-made lakes. (ER 5—Dinaric Western Balkan: OP—Opsenica, RI—Ričice-Zeleno Lake, PB—Prološko Blato, LO—Lokve, VL—Vlačine, RA—Razovac, PR—Pranjčević, PE—Peruča, ST—Štikada, TR—Tribalj, SA—Sabljaci, LE—Lešće, BR—Brljan, GO—Golubić, GP—Gusić Polje; ER 11—Pannonian Lowland: VZ—Varaždin, CK—Čakovec, DU—Dubrava, RK—Rakitje, NC—Novo Čiče, JA—Jarun, KS—Koprivnica Šoderica, PP—Popovac, PA—Pakra, BL—Biljsko Lake, SL—Sakadaš Lake, JO—Jošava, LA—Lapovac).

Ecoregion	Study Site	N	E	Altitude (m asl)	Surface Area (km^2)	Depth Max (m)	Depth Mean (m)
ER 5	OP	45.36745	15.66164	575	0.89	6.7	2.9
	RI	43.49671	17.13342	387	1.9	39.6	16
	PB	43.47466	17.12161	269	2.11	5.3	4
	LO	45.36859	14.70618	767	2.1	44.7	33.8
	VL	44.15676	15.42684	122	0.28	10	4
	RA	45.20495	15.74683	9	0.65	7.5	5.8
	PR	43.56273	16.71913	284	0.65	20	6.3
	PE	43.82191	16.55311	330	20.09	31	20.8
	ST	44.29232	15.81408	553	3.34	6.5	4.3
	TR	45.22876	14.66736	60	0.41	4	2.7

Table 1. *Cont.*

Ecoregion	Study Site	N	E	Altitude (m asl)	Surface Area (km^2)	Depth Max (m)	Depth Mean (m)
ER 5	SA	45.22878	15.22602	320	1.7	6.2	3
	LE	45.35769	15.30444	182	1.46	42.5	21
	BR	44.00897	16.03684	187	0.03	18	10
	GO	44.09888	16.22131	307	0.17	6	3
	GP	44.94508	15.11866	430	0.43	6.5	5.4
ER 11	VZ	46.39212	16.16303	191	2.85	8.7	5
	CK	46.31263	16.37021	168	10.5	13.2	7
	DU	46.32488	16.60154	150	16.6	13.4	8
	RK	45.78997	15.8441	120	2	5	3
	NC	45.71174	16.10337	103	0.9	40	15
	JA	45.7816	15.92521	115	6.4	7	4
	KS	46.23603	16.90369	128	1.5	20	8
	PP	45.6377	16.87405	96	0.75	6.5	2
	PA	45.43809	16.89871	104	2.73	6.3	3
	BL	45.59153	18.73839	80	1.25	5	3
	SL	45.60828	18.80041	79	0.12	7	4
	JO	45.32281	18.45246	93	0.79	1.4	1
	LA	45.48027	18.11297	123	0.5	11	5

2.2. Sampling Protocol

Sampling was carried out from July to September 2016 and 2017 on 28 locations. The samples were taken from two ecoregions in Croatia [44]: Dinaric Western Balkan (15 samples) and Pannonian Lowlands (13 samples) (Table 1). The lake surface was sampled once, using a modified Chironomid Pupal Exuvial Technique (CPET) [36] to collect individuals from the studied artificial lakes. Samples were collected using an exuviae hand net with a mesh size of 300 μm along the lake edge with a transect length of 10 m in the area of accumulation of debris and pupal skins. Each sample was preserved in a 4% formaldehyde solution. Further analysis included the separation of the exuviae in the laboratory and the preparation of permanent slides by mounting them in the Berlese mounting medium. The Chironomidae were identified to the lowest taxonomic level using the identification keys of Wiederholm [45] and Langton [46], based on their morphological characteristics.

2.3. Environmental Parameters

The environmental parameters measured at each site with WTW probes were oxygen concentration (oximeter WTW Oxi 330/SET), conductivity (WTW LF 330), and pH (WTW pH 330); water temperature was also measured. Altitude, water level fluctuations, depth of the lake, and consumption of 0.1 M HCl for water hardness were assessed for each lake. Analyses of the water for biological oxygen demand (BOD), chemical oxygen demand (COD_{Mn}), total organic carbon (TOC), total nitrogen (TN), nitrate (NO_3^-), and orthophosphate (PO_4^{3-}) were carried out according to APHA standard methods (1992) in the Central Water Management Laboratory of Hrvatske Vode (legal entity for water management in Croatia). Land use in the vicinity of the man-made lakes was calculated using the Corine Land Cover classification in the GIS tool [47].

2.4. Data Analyses

Prior to data analysis, identified taxa with only one or two collected individuals (exuviae) were excluded from further calculations, data were transformed using square root transformation (both species and environmental data), and analyses considered the relative abundance of each taxon [48]. Only environmental data matrices were normalized [49].

The similarity of chironomid assemblages between different study sites was calculated using cluster analysis and non-metric multidimensional scaling (nMDS) ordination based on the Bray-Curtis similarity matrix. In this ecological study, nMDS was used for the visualization of multidimensional data of identified species and presenting it in two-dimensional space to identify the gradients and patterns between samples. Euclidean distance was used in the resemblance matrix for environmental data. The statistical software used for data analysis was Primer 6 with Permanova [50]. The same software was used to visualize the data–plots from PCA and dbRDA. PCA was applied to obtain the underlying data structure, and dbRDA was applied to find the connections between species composition with investigated environmental variables to determine key factors that are responsible for shaping community structure or how each environmental variable influences the community. For the ordination of species occurrence in relation to environmental variables, Canonical correspondence analysis (CCA) was performed using CANOCO 5.00 software [51]. A total of 20 environmental variables were tested using the Monte Carlo permutation test with 999 permutations to test the significance of the identified community. Only statistically significant ($p < 0.05$) environmental variables were considered for data visualization. All recorded taxa were considered for the calculation of the diversity indices.

3. Results

3.1. Chironomidae Assemblages

A total of 5698 chironomid pupal exuviae were collected during the sampling campaign. One hundred forty-one taxa, of which 97 species from five subfamilies, were identified: Tanypodinae (10 taxa), Buchonomyiinae (1 taxon), Diamesinae (1 taxon), Orthocladiinae (40 taxa) and Chironominae–Chironomini (51 taxa), Tanytarsini (38 taxa) (Tables S1 and S2). Tanytarsini accounted for about 40% of all identified taxa, with *Paratanytarsus* spp. being the most abundant (13.55% of all individuals collected). The most common Tanypodinae species was *Procladius choreus* (2.51%). *Psectrocladius brehmi* (4.07%) and *Cricotopus sylvestris* (2.23%) were the most common Orthocladiinae species, accounting for 26.55% of all individuals collected. The most abundant Chironomini species were *Microchironomus tener* (9.92%), Tanytarsini *Paratanytarsus* spp., and *Constempellina* spp., which accounted for 13.55% and 4.95% of all identified taxa, respectively. The most abundant species was *Procladius choreus*, which was found at 18 surveyed sites in both ecoregions. Of the 141 taxa, 57 were collected in both ecoregions.

In the Dinaric Western Balkan ecoregion, 3470 exuviae were collected, and 106 taxa were identified, with *Paratanytarsus bituberculatus* being the most common species. In the Pannonian Lowlands ecoregion, 2228 individuals were collected, and 94 taxa were identified. The most abundant species was *Microchironomus tener*, followed by the genus *Cricotopus*. The sites with the highest Shannon-Wiener index were Lokve, Štikada, and Gusić Polje (2.97; 2.89; 2.85) in the Dinaric ecoregion and Čakovec, Varaždin, and Dubrava (3.03; 2.85; 2.84) in the Pannonian ecoregion. The lowest index values were at Ričice (0.63) in the Dinaric and Sakadaš Lake (0.69) in the Pannonian ecoregion.

The nMDS analysis revealed similarities between the sites that are strongly dependent on the ecoregion (Figure 2).

In the Dinaric ecoregion, Pološko Blato had the highest Chironomidae exuviae abundance, while Rakitje was the site where we sampled the highest number of exuviae in the Pannonian ecoregion. Biljsko, Sakadaš, Jošava, Popovac Lake, and Sabljaci were the sites with the lowest abundance of exuviae.

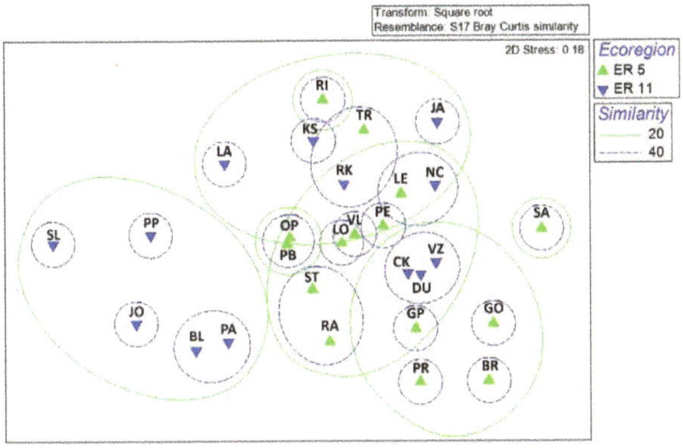

Figure 2. Non-metric multidimensional scaling (nMDS) plot shows the distribution of Chironomidae species based on Bray-Curtis resemblance and group-average hierarchical cluster analysis. Samples from two different ecoregions are shown in different colors. The abbreviations of the study sites correspond to those in Table 1.

3.2. *Environmental and Spatial Drivers*

The results of the CCA ordination analysis of the samples and environmental data are shown in Figure 3. Three of the 20 environmental variables tested were statistically significant ($p < 0.05$). All four axes (eigenvalues of 0.664, 0.588, 0.434, and 0.382) were able to explain a species-environmental correlation of 39.1%. The Monte Carlo permutation test showed that the ordination between exuviae (101 taxa) and environmental variables was significant (first axis F-ratio = 2.26, $p = 0.002$; overall: trace = 5.295, $p = 0.002$). Axis 1 was strongly correlated with ecoregion and TOC (total organic carbon), suggesting that these environmental variables are essential for structuring the Chironomidae assemblages (Figure 3).

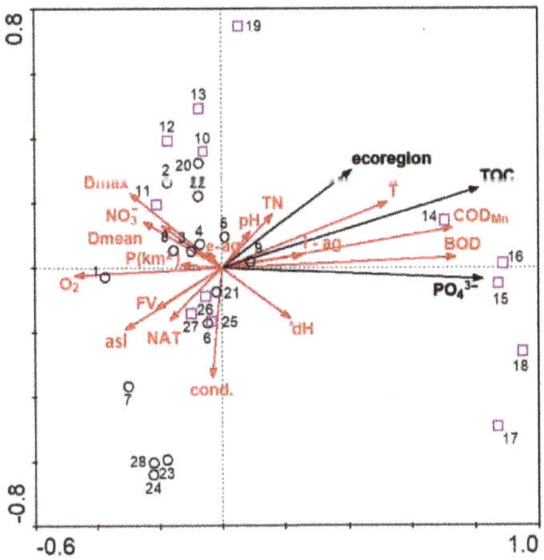

Figure 3. F1 × F2 plane of Canonical correspondence analysis (CCA) based on 28 samples and 20 environmental variables. Legend: T—water temperature; asl—altitude; FV—water level fluctuation; depth of lake (Dmax and Dmean); P (km^2)—surface area of a reservoir; O_2—dissolved oxygen concentration; cond—conductivity; °dH—water hardness; BOD—biological oxygen demand; COD_{Mn}—chemical oxygen demand; TOC—total organic carbon; TN—total nitrogen; NO_3^-—nitrate; PO_4^{3-}—orthophosphate concentration; i-ag—intensive agricultural land use; e-ag—extensive agricultural land use; NAT—natural and near-natural land use. The different ecoregions are labeled

with different symbols (ER 5—black circle; ER 11—purple square), and three black arrows represent three statistically significant environmental variables. The abbreviations are as in Figure 1.

Conductivity was the only statistically significant environmental parameter influencing Chironomidae assemblages in the Dinaric Western Balkan ecoregion with *Cricotopus* spp., *Psectrocladius sordidelus*, and *Chironomus plumosus* (Figure 4). In the Pannonian ecoregion, the community was structured under the influence of TOC, PO_4^{3-} (orthophosphate), and altitude. The Jošava, Sakadaš, and Pakra sites correlated most strongly with PO_4^{3-} values, while TOC had the greatest influence on the community in Popovac and Biljsko Lake (Figure 5a). The species that contributed to the environmental status of these reservoirs are highlighted in Figure 5b.

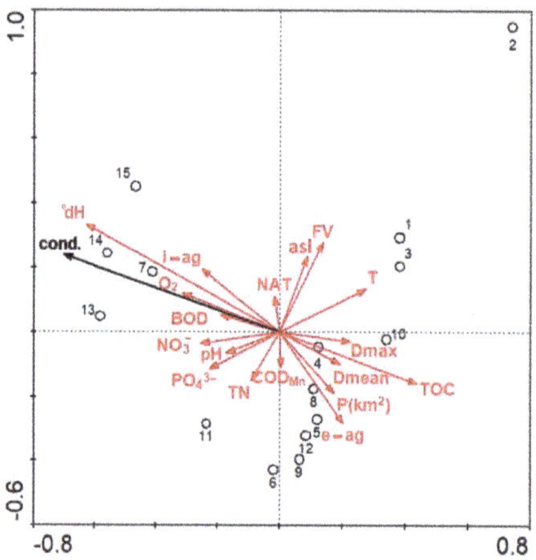

Figure 4. The CCA analysis shows the effects of environmental data on community structure in the Dinaric Western Balkan ecoregion; the statistically significant variable is marked with a black arrow. The circles represent samples: 1—Opsenica, 2—Ričice-Zeleno Lake, 3—Pološko Blato, 4—Lokve, 5—Vlačine, 6—Razovac, 7—Pranjčević, 8—Peruča, 9—Štikada, 10—Tribalj, 11—Sabljaci, 12—Lešće, 13—Brljan, 14—Golubić, 15—Gusić Polje.

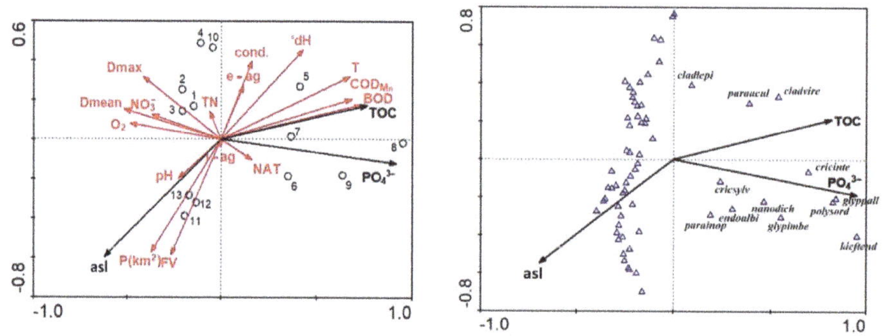

(a) (b)

Figure 5. The CCA analysis shows the effects of environmental data on community structure in the Pannonian Lowlands ecoregion: (**a**) shows the distribution of sampling sites and environmental parameters; (**b**) shows the distribution of Chironomidae taxa with the highest correlated environmental parameters. The statistically significant variables are marked with a black arrow. The circles represent samples: 1—Rakitje, 2—Novo Čiče, 3—Jarun, 4—Koprivnica Šoderica, 5—Popovac, 6—Pakra, 7—Biljsko Lake, 8—Sakadaš Lake, 9—Jošava, 10—Lapovac, 11—Varaždin, 12—Čakovec, 13—Dubrava). Abbreviation of species: cladlepi—*Cladotanytarsus lepidocalcar*, cladvire—*Cladopelma virescens*, cricinte—*Cricotopus intersectus*, cricsylv—*Cricotopus sylvestris*, endoalbi–*Endochironomus albipennis*, glypimbe—*Glyptotendipes imbecilis*, glyppall—*Glyptotendipes pallens*, kieftend—*Kiefferulus tendipediformis*, nanodich—*Nanocladius dichromus*, paraacul—*Parachironomus arcuatus*, parainop—*Paratanytarsus inopertus*, polysord—*Polypedilum sordens*.

The Spearman correlation test showed a correlation between O_2 (oxygen concentration), conductivity, BOD (biological oxygen demand), TOC, and PO_4^{3-} with the highest value of 0.568 ($p < 0.05$). dbRDA results are shown in Figure 6. Together, these five variables could explain 79.12% of the species community data, with TOC being the most significant, with 21.56% of the explained data.

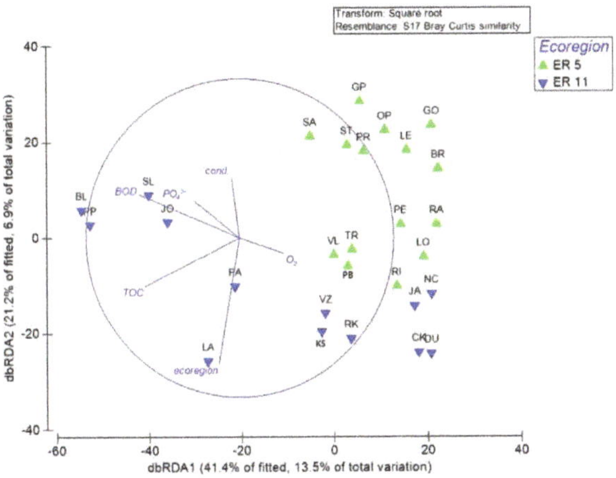

Figure 6. The dbRDA plot shows environmental variables which had the strongest correlation with species composition. Abbreviations are as in Table 1.

Principal component analysis (PCA) (Figure 7) shows that the ordination of samples in the Dinaric Western Balkan ecoregion correlates negatively with total organic carbon and orthophosphate content.

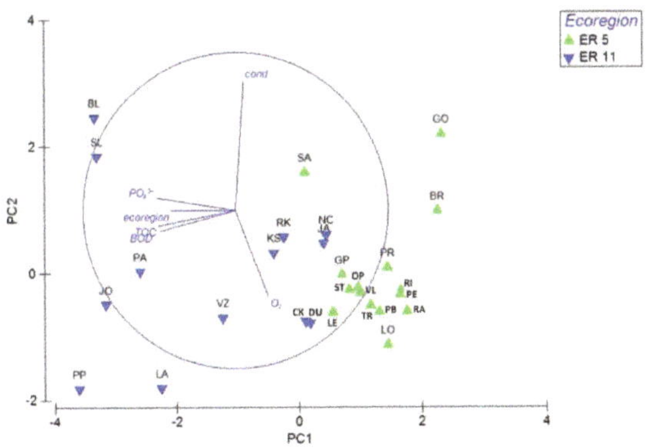

Figure 7. Principal component analysis (PCA) plot with the environmental variables ($p < 0.05$) with the highest correlation value. Abbreviations are as in Table 1.

4. Discussion

The research included 28 artificial water bodies in which 141 Chironomidae taxa, including 97 species, were found. The tribe Tanytarsini comprised about 40% of all identified taxa, mainly *Paratanytarsus* spp., with a relative abundance of 13.55%. According to Mihaljević et al. [52], the most common chironomid taxa identified in macrozoobenthos samples of studied man-made water bodies belonged to the Tanytarsini (*Tanytarsus, Paratanytarsus,* and *Cladotanytarsus*), followed by Tanypodinae and Chironomini. In previous studies on the chironomid assemblages in submerged macrophyte stands in one of the studied water bodies—Sakadaš Lake, Čerba et al. [53,54] recorded 22 and 29 chironomid taxa, respectively, while only two taxa were found in the present study of Sakadaš Lake: *Glyptotendipes pallens* and *Kiefferulus tendipediformis*. This mentioned study was focused mainly on the diversity of Chironomidae larvae in connection to the submerged plant species (*Myriophyllum spicatum* and *Ceratophyllum demersum*) in Lake Sakadaš, but in recent three-year studies [55], there is no, or very scarce macrophyte community in that lake. Macrophytes are known to represent an important habitat for macroinvertebrates [56], and their absence can potentially be directly linked to the lower number of recorded Chironomidae species in this study.

Species richness of aquatic invertebrates in man-made lakes in Croatia is relatively diverse. In a sampling campaign, Vilenica et al. [15,16] collected 32% of all Croatian Odonata species and 25% of the recorded Croatian mayfly species in man-made lakes. According to Čerba et al. [57], 41% of the total identified Chironomidae species have been recorded in reservoirs. In a study of Chironomidae in European reservoirs conducted on the basis of exuviae samples, the dominant representatives belonged to the Chironominae subfamily [38], with *Paratanytarsus* spp. being the most abundant taxa. *Paratanytarsus* is a fairly widespread genus. They are known to favor lentic habitats where they build living tubes and feed on fine organic particles [17,58]. *Paratanytarsus bituberculatus*, the dominant species in our study, is a species characteristic of stagnant waters [46] but can also be found in slow-flowing streams [59]. This corresponds to the fact that some reservoirs have a short water retention time, which makes them a suitable habitat for this species. The most common Tanypodinae species in our study was *Procladius choreus*, a species that inhabits lakes and reservoirs in a wide range of trophic states, from oligotrophic to hypertrophic. This highly adaptable species can live in different water quality ranges [60,61]. Interestingly, in samples of macrozoobenthos from man-made lakes, *P. choreus* was found predominately in littoral sediments at depths between 0.5 and 1 m [52], and the most abundant among Tanypodinae species in the mentioned macrozoobenthos study was *Ablablesmyia monilis*. Since the sampling of larvae that mainly live in greater depths of lakes rather than littoral zones can be challenging, e.g., *Microchironomus tener* [62], the CPET method and the collection of pupal exuviae has proven to be a more favorable method for ecological studies [63,64]. Many studies indicate that the CPET method can be used to obtain a relatively high number of species and genera in contrast to the hand-net sampling method of littoral macroinvertebrate fauna [39,41,65]. According to Anderson and Ferrington [40], the CPET method requires 1/3 less time compared to the macrozoobenthos sampling. Another advantage of the CPET method in biomonitoring is that it is a non-invasive method for the environment [66]. Furthermore, it is a cost-effective and more precise method of species identification, resulting in higher species richness in comparison to hand-net sampling [39,66]. An illustration of some of the advantages of this method is given in Table S3.

Croatia is divided into two ecoregions, with each ecoregion having different environmental characteristics, e.g., climate, substrate, water chemistry, precipitation, etc. [44,67,68]. The clustering of reservoirs based on the Chironomidae assemblages is directly related to the spatial drivers defined by the ecoregion. In man-made lakes of the Pannonian Lowlands ecoregion, the species *Polypedilum sordens, Glyptotendipes imbecilis, G. pallens, Endochironomus albipennis, Cricotopus intersectus, C. sylvestris, Paratanytarsus inopertus,* and *Kiefferulus tendipediformis* were positively correlated with the concentration of orthophosphates and negatively correlated with increasing depth of the reservoir. These species are typically found in sediments with decaying plants and nutrient-rich waters with good oxygen content; some of them can even live in highly eutrophic waters [28]. Based on the orthophosphate content, the Pannonian ecoregion is exposed to greater anthropogenic pressure than the Dinaric ecoregion. Orthophosphate content alters water quality and ecological status, leading to a decline in biodiversity and assemblage structure [19]. The sites with the highest orthophosphate levels are under greater anthropological pressure; these include Jošava, Biljsko, Čakovec, Varaždin, Koprivnica Šoderica, Dubrava, and Pakra. The first two sampling localities (Jošava and Biljsko Lake) showed poor species diversity. One of the environmental variables that has significantly influenced the Chironomidae assemblage is the maximum depth of lakes. Deeper lakes or reservoirs tend to have smaller littoral areas, with less light passing through along the shoreline. Basin shape changes with connectivity, and scaling relationships can be used to estimate the relative size of the littoral region [69]. In the littoral zone of the lake, the water column is more easily mixed under the influence of the wind and, as such, is, therefore, directly related to the oxygen content in the water [70]. Conductivity and, consequently, water hardness were the only two significant environmental parameters influencing Chironomidae assemblages in the Dinaric ecoregion, reflecting the conditions typical of the karst region [71,72].

Almost all major European rivers are dammed or regulated in some parts [73]. The benefits of damming a river include the utilization of the upstream areas for hydropower and water supply, while the downstream areas are flood protection areas [74]. Such modifications also alter the river flow, which in turn leads to changes in the river bed. These changes affect the physical and chemical properties of the water body, can have a negative impact on water quality, and can potentially cause changes in the biotic communities in the newly created ecosystem (e.g., the structure of the benthic community, preventing fish mobility) [75,76]. Additional high anthropogenic pressure on these habitats (e.g., high organic pollutant or nutrient loads) and differences in water levels, together with low habitat diversity and low nutrient availability, contribute to a lower diversity of biota [15,77,78].

5. Conclusions

Chironomidae pupal exuviae are a good tool for ecological studies on man-made or heavily modified water bodies. In general, CPET assesses the "health" of the aquatic ecosystem and offers advantages over standard macroinvertebrate sampling, such as reduced habitat disruption and more precise species identification. In this case, unlike benthic samples from reservoirs, exuviae provided 69 new species for Croatian chironomid fauna, as a great contribution to the checklist. Even though the sampling in this study was possible only on one occasion during the summer months, this research has provided valuable and useful information on the diversity of the Chironomidae community in the man-made and highly altered water bodies, as some of the most vulnerable aquatic habitats to the anthropogenic influence. Furthermore, clear responses from the "CPET chironomid community" to the environmental parameters indicate that the CPET method could be a useful tool for the ecological studies and bioassessment of water quality in Croatia.

Supplementary Materials: The following supporting information can be downloaded at https://www.mdpi.com/article/10.3390/w16202917/s1, Table S1: Chironomidae taxa recorded in Pannonian Lowlands Ecoregion; Table S2: Chironomidae taxa recorded in Dinaric Western Balkan Ecoregion. Table S3: An overview and comparison of sampling efforts and results between benthic and exuviae methodology.

Author Contributions: Conceptualization, Z.M., V.E. and D.Č.; methodology, Z.M., V.E. and D.Č.; validation, Z.M.; formal analysis, V.E.; investigation, Z.M., V.E., N.V. and D.Č.; resources, Z.M. and D.Č.; writing—original draft preparation, V.E., D.Č. and Z.M.; writing—review and editing, Z.M., V.E., N.V. and D.Č.; supervision, Z.M. and D.Č.; project administration, Z.M.; funding acquisition, Z.M. All authors have read and agreed to the published version of the manuscript.

Funding: This research was part of the project: "Development of a classification system for assessing the ecological potential of artificial and heavily modified surface waters", funded by Central Water Management Laboratory of Hrvatske Vode (legal entity for water management in Croatia). Project number: 10-023/16.

Data Availability Statement: The data presented in this study are available on request from the corresponding author.

Acknowledgments: Authors would like to thank the Central Water Management Laboratory of Hrvatske Vode (legal entity for water management in Croatia) for providing land use and water physicochemical data and Nera Vuić, mag. biol. for assisting with the graphical illustrations.

Conflicts of Interest: The authors declare no conflicts of interest. The funders had no role in the design of the study; in the collection, analyses, or interpretation of data; in the writing of the manuscript; or in the decision to publish the results.

References

1. Seaman, W., Jr.; Sprague, L.M. *Artificial Habitats for Marine and Freshwater Fisheries*, 1st ed.; Academic Press: San Diego, CA, USA, 1991; pp. 1–267, ISBN 9780080571171.
2. Petrere, M., Jr. Fisheries in Large Tropical Reservoirs in South America. *Lakes Reserv. Res. Manag.* **1996**, *2*, 111–133. [CrossRef]
3. Habdija, I.; Primc-Habdija, B. *Limnologija—Ekologija Slatkih Voda*; Alfa: Zagreb, Croatia, 2019; pp. 1–352, ISBN 9789533641683.
4. Nilsson, C.; Reidy, C.A.; Dynesius, M.; Revenga, C. Fragmentation and Flow Regulation of the World's Large River Systems. *Science* **2005**, *308*, 405–408. [CrossRef] [PubMed]
5. Petts, G.E. *Impounded Rivers: Perspectives for Ecological Management*; John Wiley: Chichester, UK, 1984; pp. 1–326, ISBN 0471103063.
6. Wang, Z.; Hu, C. Strategies for Managing Reservoir Sedimentation. *Int. J. Sediment Res.* **2009**, *24*, 369–384. [CrossRef]
7. Carmignani, J.R.; Roy, A.H. Ecological Impacts of Winter Water Level Drawdowns on Lake Littoral Zones: A Review. *Aquat. Sci.* **2017**, *79*, 803–824. [CrossRef]
8. Wiatkowski, M. Influence of Słup Dam Reservoir on Flow and Quality of Water in the Nysa Szalona River. *Pol. J. Environ. Stud.* **2011**, *20*, 469–478.
9. Feng, L.; Hou, X.; Zheng, Y. Monitoring and Understanding the Water Transparency Changes of Fifty Large Lakes on the Yangtze Plain Based on Long-Term MODIS Observations. *Remote Sens. Environ.* **2019**, *221*, 675–686. [CrossRef]
10. Tundisi, J.G.; Matsumura-Tundisi, T. Integration of Research and Management in Optimizing Multiple Uses of Reservoirs: The Experience in South America and Brazilian Case Studies. *Hydrobiologia* **2003**, *500*, 231–242. [CrossRef]
11. Xu, Y.; Wang, L.; Tang, Q.; Naselli-Flores, L.; Jeppesen, E.; Han, B.P. The Relationship Between Phytoplankton Diversity and Ecosystem Functioning Changes with Disturbance Regimes in Tropical Reservoirs. *Ecosystems* **2023**, *26*, 752–767. [CrossRef]
12. Davies, S.P.; Jackson, S.K. The Biological Condition Gradient: A Descriptive Model for Interpreting Change in Aquatic Ecosystems. *Ecol. Appl.* **2006**, *16*, 1251–1266. [CrossRef]
13. King, R.S.; Baker, M.E. Use, Misuse, and Limitations of Threshold Indicator Taxa Analysis (TITAN) for Natural Resource Management. In *Application of Threshold Concepts in Natural Resource Decision Making*; Springer: New York, NY, USA, 2014; pp. 231–254, ISBN 978-1-4899-8040-3.
14. Vilenica, M.; Previšić, A.; Ivković, M.; Popijač, A.; Vučković, I.; Kučinić, M.; Kerovec, M.; Gattolliat, J.L.; Sartori, M.; Mihaljević, Z. Mayfly (Insecta: Ephemeroptera) Assemblages of a Regulated Perennial Mediterranean River System in the Western Balkans. *Biologia* **2016**, *71*, 1038–1048. [CrossRef]
15. Vilenica, M.; Vučković, N.; Mihaljević, Z. Littoral Mayfly Assemblages in South-East European Man-Made Lakes. *J. Limnol.* **2019**, *78*, 47–59. [CrossRef]
16. Vilenica, M.; Pozojević, I.; Vučković, N.; Mihaljević, Z. How Suitable Are Man-Made Water Bodies as Habitats for Odonata? *Knowl. Manag. Aquat. Ecosyst.* **2020**, *421*, 13. [CrossRef]
17. Armitage, P.D.; Cranston, P.S.; Pinder, L.C.V. *The Chironomidae: Biology and Ecology of Non-biting Midges*; Chapman and Hall: London, UK, 1995; pp. 1–572, ISBN 041245260X.
18. Moog, O. *Fauna Aquatica Austriaca. Katalog zur Autökologischen Einstufung Aquatischer Organismen Österreichs*, 2nd ed.; Bundesministerium für Land- und Forstwirtschaft, Umwelt und Wasserwirtschaft Wasserwirtschaftskataster: Wien, Austria, 2002; pp. 1–670, ISBN 3-85-174-044-0.
19. Fraterrigo, J.M.; Downing, J.A.; Fraterrigo, J.M.; Downing, J.A. The Influence of Land Use on Lake Nutrients Varies with Watershed Transport Capacity. *Ecosystems* **2008**, *11*, 1021–1034. [CrossRef]
20. Motta, L.; Massaferro, J. Climate and Site-Specific Factors Shape Chironomid Taxonomic and Functional Diversity Patterns in Northern Patagonia. *Hydrobiologia* **2019**, *839*, 131–143. [CrossRef]
21. Belle, S.; Goedkoop, W. Functional Diversity of Chironomid Communities in Subarctic Lakes across Gradients in Temperature and Catchment Characteristics. *Limnology* **2021**, *22*, 5–16. [CrossRef]
22. Cranston, P.S. Introduction. In *The Chironomidae: Biology and Ecology of Nonbiting Midges*; Armitage, P.D., Cranston, P.S., Pinder, L.C.V., Eds.; Chapman and Hall: London, UK, 1995; pp. 1–5.

23. Hazra, N.; Saha, G.K.; Mazumdar, A.; Chaudhuri, P.K. Records of Chironomids of the Tribe Pentaneurini (Diptera: Chironomidae) in the Eastern Himalayas of India. *Ann. Soc. Entomol. Fr.* **2011**, *47*, 330–339. [CrossRef]
24. Lindegaard, C. Classification of Water-Bodies and Pollution. In *The Chironomidae: Biology and Ecology of Non-Biting Midges*; Armitage, P.D., Cranston, P.S., Pinder, L.C.V., Eds.; Chapman and Hall: London, UK, 1995; pp. 385–404, ISBN 041245260X.
25. Schiffels, S. Commensal and Parasitic Chironomidae. *Denisia* **2014**, *33*, 393–407.
26. Vilenica, M.; Ergović, V.; Mihaljević, Z. Mayfly (Ephemeroptera) Assemblages of a Pannonian Lowland Mountain, with First Records of the Parasite Symbiocladius Rhithrogenae (Zavrel, 1924) (Diptera: Chironomidae). *Ann. Limnol.* **2018**, *54*, 31. [CrossRef]
27. Berg, M.B. Larval Food and Feeding Behaviour. In *The Chironomidae. Biology and Ecology of Nonbiting Midges*; Armitage, P.D., Cranston, P.S., Pinder, L.C.V., Eds.; Chapman and Hall: London, UK, 1995; pp. 136–168, ISBN 041245260X.
28. Vallenduuk, H.J. Chironomini Larvae of Western European Lowlands (Diptera: Chironomidae). Keys with Notes to the Species. In *With a Redescription of Glyptotendipes (Caulochironomus) Nagorskayae and a First Description of Glyptotendipes (Caulochironomus) Kaluginae New Species*, 1st ed.; Erik Mauch Verlag: Dinkelscherben, Germany, 2017; pp. 1–216, ISBN 0935-333X.
29. EC Parliament and Council. Directive 2000/60/EC of the European Parliament and of the Council Establishing a Framework for Community Action in the Field of Water Policy. *OJEU*, 22 December 2000; pp. 1–72.
30. Milošević, D.; Čerba, D.; Szekeres, J.; Csányi, B.; Tubić, B.; Simić, V.; Paunović, M. Artificial Neural Networks as an Indicator Search Engine: The Visualization of Natural and Man-Caused Taxa Variability. *Ecol. Indic.* **2016**, *61*, 777–789. [CrossRef]
31. Ruse, L. Chironomid Pupal Exuviae as Indicators of Lake Status. *Arch. Hydrobiol.* **2002**, *3*, 367–390. [CrossRef]
32. Raunio, J.; Muotka, T. The Use of Chironomid Pupal Exuviae in River Biomonitoring: The Importance of Sampling Strategy. *Arch. Hydrobiol.* **2005**, *164*, 529–545. [CrossRef]
33. Calle-Martínez, D.; Casas, J.J. Chironomid Species, Stream Classification, and Water-Quality Assessment: The Case of 2 Iberian Mediterranean Mountain Regions. *J. N. Am. Benthol. Soc.* **2006**, *25*, 465–476. [CrossRef]
34. Raunio, J.; Paasivirta, L. Emergence Patterns of Lotic Chironomidae (Diptera: Nematocera) in Southern Finland and the Use of Their Pupal Exuviae in River Biomonitoring. *Fundam. Appl. Limnol.* **2008**, *170*, 291–301. [CrossRef]
35. Wilson, R.S.; Bright, P.L. The Use of Chironomid Pupal Exuviae for Characterising Streams. *Freshw. Biol.* **1973**, *3*, 283–302. [CrossRef]
36. Wilson, R.S.; Ruse, L.P. *A Guide to the Identification of Genera of Chironomid Pupal Exuviae Occurring in Britain and Ireland (Including Common Genera from Northern Europe) and Their Use in Monitoring Lotic and Lentic Fresh Waters*; Freshwater Biological Association: Ambleside, UK, 2005; pp. 1–176, ISBN 978-0-900386-73-2.
37. Wilson, R.S.; McGill, J.D. A New Method of Monitoring Water Quality in a Stream Receiving Sewage Effluent, Using Chironomid Pupal Exuvia. *Water Res.* **1977**, *11*, 959–962. [CrossRef]
38. Bitušík, P.; Svitok, M.; Kološta, P.; Hubková, M. Classification of the Tatra Mountain Lakes (Slovakia) Using Chironomids (Diptera, Chironomidae). *Biologia* **2006**, *61*, 191–201. [CrossRef]
39. Ferrington, L.C.; Blackwood, M.A.; Wright, C.A.; Crisp, N.H.; Kavanaugh, J.L.; Schmidt, F.J. *A Protocol for Using Surface-Floating Pupal Exuviae of Chironomidae for Rapid Bioassessment of Changing Water Quality*; International Association of Hydrological Sciences: Wien, Austria, 1991; pp. 181–190, ISBN 0947571086.
40. Anderson, A.M.; Ferrington, L.C. Resistance and Resilience of Winter-Emerging Chironomidae (Diptera) to a Flood Event: Implications for Minnesota Trout Streams. *Hydrobiologia* **2013**, *707*, 59–71. [CrossRef]
41. Saulino, H.H.L.; Cañedo-Argüelles, M.; Trivinho-Strixino, S.; Gorni, G.R.; Corbi, J.J. Chironomid Pupal Exuviae Communities Support the "Field of Dreams" Hypothesis after the Riparian Vegetation Recovery in Headwater Urban Streams. *Ecol. Indic.* **2021**, *127*, 107766. [CrossRef]
42. Downes, B.; Barmuta, L.A.; Fairweather, P.G.; Faith, D.P.; Keough, M.J.; Lake, P.S.; Mapstone, B.D.; Quinn, G.P. *Monitoring Ecological Impacts. Concepts and Practice in Flowing Waters*; Cambridge University Press: Cambridge, UK, 2002; pp. 1–421, ISBN 9780521771573.
43. Ruse, L.P. Chironomid (Diptera) species recorded from UK lakes as pupal exuviae. *J. Entomol. Acarol. Res.* **2013**, *45*, e13. [CrossRef]
44. Illies, J. *Limnofauna Europaea. Eine Zusammenstellung Aller Die Europäische Binnengewässer Bewohnenden Mehrzelligen Tierarten Mit Angaben Über Ihre Verbreitung Und Ökologie*, 2nd ed.; Gustav Fischer Verlag: Stuttgart, Germany, 1978; pp. 1–532, ISBN 0013-2969.
45. Wiederholm, T. *Chironomidae of the Holarctic Region, Keys and Diagnoses*; Publishing House of the Swedish Research Councils: Stockholm, Sweden, 1983; pp. 1–457, ISBN 25459.
46. Langton, P.H. *A Key to Pupal Exuviae of West Palearctic Chironomidae*; Private Publishing: Huntingdon, UK, 1991; pp. 1–386.
47. CLC Hrvatska. *CORINE Land Cover Hrvatska*; Hrvatska Agencija za Okoliš i Prirodu: Zagreb, Croatia, 2013.
48. McCune, B.; Grace, J.B. *Analysis of Ecological Communities*, 2nd ed.; MjM Software Design: Gleneden Beach, OR, USA, 2002; pp. 1–300, ISBN 0-9721290-0-6.
49. Pollesch, N.L.; Dale, V.H. Normalization in Sustainability Assessment: Methods and Implications. *Ecol. Econ.* **2016**, *130*, 195–208. [CrossRef]
50. Clarke, K.R.; Gorley, R.N. *Primer V6: User Manual—Tutorial*; PRIMER-E Limited: Plymouth, UK, 2006; pp. 1–190.
51. ter Braak, C.J.F.; Šmilauer, P. *CANOCO Reference Manual and User's Guide to Canoco for Windows: Software for Canonical Community Ordination (Version 4)*; Microcomputer Power: Ithaca, NY, USA, 1998; pp. 1–500.

52. Mihaljević, Z.; Vučković, N.; Čerba, D. Littoral Chironomid Assemblages in the Reservoirs along the Eastern Adriatic Coast. In *Book of Abstracts, Proceedings of the 20th International Symposium on Chironomidae, Trento, Italy, 2–8 July 2017*; Trident: Barnala, India, 2017.
53. Čerba, D.; Mihaljević, Z.; Vidaković, J. Colonisation of Temporary Macrophyte Substratum by Midges (Chironomidae: Diptera). *Ann. Limnol.-Int. J. Lim.* **2010**, *46*, 181–190. [CrossRef]
54. Čerba, D.; Mihaljević, Z.; Vidaković, J. Colonisation Trends, Community and Trophic Structure of Chironomid Larvae (Chironomidae: Diptera) in a Temporal Phytophylous Assemblage. *Fundam. Appl. Limnol.* **2011**, *179*, 203–214. [CrossRef]
55. Vilenica, M.; Brigić, A.; Ergović, V.; Koh, M.; Alegro, A.; Šegota, V.; Rimac, A.; Rumišek, M.; Mihaljević, Z. Taxonomic and Functional Odonata Assemblage Metrics: Macrophyte–Driven Changes in Anthropogenically Disturbed Floodplain Habitats. *Hydrobiologia* **2024**, *851*, 3787–3807. [CrossRef]
56. Tarkowska-Kukuryk, M. Water Soldier *Stratiotes aloides* L. (Hydrocharitaceae) as a Substratum for Macroinvertebrates in a Shallow Eutrophic Lake. *Pol. J. Ecol.* **2006**, *3*, 441–451.
57. Čerba, D.; Koh, M.; Ergović, V.; Mihaljević, Z.; Milošević, D.; Hamerlík, L. Chironomidae (Diptera) of Croatia with Notes on the Diversity and Distribution in Various Habitat Types. *Zootaxa* **2020**, *4780*, 259–274. [CrossRef]
58. Grzybkowska, M. Diel Drift of Chironomidae in a Large Lowland River (Central Poland). *Netherland J. Aquat. Ecol.* **1992**, *26*, 355–360. [CrossRef]
59. Malo, J.; Morais, M.; Pinto, P.A. Contribution to the Knowledge of the Chironomid Fauna (Diptera) in Southern Portugal. *Ann. Limnol.-Int. J. Lim.* **1998**, *34*, 165–170. [CrossRef]
60. Saether, O.A. Chironomid Communities as Water Quality Indicators. *Holarct. Ecol.* **1979**, *2*, 65–74. [CrossRef]
61. Čerba, D.; Vlaičević, B.; Davidović, R.A.; Koh, M.; Ergović, V.; Turković Čakalić, I. Chironomidae in Shallow Water Bodies of a Protected Lowland Freshwater Floodplain Ecosystem. *Sci. Prog.* **2023**, *106*, 00368504231172653. [CrossRef]
62. Moller Pillot, H.K.M. *Chironomidae Larvae, Biology and Ecology of the Chironomini*; KNNV Publishing: Zeist, The Netherlands, 2009; pp. 1–270, ISBN 9789004284418.
63. Raunio, J.; Paavola, R.; Muotka, T. Effects of Emergence Phenology, Taxa Tolerances and Taxonomic Resolution on the Use of the Chironomid Pupal Exuvial Technique in River Biomonitoring. *Freshw. Biol.* **2007**, *52*, 165–176. [CrossRef]
64. Ruse, L. Classification of Nutrient Impact on Lakes Using the Chironomid Pupal Exuvial Technique. *Ecol. Indic.* **2010**, *10*, 594–601. [CrossRef]
65. Kranzfelder, P.; Anderson, A.M.; Egan, A.T.; Mazack, J.E.; Bouchard, R.W.; Rufer, M.M.; Ferrington, L.C. Use of Chironomidae (Diptera) Surface-Floating Pupal Exuviae as a Rapid Bioassessment Protocol for Water Bodies. *J. Vis. Exp.* **2015**, *101*, e52558. [CrossRef]
66. Bouchard, R.W.; Ferrington, L.C. The Effects of Subsampling and Sampling Frequency on the Use of Surface-Floating Pupal Exuviae to Measure Chironomidae (Diptera) Communities in Wadeable Temperate Streams. *Environ. Monit. Assess.* **2011**, *181*, 205–223. [CrossRef]
67. Kottek, M.; Grieser, J.; Beck, C.; Rudolf, B.; Rubel, F. World Map of the Köppen-Geiger Climate Classification Updated. *Meteorol. Z.* **2006**, *15*, 259–263. [CrossRef] [PubMed]
68. Bonacci, O.; Roje-Bonacci, T. Drastic Hydrological Changes Caused by Hydroelectrical Development in Karst: A Case of the Karst River Zrmanja (Croatia). *Environ. Earth Sci.* **2015**, *74*, 6767–6777. [CrossRef]
69. Seekell, D.; Cael, B.; Norman, S.; Byström, P. Patterns and Variation of Littoral Habitat Size Among Lakes. *Geophys. Res. Lett.* **2021**, *48*, e2021GL095046. [CrossRef]
70. Thorp, J.H. Functional Relationships of Freshwater Invertebrates. In *Thorp and Covich's Freshwater Invertebrates*, 4th ed.; Elsevier: Amsterdam, The Netherlands, 2015; pp. 65–82, ISBN 978-0-12-385026-3.
71. Griffiths, H.I.; Krystufek, B.; Reed, J.M. *Balkan Biodiversity: Pattern and Process in the European Hotspot*; Kluwer Academic Publishers: London, UK, 2004; pp. 1–332, ISBN 978-90-481-6732-6.
72. Popijač, A.; Sivec, I. First Records of the Alpine Stonefly Species *Protonemura julia* Nicolai, *1983* (Insecta, Plecoptera) in Croatia. *Nat. Croat.* **2009**, *18*, 83–89.
73. Tockner, K.; Zarfl, C.; Robinson, C.T. *Rivers of Europe*, 2nd ed.; Elsevier: Amsterdam, The Netherlands, 2022; pp. 1–942, ISBN 978-0-08-102612-0.
74. Kron, W.; Müller, O. Efficiency of Flood Protection Measures: Selected Examples. *Water Policy* **2019**, *21*, 449–467. [CrossRef]
75. Moss, B. *Ecology of Freshwater: Man and Medium, Past to Future*; Blackwell Science: Oxford, UK, 1998; pp. 1–584, ISBN 0632035129.
76. Sternberg, R. Damming the River: A Changing Perspective on Altering Nature. *Renew. Sustain. Energy Rev.* **2006**, *10*, 165–197. [CrossRef]
77. Allan, J.D. Landscapes and Riverscapes: The Influence of Land Use on Stream Ecosystems. *Source Annu. Rev. Ecol. Evol. Syst.* **2004**, *35*, 257–284. [CrossRef]
78. Brasil, L.S.; Juen, L.; Cabette, H.S.R. The Effects of Environmental Integrity on the Diversity of Mayflies, Leptophlebiidae (Ephemeroptera), in Tropical Streams of the Brazilian Cerrado. *Ann. Limnol.-Int. J. Lim.* **2014**, *50*, 325–334. [CrossRef]

water

MDPI

Article

Diversity and Ecological Functions of Fungal Communities in Tangchi Hot Spring in Lujiang (China)

Feng-Qin Zhang [1,*], Jun Liu [2] and Xiao-Ju Chen [1]

1 College of Biological and Environmental Engineering, Chaohu University, Chaohu 238000, China
2 School of Food and Biological Engineering, Hefei University of Technology, Hefei 230601, China
* Correspondence: 053042@chu.edu.cn

Abstract: Hot springs contain many microorganisms, and a great deal of research has been conducted on the bacteria in hot springs but less on the fungi. Microbial diversity research in hot springs will help to further develop and utilize the microbial resources from hot springs. This is the first study to analyze the fungal diversity and ecological functions in Tangchi hot spring (China). In this paper, the fungal ITS sequences in the water samples of Tangchi hot spring were sequenced by Illumina MiSeq high-throughput sequencing technology and bioinformatically analyzed to explore the fungal diversity and ecological functions in the hot spring. The results revealed that the dominant fungal phylum was Ascomycota, followed by Basidiomycota, Chytridiomycota, and Olpidiomycota, and the dominant fungal genera were *Rhizophydium*, *Aureobasidium*, *Rhodotorula,* and *Sclerotinia* in Tangchi hot spring. The FUNGuild functional analysis results revealed that the dominant guilds in Tangchi hot spring were plant pathogens followed by the undefined saprotroph and the animal endosymbiont–animal pathogen–endophyte–plant pathogen–undefined saprotroph. This study can fill the gap in people's understanding of fungal resources and offer theoretical support for understanding the fungal diversity and fungal resource development and utilization in Tangchi hot spring.

Keywords: Tangchi hot spring; fungi; high-throughput sequencing technology; FUNGuild; diversity

1. Introduction

Tangchi hot spring is located in Lujiang County, Hefei City, China. The Tangchi hot spring water is mainly used for seismic monitoring, and the hot water flowing from the well is supplied to the local people and tourists for bathing. The water chemistry type of Tangchi hot spring is HCO_3 Ca·Mg [1].

In this study, we used high-throughput sequencing technology to analyze the diversity of the fungi in Tangchi hot spring. High-throughput sequencing is widely used in many fields of scientific research and precision medical testing, such as genome sequencing [2], transcriptome sequencing [3], metagenomic sequencing [4], epigenetic sequencing [5], exon capture sequencing [6], and target region sequencing [7]. In recent years, high-throughput sequencing has been used for analyzing the community structure of microbes due to its low cost, high throughput, and rapid generation of large amounts of data, avoiding the limitation of the culture conditions and the objective reflection of the microbial diversity [8–10]. Illumina MiSeq runs PE300 mode and can undertake microbial taxonomic sequencing, amplicon sequencing, and nucleic acid aptamer sequencing. The bacterial communities of Tangchi hot spring have already been analyzed by Illumina MiSeq PE300 (Illumina, CA, USA) in the early stage [11]. In this study, Illumina MiSeq PE300 (Illumina, CA, USA) was also used to classify and sequence the fungi from Tangchi hot spring.

The products produced by fungi have characteristics comparable to or even better than classically derived products, and their enormous potential has been recognized [12]. Thermophilic fungi have more advantages and more significant potential regarding the production processes. Thermophiles are a class of microorganisms that live in high-temperature environments, belong to the category of extremophiles, and are essential microbial resources that can be developed and utilized. In addition to thermophilic bacteria, thermophilic fungi are also found in hot springs [13–15]. Moreover, thermophilic fungi producing cellulase were isolated and identified from Sungai Pinang hot spring, Riau Province, Indonesia [16]. Twenty-eight out of the thirty-five thermophilic fungi isolated in caves and hot springs (Cardonal region of Hidalgo, Mexico) can produce proteases [17].

FUNGuild (Fungi Functional Guild) can classify and analyze fungal communities through microecological guild [18]. Guild is a concept in microecology. It involves a class of species (regardless of their close genetic relationship) that can similarly use the same kinds of environmental resources. The functional categorization of fungi can be accomplished by combining fungal species categorization with functional guild categorization through bioinformatics methods. According to FUNGuild, the ecological functions (pathotroph, saprotroph, and symbiotroph) of the fungi in the Tangchi hot spring water can be predicted. The ecological function analysis of the fungi in hot springs can provide reference value for the development and utilization of hot springs.

This is the first analysis of the fungal communities and ecological functions in Tangchi hot spring. This study can fill the gap in people's understanding of fungal resources and provide theoretical support for developing and utilizing the fungal resources in Tangchi hot spring.

2. Materials and Methods

2.1. Sampling

Water samples were taken from Tangchi hot spring in Lujiang (China). Samples collected were packed into sterilized containers. The pH and temperature of the samples were detected on site. Then, water samples were filtered in the laboratory with 0.22 μM sterile microporous filter membranes to concentrate the microorganisms, and the filtered membranes were stored at −80 °C for subsequent DNA extraction.

2.2. Fungal ITS High-Throughput Sequencing

Total genomic DNA of microbes was extracted from Tangchi hot spring samples with E.Z.N.A™ Mag-Bind Soil DNA Kit (OMEGA, Norcross, GA, USA). The integrity and concentration of DNA samples were detected by agarose gel electrophoresis and Qubit 4.0 (Thermo, Waltham, MA, USA), respectively. The endogenous transcriptional spacer ITS region was used as the target sequence and amplified with universal primers ITS1F (CTTGGTCATTTAGAGGAAGTAA)/ITS2R (GCTGCGTTCTTCATCGATGC) [19]. The PCR amplification products were purified and homogenized, and a small fragment sequencing library was constructed using Paired-End method. The constructed library underwent library quality inspection, and then the qualified library was sequenced three times using Illumina MiSeq PE300 (Illumina, CA, USA).

Following the barcode and amplification primer sequence, every sample's data were split from the off-machine data, the barcode sequence and primer sequence were cut off, and the reads of the sample were spliced [20]. Raw tags obtained by splicing were filtered to remove low-quality tags and chimeras that did not meet the length required to generate clean tags [21].

2.3. Bioinformatics Analysis

Using the Usearch software (version 11.0.667) the obtained high-quality sequences were merged and clustered into Operational Taxonomic Units (OTUs) with a similarity higher than 97%. After chimeric sequences and singleton OTU were removed, the fungal OTU representative sequences were blasted to UNITE database [22] by Blast classification algorithm. Four α-diversity indices consisting of Ace, Chao1, Shannon, and Simpson indices were figured for each sample by Mothur software (version 3.8.31).

2.4. Function Prediction

The fungi in Tangchi hot spring were classified into ecological functional groups according to the different ways of absorbing and using environmental resources by FUNGuild (Version 1.0) [23].

2.5. Data Accession

ITS gene sequencing data were accepted at Sequence Read Archive (SRA) under accession No. SRR20831103.

3. Results

3.1. Water Sample Parameters of Tangchi Hot Spring

The temperature and pH of the water samples from Tangchi hot spring were detected on site as 63.9 °C and pH 8.2, respectively. It can be seen that Tangchi hot spring is alkaline.

3.2. Sequencing Results of Hot Spring Samples

There were 56,052 reads and 56,039 valid tags from Tangchi hot spring (Table 1). After splitting and removing redundancy, the hot spring sample sequences were clustered with OTUs under a similarity of 97%, and 541 OTUs were obtained from Tangchi hot spring. The OTU information of the fungi with relative abundance values above 1% is displayed in Figure 1 and Table 2.

Table 1. Raw and clean data of fungal communities in Tangchi hot spring.

	Barcode	SeqNum	BaseNum	MeanLen	MinLen	MaxLen
Raw Data	ACGAGTG	56,052	18,071,057	322.4	44	493
Clean Data	ACGAGTG	56,039	15,711,131	280.36	101	451

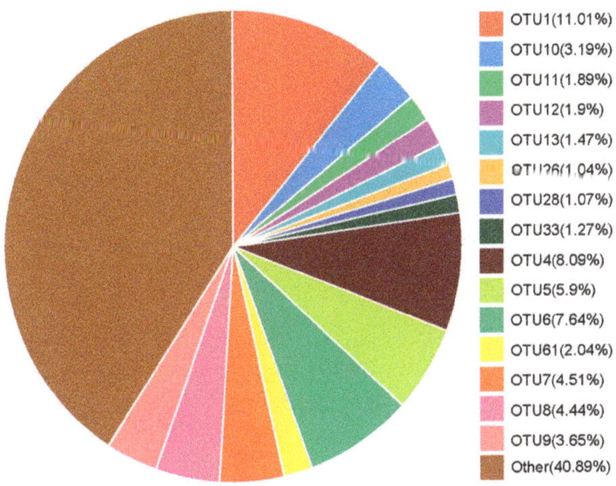

Figure 1. OTU relative abundance of fungal communities in Tangchi hot spring.

Table 2. OTU taxonomy information of fungal communities in Tangchi hot spring.

OTU ID	Taxonomy
OTU1	d__Fungi; p__Chytridiomycota; c__Rhizophydiomycetes; o__Rhizophydiales; f__Rhizophydiaceae; g__Rhizophydium; s__unclassified_Rhizophydium
OTU4	d__Fungi; p__Ascomycota; c__Leotiomycetes; o__Helotiales; f__Sclerotiniaceae; g__Sclerotinia; s__Sclerotinia_sclerotiorum_SH1152486.08FU
OTU6	d__Fungi; p__Ascomycota; c__Dothideomycetes; o__Dothideales; f__Aureobasidiaceae; g__Aureobasidium; s__unclassified_Aureobasidium
OTU5	d__Fungi; p__Basidiomycota; c__Tremellomycetes; o__Cystofilobasidiales; f__Mrakiaceae; g__Tausonia; s__Tausonia_pullulans_SH1650607.08FU
OTU7	d__Fungi; p__Olpidiomycota; c__GS17; o__GS17; f__unclassified_GS17;g__unclassified_GS17; s__unclassified_GS17
OTU8	d__Fungi; p__Basidiomycota; c__Microbotryomycetes; o__Sporidiobolales; f__Sporidiobolaceae; g__Rhodotorula; s__Rhodotorula_babjevae_SH2272266.08FU
OTU9	d__Fungi; p__Ascomycota; c__Dothideomycetes; o__Pleosporales; f__Phaeosphaeriaceae; g__Sclerostagonospora; s__Sclerostagonospora_lathyri_SH1525096.08FU
OTU10	d__Fungi; p__Basidiomycota; c__Microbotryomycetes; o__Sporidiobolales; f__Sporidiobolaceae; g__Rhodotorula; s__Rhodotorula_mucilaginosa_SH1181860.08FU
OTU61	d__Fungi; p__Ascomycota; c__Saccharomycetes; o__Saccharomycetales; f__Saccharomycetaceae; g__Saccharomyces; s__Saccharomyces_cerevisiae_SH1613178.08FU
OTU12	d__Fungi; p__Ascomycota; c__Dothideomycetes; o__Pleosporales; f__Pleosporaceae; g__Alternaria; s__unclassified_Alternaria
OTU11	d__Fungi; p__Ascomycota; c__Dothideomycetes; o__Capnodiales; f__Cladosporiaceae; g__Cladosporium; s__Cladosporium_exasperatum_SH2320217.08FU
OTU13	d__Fungi;p__Basidiomycota; c__Tremellomycetes; o__Cystofilobasidiales; f__Mrakiaceae;g__Mrakia; s__unclassified_Mrakia
OTU33	d__Fungi; p__Basidiomycota; c__Agaricostilbomycetes; o__Agaricostilbales; f__Ruineniaceae; g__Ruinenia; s__Ruinenia_clavata_SH1575142.08FU
OTU28	d__Fungi; p__Ascomycota; c__Dothideomycetes; o__Dothideales; f__Aureobasidiaceae; g__Aureobasidium; s__Aureobasidium_pullulans_SH1515141.08FU
OTU26	d__Fungi; p__Chytridiomycota; c__unclassified_Chytridiomycota; o__unclassified_Chytridiomycota; f__unclassified_Chytridiomycota; g__unclassified_Chytridiomycota; s__unclassified_Chytridiomycota

3.3. *Composition of Fungal Communities*

The OTU representative sequences from Tangchi hot spring were aligned. There were 10 phyla, 35 classes, 83 orders, 180 families, 293 genera, and 414 species in Tangchi hot spring.

At the phylum level, the fungi with relative abundance below 1% were assigned to Other. The fungi communities of Tangchi hot spring were mainly classified into four phyla (Figure 2a). There were Ascomycota (43.82%), Basidiomycota (32.38%), Chytridiomycota (16.55%), and Olpidiomycota (4.51%).

At the genus level, the fungi with relative abundance below 1% were assigned to Other. The fungal communities of Tangchi hot spring were classified into 18 genera (Figure 2b), which were *Rhizophydium* (11.68%), *Aureobasidium* (8.71%), *Rhodotorula* (8.56%), *Sclerotinia* (8.26%), *Tausonia* (5.90%), unclassified_GS17 (4.51%), *Sclerostagonospora* (3.65%), *Cladosporium* (2.12%), *Saccharomyces* (2.04%), *Alternaria* (1.92%), *Cystofilobasidium* (1.63%), *Mrakia* (1.47%), *Powellomyces* (1.43%), *Ruinenia* (1.27%), *Sporobolomyces* (1.26%), *Boothiomyces* (1.16%), *Aspergillus* (1.05%), and unclassified_Chytridiomycota (1.04%). There were unclassified fungi in Tangchi hot spring, indicating that the fungal genera in the hot spring have certain genera novelty.

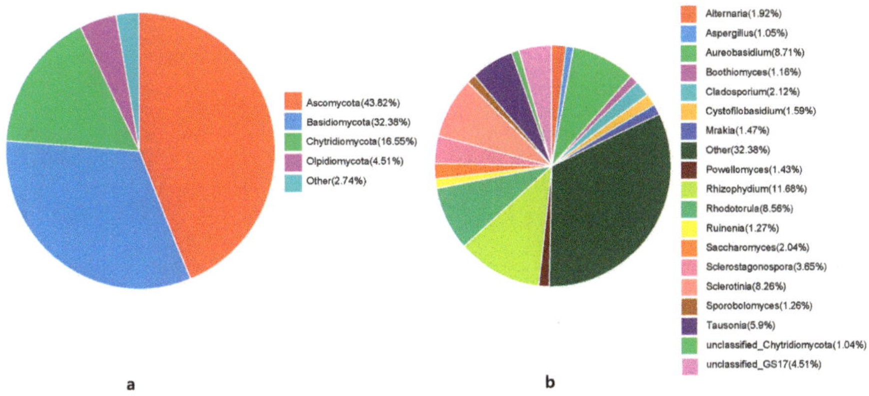

Figure 2. Phylum and genus level distributions of fungal communities in Tangchi hot spring. (**a**) Phylum level; (**b**) genus level.

Based on the taxonomic comparison of the sample, the classification of dominant species (relative abundance $\geq$ 1%) was selected to understand the dominant microorganisms' evolutionary relationships and abundance differences in the sequenced environmental samples in terms of the entire taxonomic system. The taxonomic systematic composition (including domain, phylum, class, order, family, genus, species, and OTU) of the fungi in Tangchi hot spring was analyzed (Figure 3).

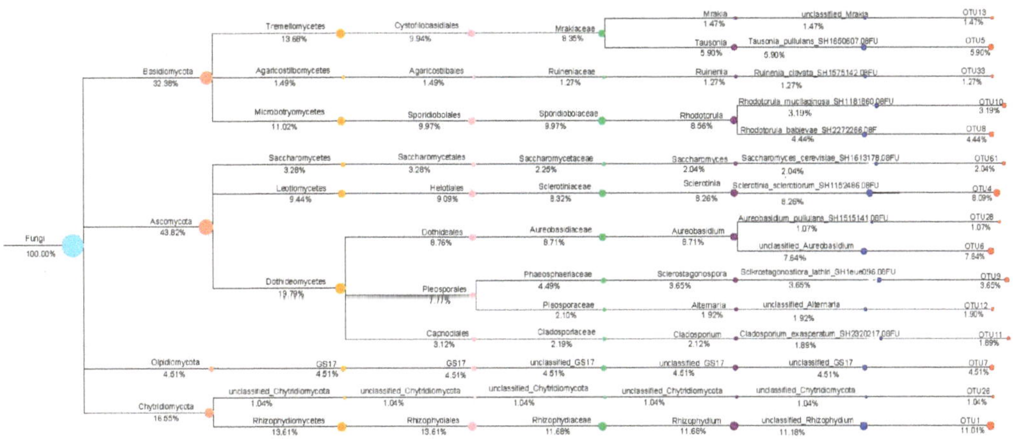

Figure 3. Taxonomic systematic compositional tree of fungi in Tangchi hot spring.

3.4. Fungal Alpha Diversity Analysis

Alpha diversity can represent the richness, diversity, and evenness of a microbial community. The Shannon index, Simpson index, and Good's coverage are used to estimate the community diversity. The Chao1 and ACE estimators can estimate the community richness, and the Shannon even index can assess the community evenness [24]. The alpha diversity indices of the fungi in Tangchi hot spring and single factor analysis of variance were calculated (Table 3). The results suggested that the Shannon, Chao1, and ACE indices of the fungi in Tangchi hot spring were high; in addition, the Simpson index of the fungi was small. They indicated that the fungi in Tangchi hot spring were rich and diverse. The coverage indices of Tangchi hot spring were close to 1, reflecting that the sequencing results represented the real conditions of the water samples.

Table 3. Fungal alpha diversity indexes of Tangchi hot spring.

Sample	Shannon	Chao	Ace	Simpson	Shannoneven	Coverage
Tangchi	4.339195	541.965517	542.70909	0.037157	0.689481	0.99976

3.5. Fungal Function Prediction

FUNGuild analyzed the ecological function category of the fungi in Tangchi hot spring. Only OTUs and taxons with credible levels of 'Probable' and 'Highly Probable' were selected for analysis.

According to the functional prediction of the utilization pathways of the same environmental resources based on the fungal community, the definable functional trophic modes can be detected, which can be classified into three trophic modes (pathotroph, saprotroph, and symbiotroph) and four mutual cross-trophic modes (pathotroph–saprotroph, pathotroph–symbiotroph, saprotroph–symbiotroph, and pathotroph–saprotroph–symbiotroph) according to the absorption and utilization of environmental resources (Figure 4). Pathotroph (35.4%) dominates in Tangchi hot spring, followed by pathotroph–saprotroph (29.3%) and saprotroph (28.7%).

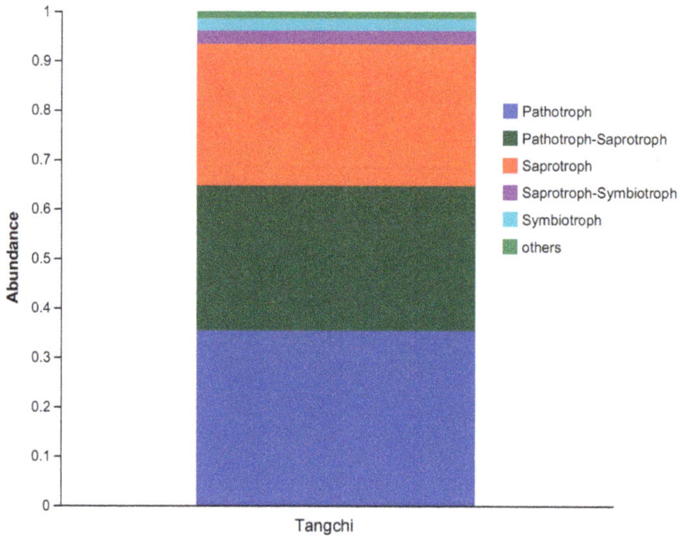

Figure 4. Trophic modes of fungi in Tangchi hot spring.

The above seven trophic modes were further assigned, among which the guilds with the relative abundance of dominant OTUs greater than 1% belonged to 11 guilds, and the guilds with dominant OTUs less than 1% were combined into others (Figure 5). The dominant guilds in Tangchi hot spring were plant pathogen (27.3%), undefined saprotroph (22.2%), animal endosymbiont–animal pathogen–endophyte–plant pathogen–undefined saprotroph (17.7%), animal pathogen–undefined saprotroph (7.3%), animal pathogen (6.8%), wood saprotroph (2.8%), endophyte–litter saprotroph–soil saprotroph–undefined saprotroph (2.1%), soil saprotroph–undefined saprotroph (1.8%), animal pathogen–plant pathogen–undefined saprotroph (1.5%), endophyte (1.2%), and plant saprotroph (1.0%).

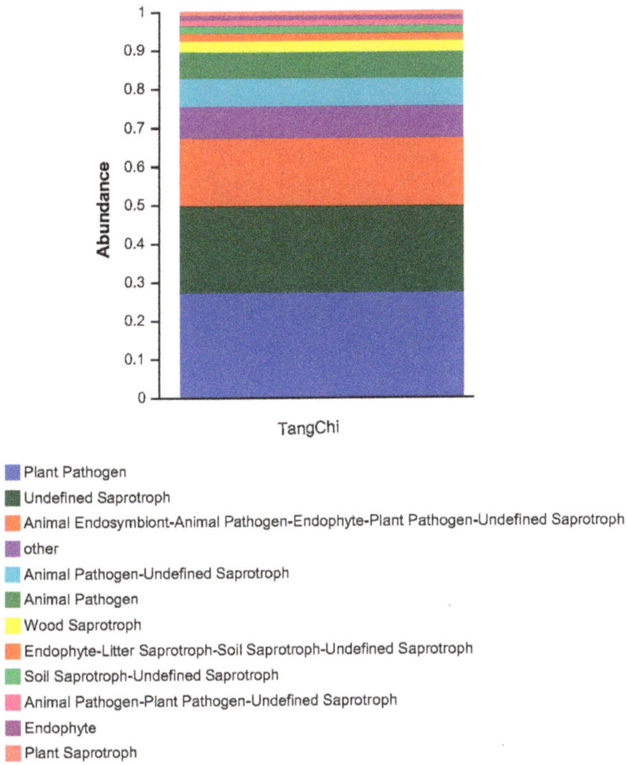

Figure 5. Guilds of fungi in Tangchi hot spring.

4. Discussion

In this study, the diversity of the fungi in Tangchi hot spring was studied by high-throughput sequencing technology, which broke the technical bottleneck that a large number of microorganisms could not be isolated and cultured. At the phylum taxonomic level, the prevalent fungal phyla in Tangchi hot spring were Ascomycota (43.82%), Basidiomycota (32.39%), and Chytridiomycota (16.5%). Ascomycota and Basidiomycota have also been found to be the dominant phyla in the hot springs of the soda lakes in the Kenyan rift valley (83.3% and 15.8%, respectively) [4], Boiling Springs Lake (63.9% and 8.20%) [25], and Haikou hot spring (43.88% and 18.37%, respectively) [26]. Ascomycota and Basidiomycota have a good heat-resistant mechanism. Further, 89 species of heat-resistant fungi are reported worldwide, of which more than half (49 species) belong to Ascomycota and Basidiomycota [27,28]. Ascomycota is the largest fungal phylum and an extremely diverse group, with an estimated global species richness of approximately 154,500 species [29]. Ascomycota has two types: saprophytic and parasitic. Saprophytic Ascomycetes grow on various substrates, such as soil, decaying wood, animal and plant residues, etc. Parasitic species can cause plant diseases, with a few parasitizing on humans, animals, or insects. Some Ascomycetes form lichens in symbiosis with algae, and mycorrhizal fungi in symbiosis with higher plants [30]. Ascomycetes are used for industrial applications, food production, and flavoring, and the sporophores of morels and truffles are precious edible mushrooms. Basidiomycota can degrade lignin and plant polysaccharides [31], and they also produce secondary metabolites such as terpenoids and meroterpenoids [32]. To our knowledge, Chytridiomycota, 16.55% in Tangchi hot spring, has not yet been found in any other hot springs. Chytridiomycota is the most abundant basal fungal phylum involved in the vital processes in both terrestrial and aquatic ecosystems. However, its diversity and richness remain mysterious [33]. Chytridiomycota is a low-level fungus that parasitizes the larvae stage of black flies, mosquitoes, scale insects, and ants [34].

At the genus taxonomic level, the dominant fungal genera in Tangchi hot spring were Rhizophydium, Aureobasidium, Rhodotorula, Sclerotinia, Tausonia, unclassified_GS17, Sclerostagonospora, Cladosporium, Saccharomyces, Alternaria, Cystofilobasidium, Mrakia, Powellomyces, Ruinenia, Sporobolomyces, Boothiomyces, Aspergillus, and unclassified_Chytridiomycota. We know that there were several unclassified fungi genera in Tangchi hot spring, demonstrating a certain novelty in the fungi of Tangchi hot spring. As far as we know, Rhizophydium has not been reported in other hot springs but was detected in Tangchi hot spring. So far, Aureobasidium, existing in Tangchi hot spring, has also been discovered in the hot springs of the soda lakes in the Kenyan rift valley [4]. Aureobasidium is a dimorphic fungal genus with both primary yeast and secondary filamentous cells and an extensive distribution. The Aureobasidium genus can utilize various waste products and synthesize natural products such as pullulan, fructooligosaccharides, and melanin [35]. Cystobasidium, discovered in Tangchi hot spring, is also found in the hot springs in Haikou (China) (4.15%) [26], while no other common genera were found.

This study predicted the ecological functions of Tangchi hot spring by FUNGuild. From the perspective of trophic type, pathotroph, pathotroph–saprotroph, and saprotroph are the major trophic types, which may be related to Ascomycota and Basidiomycota being the dominant fungal phyla. Some Ascomycota are saprotrophs, which are important decomposers and can decompose refractory material. They play an important role in nutrient cycling [36]. However, some Ascomycota are animal pathogens, plant pathogens, or animal–plant pathogens, which are a few parasites, poultry and insects, and many parasitic plants. These aspects are consistent with the analysis results that the dominant guilds of Tangchi hot spring are plant pathogens and undefined saprotrophs. The dominant fungal genera, *Rhizophydium* and *Aureobasidium*, in Tangchi hot spring are not annotated in FUNGuild, which indicates that the annotations in FUNGuild are not comprehensive and its database needs to be supplemented and improved [23]. *Sclerotinia sclerotiorum* is a plant pathogen that can cause disease in more than 400 broad-leaved plant species, including vegetable crops, resulting in major yield losses [37,38]. *Rhodotorula*, 8.54% in Tangchi hot spring, is a common saprophytic yeast and an opportunistic (when animal immune function is impaired) pathogen in animals (including humans). It can induce skin infections in chickens and marine animals, lung infections and otitis media in sheep and cattle [39], and infections such as folliculitis and keratitis in humans [40,41]. According to reports, *Rhodotorula* has also been found in the hot springs of the soda lakes in the Kenyan rift valley and Saudi springs [4,42]. This study demonstrates the potential pathogens in hot springs that pose safety risks to human health, and some populations are not suitable for soaking in these hot springs. It can be seen that the FUNGuild function prediction results are consistent with the composition of the fungal communities of Tangchi hot spring.

5. Conclusions

This paper analyzed the fungal diversity of Tangchi hot spring in Lujiang using high-throughput sequencing technology for the first time. The results showed that there are rich fungal resources in Tangchi hot spring. Moreover, there are unclassified fungi in this hot spring, which shows that the fungi in Tangchi hot spring have a certain novelty. Pathogenic fungi account for a large proportion in Tangchi hot spring. There are still a large number of unknown fungal functions in Tangchi hot spring, and further research will be conducted in this area in the future. This study offers support for understanding the fungal diversity, resource development, and utilization in Tangchi hot spring, and it provides health guidance for soaking in this hot spring.

Author Contributions: Conceptualization, F.-Q.Z.; data preparation, F.-Q.Z.; methodology, F.-Q.Z.; analysis, F.-Q.Z. and J.L.; writing—original draft preparation, F.-Q.Z.; writing—review and editing, F.-Q.Z. and J.L.; supervision, X.-J.C.; funding acquisition X.-J.C. All authors have read and agreed to the published version of the manuscript.

Funding: This work was supported by the Projects of Chaohu University (grant numbers XLZ202308, kj21fdzy02, kj21kctd03), Key Research and Development Program Project in Anhui Province (grant number 202204c06020059), the Project of Educational Commission in Anhui province (grant number kj2021A1019), the School Level Key Construction Discipline of Food Science and Engineering in Chaohu University (grant number kj16zdjsxk02) and Outstanding Youth Research Project of Anhui Educational Committee (grant number 2023AH030092).

Data Availability Statement: Data are contained within the article.

Conflicts of Interest: The authors have no conflicts of interest.

References

1. Xia, Q.; Zhang, C.L.; Gao, Y.; Hu, C.; Zha, S.X.; Ma, M.G. Status and reliability appraisal of geothermal resources in Hefei city. *Geol. Anhui* **2015**, *25*, 222–226. (In Chinese) [CrossRef]
2. Heinze, B. Sequencing of complete Chloroplast genomes. *Methods Mol. Biol.* **2021**, *2222*, 89–105. [CrossRef]
3. Lu, T.; Zou, X.; Yang, Z.; Lei, F.; Chen, Y.; Liu, G.; Sun, B.; Li, Y. Transcriptome analysis of goat ovaries and follicles based on high-throughput sequencing. *J. South China Agric. Univ.* **2020**, *41*, 23–32. [CrossRef]
4. Won, K.H.; Kim, D.; Shin, D.; Hur, J.; Lee, H.; Heo, J.; Oh, J.D. High-throughput sequencing-based metagenomic and transcriptomic analysis of intestine in piglets infected with salmonella. *J. Anim. Sci. Technol.* **2022**, *64*, 1144–1172. [CrossRef]
5. Zhang, T.; Huang, P.D.; Qiu, C. Progresses in epigenetic studies of asthma from the perspective of high-throughput analysis technologies: A narrative review. *Ann. Transl. Med.* **2022**, *10*, 493. [CrossRef] [PubMed]
6. Mengual, X.; Mayer, C.; Burt, T.O.; Moran, K.M.; Dietz, L.; Nottebrock, G.; Pauli, T.; Young, A.D.; Brasseur, M.V.; Kukowka, S.; et al. Systematics and evolution of predatory flower flies (Diptera: Syrphidae) based on exon-capture sequencing. *Syst. Entomol.* **2022**, *48*, 250–277. [CrossRef]
7. Victorino, Í.M.M.; Berruti, A.; Orgiazzi, A.; Voyron, S.; Bianciotto, V.; Lumini, E. High-Throughput DNA Sequence-Based Analysis of AMF Communities. *Methods Mol. Biol.* **2020**, *2146*, 99–116. [CrossRef]
8. Liu, Z.; Iqbal, M.; Zeng, Z.; Lian, Y.; Zheng, A.; Zhao, M.; Li, Z.; Wang, G.; Li, Z.; Xie, J. Comparative analysis of microbial community structure in the ponds with different aquaculture model and fish by high-throughput sequencing. *Microb. Pathog.* **2020**, *142*, 104101. [CrossRef] [PubMed]
9. Sadeepa, D.; Sirisena, K.; Manage, P.M. Diversity of microbial communities in hot springs of Sri Lanka as revealed by 16S rRNA gene high-throughput sequencing analysis. *Gene* **2022**, *812*, 146103. [CrossRef] [PubMed]
10. Kioroglou, D.; Mas, A.; Portillo, M.C. High-throughput sequencing approach to analyze the effect of aging time and barrel usage on the microbial community composition of red wines. *Front. Microbiol.* **2020**, *11*, 562560. [CrossRef]
11. Zhang, F.Q.; Liu, J.; Chen, X.J. Comparative analysis of bacterial diversity in two hot springs in Hefei, China. *Sci. Rep.* **2023**, *13*, 5832. [CrossRef]
12. Katharina, A.S.; Johannes, K.; Joachim, M.; Andreas, L. Effects of Environmental and Nutritional Conditions on Mycelium Growth of Three Basidiomycota. *Mycobiology* **2024**, *52*, 124–134. [CrossRef]
13. Kumar, A.; Suman, R.; Sneha, K.; Anamika, K.; Puja, B.; Pammi, K. Potentials and applications of thermophilic fungi: A general review. *Invertis J. Renew. Energy* **2021**, *11*, 39–41. [CrossRef]
14. Salano, O.A.; Makonde, H.M.; Kasili, R.W.; Wangai, L.N.; Nawiri, M.P.; Boga, H.I. Diversity and distribution of fungal communities within the hot springs of soda lakes in the Kenyan rift valley. *Afr. J. Microb. Res.* **2017**, *11*, 764–775. [CrossRef]
15. Liu, K.H.; Ding, X.W.; Salam, N.; Zhang, B.; Tang, X.F.; Deng, B.W.; Li, W.J. Unexpected fungal communities in the Rehai thermal springs of Tengchong infuenced by abiotic factors. *Extremophiles* **2018**, *22*, 525–535. [CrossRef]
16. Saryono, S.; Novianty, R.; Suraya, N.; Piska, F.; Devi, S.; Pratiwi, N.W.; Ardhi, A. Molecular identification of cellulase-producing thermophilic fungi isolated from Sungai Pinang hot spring, Riau Province, Indonesia. *Biodiversitas* **2022**, *23*, 1457–1465. [CrossRef]
17. Jazmín, L.C.A.; Guadalupe, G.S.; Karina, G.G.; Rafael, O.H.D. Biotechnological insights into extracellular enzyme production by thermotolerant fungi from hot springs and caves: Morphology, pellets formation, and protease production. *Biotechnol. Appl. Biochem.* **2024**, *71*, 536–552. [CrossRef]
18. Yan, K.; Pei, Z.H.; Meng, L.N.; Zheng, Y.; Wang, L.; Feng, R.Z.; Li, Q.Z.; Liu, Y.; Zhao, X.M.; Wei, Q.; et al. Determination of community structure and diversity of seed-vectored endophytic fungi in Alpinia zerumbet. *Front. Microb.* **2022**, *13*, 814864. [CrossRef]
19. Wang, W.; Cai, T.; Yang, Y.T.; Guo, H.; Shang, Z.; Shahid, H.; Zhang, Y.R.; Qiu, S.R.; Zeng, X.N.; Xu, X.L.; et al. Diversity of fungal communities on diseased and healthy cinnamomum burmannii fruits and antibacterial activity of secondary metabolites. *Microbiol. Spectr.* **2023**, *11*, e0008023. [CrossRef]
20. Zhu, A.M.; Wu, Q.; Liu, H.L.; Sun, H.L.; Han, G.D. Isolation of rhizosheath and analysis of microbial community structure around roots of Stipa grandis. *Sci. Rep.* **2022**, *12*, 2707. [CrossRef]
21. Lu, J.J.; Mao, D.C.; Li, X.; Ma, Y.Q.; Luan, Y.Q.; Cao, Y.; Luan, Y.P. Changes of intestinal microflora diversity in diarrhea model of KM mice and effects of Psidium guajava L. as the treatment agent for diarrhea. *J. Infect. Public Health* **2020**, *13*, 16–26. [CrossRef]

22. Kõljalg, U.; Nilsson, R.H.; Abarenkov, K.; Tedersoo, L.; Taylor, A.F.S.; Bahram, M.; Bahram, M.; Bates, S.T.; Bruns, T.D.; Bengtsson-Palme, J.; et al. Towards a unified paradigm for sequence-based identification of fungi. *Mol. Ecol.* **2013**, *22*, 5271–5277. [CrossRef] [PubMed]
23. Nguyen, N.H.; Song, Z.W.; Bates, S.T.; Branco, S.; Tedersoo, L.; Menke, J.; Schillinge, J.S.; Kennedy, P.G. FUNGuild: An open annotation tool for parsing fungal community datasets by ecological guild. *Fungal Ecol.* **2016**, *20*, 241–248. [CrossRef]
24. Chen, D.Y.; Li, C.W.; Feng, L.; Zhang, Z.Z.; Zhang, H.M.; Cheng, G.Y.; Li, D.S.; Zhang, G.Q.; Wang, H.N.; Chen, Y.X. Analysis of the influence of living environment and age on vaginal fungal microbiome in giant pandas (Ailuropoda melanoleuca) by high throughput sequencing. *Microb. Pathog.* **2018**, *115*, 280–286. [CrossRef]
25. Wilson, M.S.; Siering, P.L.; White, C.L.; Hauser, M.E.; Bartles, A.N. Novel archaea and bacteria dominate stable microbial communities in North America's largest hot spring. *Microb. Ecol.* **2008**, *56*, 292–305. [CrossRef] [PubMed]
26. Shu, W.; Tian, X.Y.; Zhao, H.W. Diversity of fungi and bacteria in hot springs in Haikou, Hainan Province. *Acta Microbiol. Sin.* **2020**, *60*, 1999–2011. (In Chinese) [CrossRef]
27. Mouchacca, J. Heat tolerant fungi and applied research: Addition to the previously treated group of strictly thermotolerant species. *World J. Microb. Biot.* **2007**, *23*, 1755–1770. [CrossRef]
28. Mouchacca, J. Thermotolerant fungi erroneously reported in applied research work as possessing thermophilic attributes. *World J. Microb. Biot.* **2000**, *16*, 869–880. [CrossRef]
29. Bánki, O.; Roskov, Y.; Döring, M.; Ower, G.; Vandepitte, L.; Hobern, D.; Remsen, D.; Schalk, P.; DeWalt, R.E.; Ma, K.; et al. *Catalogue of Life Checklist (Version 2023-01-12)*; Catalogue of Life: Leiden, The Netherlands, 2023. [CrossRef]
30. Wanasinghe, D.N.; Nimalrathna, T.S.; Xian, L.Q.; Faraj, T.K.; Xu, J.C.; Mortimer, P.E. Taxonomic novelties and global biogeography of Montagnula (Ascomycota, Didymos- phaeriaceae). *MycoKeys* **2024**, *101*, 191–232. [CrossRef] [PubMed]
31. Manici, L.M.; Caputo, F.; Sabata, D.D.; Fornasier, F. The enzyme patterns of Ascomycota and Basidiomycota fungi reveal their different functions in soil. *Appl. Soil. Ecol.* **2024**, *196*, 105323. [CrossRef]
32. Sum, W.C.; Ebada, S.S.; Matasyoh, J.C.; Stadler, M. Recent progress in the evaluation of secondary metabolites from Basidiomycota. *Curr. Res. Biotechnol.* **2023**, *6*, 100155. [CrossRef]
33. Blaalid, R.; Khomich, M. Current knowledge of Chytridiomycota diversity in Northern Europe and future research needs. *Fungal Biol. Rev.* **2021**, *36*, 42–51. [CrossRef]
34. Kaczmarek, A.; Boguś, M.I. Fungi of entomopathogenic potential in Chytridiomycota and Blastocladiomycota, and in fungal allies of the Oomycota and Microsporidia. *IMA Fungus* **2021**, *12*, 29. [CrossRef]
35. Wang, P.; Jia, S.L.; Liu, G.L.; Chi, Z.; Chi, Z.M. *Aureobasidium* spp. and their applications in biotechnology. *Process Biochem.* **2022**, *116*, 72–83. [CrossRef]
36. Sun, Q.; Wu, H.L.; Chen, F.; Kang, J.H. Fungal community diversity and structure in rhizosphere soil of different crops in the arid zone of central Ningxia. *Microbiology* **2019**, *46*, 2963–2972. [CrossRef]
37. Lu, G. Engineering *Sclerotinia Sclerotiorum* resistance in oilseed crops. *Afr. J. Biotechnol.* **2003**, *2*, 6. [CrossRef]
38. Bueno, E.A.; Oliveira, M.B.; Andrade, R.V.; Petrofeza, L. Effect of different carbon sources on proteases secreted by the fungal pathogen *Sclerotinia sclerotiorum* during Phaseolus vulgaris infection. *Genet. Mol. Res GMR* **2012**, *11*, 2171. [CrossRef] [PubMed]
39. Wirth, F.; Goldani, L.Z. Epidemiology of *Rhodotorula*: An emerging pathogen. *Interdiscip. Perspect. Infect. Dis.* **2012**, *2012*, 465717–465723. [CrossRef] [PubMed]
40. Ring, J.; Jaeger, T.; Anliker, M.D. *Rhodotorula mucilaginosa* infection in Li-Fraumeni-like syndrome: A new pathogen in folliculitis. *Brit. J. Dermatol.* **2011**, *164*, 1120–1122. [CrossRef]
41. Giovannini, J.; Lee, R.; Zhang, S.X.; Jun, A.S.; Bower, K.S. *Rhodotorula keratitis*: A rarely encountered ocular pathogen. *Case Rep. Ophthalmol.* **2014**, *5*, 302–310. [CrossRef] [PubMed]
42. Selim, S.; Hassan, S.; Hagagy, N.; Kraková, L.; Grivalský, T.; Pangallo, D. Assessment of microbial diversity in Saudi springs by culture dependent and independent methods. *Geomicrobiol. J.* **2016**, *34*, 443–453. [CrossRef]

water

MDPI

Article

Assessment of the Divergent Influence of Natural and Non-Seasonal Hydrological Fluctuations on Functional Traits and Niche Characteristics of Plant Guilds along the Xiangxi River, China

Xiaoling Li [1], Wenxiong Yi [1,†], Shaoting Xu [1,†], Di He [1], Qifeng Min [1], Gong Chen [1], Jin Yang [1], Danli Deng [1,*], Zhengjian Yang [1], Guiyun Huang [2], Meixiang Hu [2] and Chen Ye [3,*]

1 Engineering Research Center of Eco-Environment in Three Gorges Reservoir Region, Ministry of Education, Hubei International Scientific and Technological Center of Ecological Conservation and Management in the Three Gorges Area, College of Biological and Pharmaceutical Science, China Three Gorges University, Yichang 443002, China; lixiaolinggz@126.com (X.L.); yi13381691544@163.com (W.Y.); jmyangjin@126.com (J.Y.); yangzj1984@ctgu.edu.cn (Z.Y.)

2 Rare Plants Research Institute of Yangtze River, Three Gorges Corporation, Yichang 443000, China; huangguiyun@ctg.com.cn (G.H.); humeixiang@ctg.com.cn (M.H.)

3 Key Laboratory of Aquatic Botany and Watershed Ecology, Wuhan Botanical Garden, Chinese Academy of Sciences, Wuhan 430074, China

* Correspondence: dengdanli@ctgu.edu.cn (D.D.); yechen922@wbgcas.cn (C.Y.)

† These authors contributed equally to this work.

Abstract: The reservoir water level fluctuation zones (RWLFZs) and the natural riparian zones (NRZs) are two riparian ecosystems with dramatically opposite hydrological rhythms that notably influence the plant guilds. However, little is known about the discrepancies of the functional traits and niche characteristics of plant guilds in the RWLFZs and NRZs under different flooding rhythms. The aims of this study were to assess the divergent influence of natural and non-seasonal hydrological fluctuations on functional traits and niche characteristics of plant guilds. The results showed that 78 vascular plant species were identified in the riparian zones of the Xiangxi River basin. The dominant species were annuals in the two riparian ecosystems and their percentage increased temporally from 65.79% in the NRZs to 67.34% in the RWLFZs. Compared with the NRZs, the specific leaf area, vegetation coverage and the aboveground biomass in the RWLFZs significantly increased by 74%, 27% and 19.6%, respectively, while the water-use efficiency of the RWLFZ decreased by 59.6% and there was no significant difference in the net photosynthetic rate between them, which showed that annuals in the RWLFZs adopted the R adaptation strategy of being fast-growing with a short lifespan and quickly acquiring and investing resources by altering leaf morphology, including expanding the leaf area. The Simpson dominance index of RWLFZs was significantly higher than that of NRZ. Thus, counter-seasonally hydrological alterations had significant effects on functional traits of dominant species in the RWFLZs. Moreover, the highly adaptable and widely distributed species with larger niche breadths and high important values usually had a higher niche overlap value in the RWLFZs than in the NRZs, which showed that the competition in the plant communities after experiencing anti-seasonal flooding was much more intraspecific than interspecific and would facilitate the expansion of species niches. Our findings imply that the species with large niche breadths and high important values should be prioritized in ecological restoration efforts in the newly formed hydro-fluctuation zones of the TGR.

Keywords: non-seasonally and naturally hydrological fluctuations; functional traits; niche characteristics; Xiangxi River; Three Gorges Reservoir

1. Introduction

The operation of reservoirs exerts a substantial influence on the plant community, including species composition, community structure and distribution patterns [1,2]. It is estimated that two-thirds of the world's fresh water flowing to the oceans is obstructed by more than about 50,000 large dams—defined as those higher than 15 m in operation worldwide [3]. Ali demonstrated that the plant species in the water level fluctuation zones (WLFZs) of the Nubia Reservoir in Sudan, Africa, decreased sharply with only half of the plant species existing after 30 years of water storage [4]. Moreover, dams anticipate large-scale plant population disappearance and loss of terrestrial and arboreal species in submerged lowland forest areas [1]. So far, however, few experimental studies have been conducted to test the effect of water level fluctuations on the functional traits and niche characteristics of dominant species of plant communities in the reservoir water level fluctuation zones (RWLFZs) with a large vertical height difference compared with the natural riparian zones (NRZs). Therefore, it is of great importance to study the functional traits and niche of dominant species for evaluating the effects of the reservoir impoundments on vegetation community characteristics for better prediction of RWLFZ plant community dynamics and for restoration efforts of riparian ecosystems.

The Three Gorges Dam (TGD), the world's largest hydroelectric project, was built on the Yangtze River in China. The Three Gorges Dam (TGD) began operating in June 2003, and the water level of the Three Gorges Reservoir (TGR) rose from approximately 60 m above sea level (asl) to 135 m asl above the original riverbank. The second impoundment occurred in October 2006, and the water level rose to 156 m asl. With the full operation of the TGD in 2010, the water level fluctuates from 145 m asl. in summer (May to September) to 175 m asl in winter (October to April) with a vertical height difference of 30 m in the reservoir, resulting in the formation of an unnatural RWLFZ with a total area of 350 km^2 [5]. The reversal of flooding time and prolonged submergence duration have dramatically altered the hydrological regimes and plant community in the RWLFZs [5–7]. For example, before the construction of the Three Gorges Dam (TGD), the vertical drop of the NRZs of the main stream of Yangtze River was 10~47 m asl. and the NRZ was only inundated by natural summer flooding [8–10]. Thus, the complete operation of TGD generated different types of WLFZs, including the RWLFZs and the NRZs. For the RWLFZ, it had not experienced flooding before the impoundment of the TGR, and the vegetation in the RWLFZ is subjected to the anti-seasonally hydrological rhythm (i.e., summer exposure and winter inundation). The newly formed ecosystems in the RWLFZs offer special challenges to restoration managers because of the reversal of flooding time and prolonged submergence duration, their short growing seasons, and long recovery periods.

Hydrological regime alterations anticipate a series of unfavorable vegetation changes, such as vegetation type, distribution patterns, community species composition and diversity, coverage and functional traits of plant guilds [11,12]. Riverine plant communities exhibit varied responses to diverse water level regimes [13]. The impoundment at the TGR is likely to exert distinct influences on the various vegetation types within the WLFZs. Furthermore, the duration of flooding is identified as a significant factor impacting species distribution [14]. After 20 times of flood–dry cycles in the RWLFZ of the TGR since the impoundment in 2003, some previous dominant species that had not experienced multiple impoundments gradually died out, showing a certain degree of degradation, and some new dominant species would appear and grow more and more vigorously [13]. These alterations would accelerate the disappearance of original vegetation, especially trees and shrub, the formation of permanent waters and trigger the appearance of novel plant communities, which gradually adapt to the environment of the RWLFZ in the TGR [12,15,16]. The species composition and structure of the plant community are still extremely unstable [17]. Zhang et al. [18] documented that the perennial plants in the reservoir's riparian zone exhibited a markedly higher photosynthetic capacity per unit area, leaf nitrogen concentration, and stem mass ratio compared to their upland counterparts. However, little is known about the effects of prolonged submergence (4–8 months) and the reversal of inundation time of TGR on functional traits and niche characteristics of plant guilds in different WLFZs.

Previous investigations have noted that specific plant species are capable of withstanding extended periods of winter flooding in the Three Gorges Reservoir Area, facilitated by a "physiological compensation" mechanism that enhances their photosynthetic capacity [18,19]. Perennial vegetation within the WLFZ displayed a notably enhanced area-based photosynthetic capacity when compared to upland species. In addition, these WLFZ perennials exhibited significantly greater stomatal conductance, leaf nitrogen concentration, and stem mass ratio than their upland counterparts [18]. However, it remains largely unclear whether dominant plant species in the RWLFZs altered functional traits in response to the flood–dry–flood alternate changes after 20 cycles of operation of the TGR compared with the NRZs.

Niche theory has become an important means of evaluating biodiversity mechanisms, species coexistence and competition and the status of populations in natural plant communities [16,20]. The niche may be well characterized by two important niche indices, i.e., niche breadth and niche overlap [16,21]. The size of niche breadth of species depends on the utilization of environmental resources and adaptive capacity of species to environments [22]. The changes in soil environments caused by anthropogenic activities lead to competition for limited resources among plant species and adaptability to environments [23]. Because niche breadth and overlap can be quantitatively evaluated, they can effectively reflect the range of resource characteristics across which a species exists in its specific ecosystem, and indicate the extent that species utilize different types of resources and measure potential competition between species in natural plant communities [12,16]. The optimal fundamental niche of a species can enhance our comprehension of the underlying mechanisms that drive variations in abundance patterns, dominance, coexistence among plant species, and biodiversity across flood gradients [24]. In the present study, two types of the WLFZs (i.e., RWLFZ and NRZ) with different hydrological regimes in the Xiangxi River were selected to investigate the changes in plant community structure, functional traits and niche characteristics of dominant plant populations and try to address the following specific questions:

(1) What is the divergent influence of counter-seasonally and naturally hydrological regimes on the plant community species composition and functional traits in the two types of the WLFZs (i.e., RWLFZ and NRZ) in the Xiangxi River of the TGR?
(2) What are the divergent niche characteristics of dominant species of plant communities in the two types of the WLFZs (i.e., RWLFZ and NRZ) and the mechanisms of competition of dominant plant guilds in the Xiangxi River of the TGR?
(3) What are the ecological adaptation strategies of plant guilds based on functional traits and niche characteristics in the different types of the WLFZs of the TGR?

This work will contribute significantly to identifying the most suitable plant species for ecological restoration and re-vegetation endeavors in the unique riparian forests of the TGR and analogously environmentally delicate ecosystems.

2. Study Area and Research Method

2.1. Study Area Overview

The Xiangxi River is located in the western part of Hubei Province (with a longitude from 110°25′ E to 111°06′ E and a latitude from 31°04′ N to 31°34′ N) [25], originating from the Shennongjia Forest District, with a total length of 97.3 km. The river mouth is 34.5 km upstream from the TGD, and the basin area is 3183 km^2 with a natural drop of 1054 m [26]. Since the impoundment of the TGD, the Xiangxi River has transformed from a natural river channel into a reservoir-type river. When the TGR reaches a water level of 175 m, the bay's backwater length is approximately 42 km [27]. The region has a subtropical continental monsoon climate, with distinct seasons and abundant rainfall. Precipitation within the basin decreases from northwest to southeast, with an annual rainfall ranging from 1440 to 1000 mm. The rainy season is mostly concentrated between June and September. The main types of soil are yellow brown soil and lime soil. The undulating topography and high hills account for more than eighty percent of the total area [28]. Before submergence, vegetation in the riparian forests was dominated by annuals, i.e., *Setaria viridis*, *Digitaria ciliaris*, and *Leptochloa chinensis*, perennials including *Cynodon dactylon*, *Hemarthria altissima*, and *Capillipedium assimile*, and trees such as *Ficus tikoua*, *Pterocarya stenoptera*, and *Vitex negundo* [6]. However, after the prolonged submergence, the number of plant species is strongly reduced and annual plants, i.e., *Echinochloa crusgalli* and *Bidens tripartita*, and perennials including *C. dactylon* are dominant species [6].

2.2. Field Investigation and Sampling

Vegetation surveys were conducted across nine transects from June to September in 2023 (Figure 1). Of these, there were six transects in the RWLFZs and three in the NRZs. The surveys were carried out during the flood-regulation periods, with each transect spanning widths between 40 and 60 m. The water level fluctuated between 145 m and 175 m. Therefore, we divided each transect into three elevation zones (145–155 m, flooding depth 20–30 m; 155–165 m, flooding depth 10–20 m and 165–175 m, flooding depth 0–10 m) in the RWLFZs and one elevation zone (175–180 m, no flooding) in the NRZs. For the elevation zone of each transect, five 1 × 1 m square sampling plots were established with 20 m spacing between each plot. Soil density (0–20 cm) was measured by the cutting ring method. At each transect's elevation zone, three topsoil samples from 0 to 20 cm depth were randomly gathered and homogeneously combined to create a single composite sample for analysis. A total of 105 quadrats were sampled and five samples were mixed in each quadrat, sealed in plastic bags and brought to the laboratory. The name, number, height, coverage, frequency and above-ground biomass of each plant species in each quadrat were recorded, and finally, all plants were harvested to measure the above-ground biomass of vegetation in each quadrat. The sketch maps of the study area and sampling sections are shown in Figure 1, and the basic characteristics of the quadrats are shown in Table 1.

Table 1. Basic characteristics of the plots.

Sample Code	Sample Type	Elevation (m)	Slope (°)	Aspect	Longitude (°)	Latitude (°)	Locality
R1	RWLFZ	145–175	22–27	Northeast	110°45′ E	30°58′ N	Bazhi gate in Zigui
R2	RWLFZ	145–175	32–39	Northwest	110°45′ E	31°00′ N	Wangu temple in Zigui
R3	RWLFZ	145–175	30–35	East	110°44′ E	31°02′ N	Xiangjia dam in Zigui
R4	RWLFZ	170–175	18–29	Northwest	110°47′ E	31°07′ N	Diaoyang rock in Zigui
R5	RWLFZ	160–175	16–20	Northwest	110°48′ E	31°07′ N	Yufang flat in Zigui
R6	RWLFZ	165–175	10–18	Southwest	110°51′ E	31°11′ N	Jianyangping bridge in Zigui
N7	NRZ	175–180	18–20	Southwest	110°50′ E	31°08 N	Lijia mountain in Zigui
N8	NRZ	175–180	6–13	Southeast	110°49′ E	31°07′ N	Shijia dam in Zigui
N9	NRZ	175–180	10–18	Southwest	110°51′ E	31°11′ N	Yangdao river in Zigui

Notes: The sample type in the table represents the type of hydrological fluctuation zone where the sampling point is located; elevation represents the collection of samples in that altitude range; slope represents the slope range of this area; aspect represents the orientation of the sampling point.

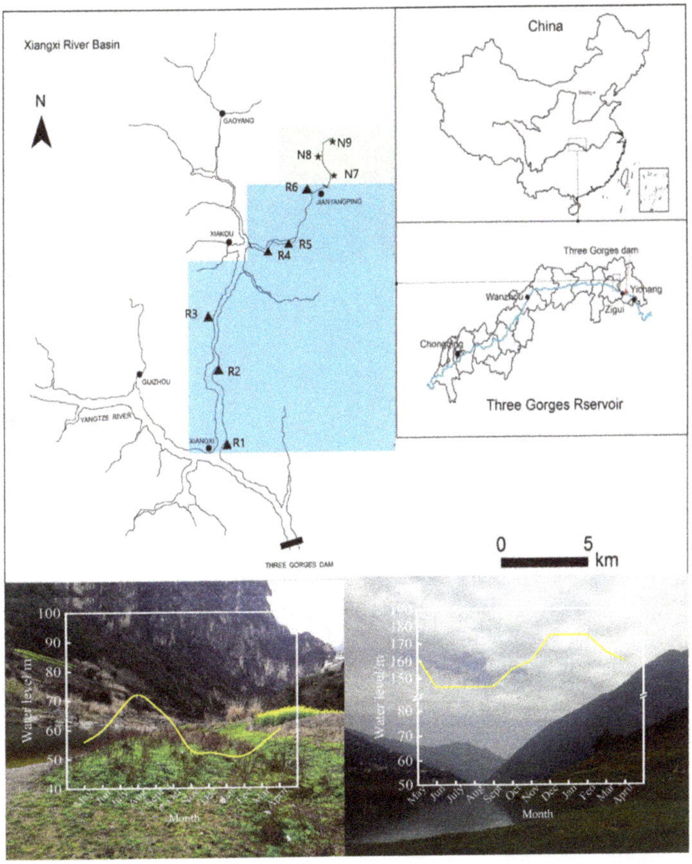

Figure 1. Sketch map of study area and sampling sections. Notes: R1–R6 represent six transects in the RWLFZs, and N7–N9 represent three transects in the NRZs. Please refer to Table 1 for details. The lower half of the figure illustrates the hydrological variation patterns, with the left side depicting the natural hydrological fluctuation zones and the right side representing the counter-seasonally hydrological fluctuations zones.

2.3. Leaf Functional Traits

The Li-6400 portable photosynthetic equipment (Li-cor, Inc., Lincoln, NE, USA) was utilized to measure photosynthesis parameters in the field at 9:00–11:00 or 15:00–17:00 on a clear day to avoid midday depression of photosynthesis. Following the measurement, the measured blades' markers (1–9) were collected and stored in an envelope bag (the envelope bag was labeled with sampling point information), and the labels were sorted and mailed back to the laboratory for measurement of further indicators. Photosynthesis parameters comprise the net photosynthesis rate (P_n), transpiration rate (T_r), stomatal conductance (C_s), inter-cellular CO_2 concentration (C_i), leaf temperature (T_l), atmosphere temperature (T_a), atmosphere CO_2 concentration (C_a), water-use efficiency (WUE, P_n/T_r), etc. The unfolded leaves were collected in the laboratory in time, scanned, and saved by the Canon Scan Li DE 110 scanner (Canon, Tokyo, Japan), the leaf area of each leaf was calculated by Image J software (v1.8.0.345) (cm^2), and the blade thickness was measured by avoiding the main vein with a spiral micrometer [29]. The scanned leaves were placed in the oven and dried to a constant weight at 75 °C. The dry matter weight of each leaf (g) was weighed and recorded (precision: 0.001 g). The specific leaf area (leaf area/leaf dry weight, cm^2/g) of each plant species in various quadrats was calculated and recorded [30].

2.4. Environmental Factors

Soil samples were prepared for analysis by air-drying and passing them through a sieve with a mesh size of less than 2 mm. The pH value of soil was determined by the potentiometric method [31]. The determination of soil organic matter (OM) content was conducted using the dichromate volumetric method [32]. Total phosphorus (TP) and available phosphorus (AP) were measured using the molybdenum-antimony anti-spectrophotometric method [33]. Total phosphorus (TK) and available potassium (AK) were determined using the flame photometric method [34]. Ammonium (NH_4^+-N) and nitrate (NO_3^--N) were determined using spectrophotometry [35]. The soil moisture (SM) was measured using the oven-drying method. The heterogeneity of soil environmental factors in different elevations in the TGR is shown in Figure 2.

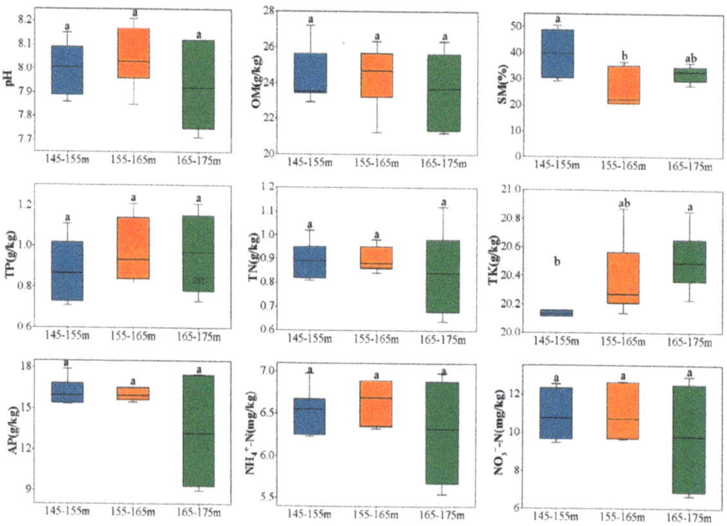

Figure 2. Characteristics of soil factors at different altitudes. Different letters represent significant differences ($p < 0.05$). SM: soil moisture, OM: organic matter, TN: total nitrogen, TP: total phosphorus, TK: total potassium, AP: available phosphorus, AK: available potassium, NO_3^--N: nitrate, NH_4^+-N: ammonium.

2.5. Data Analysis

2.5.1. Importance Value

Within each vegetation assessment, individual samples were categorized to represent distinct resource states. The importance value (P_i) for each species was derived from a combination of their relative coverage (RC), relative frequency (RF), relative height (RH), and relative density (RD) across various quadrats. These metrics were then utilized as parameters to identify the predominant herbaceous species within the water level fluctuation zones and to conduct ecological niche analysis [36]. The computation is articulated by the following formula:

$$P_i = (RC + RF + RH + RD)/4$$

where P_i is the importance value of the i-th species; RC is calculated as the projective coverage of a species divided by the sum of coverages of all species across all quadrats. RF is the count of occurrences of a species divided by the total occurrences of all species in all quadrats. RH is the average height of a species divided by the total height of all species across all quadrats. *RD* is the total count of individuals of a species divided by the total individual count of all species in all quadrats.

2.5.2. Species Diversity Index

Community alpha diversity was assessed through a suite of indices, including species richness (*S*), the Shannon diversity index (*H*), Pielou's evenness index (*E*), and the Simpson dominance index (*D*). The Shannon diversity index represents the species diversity of the community, including species richness (number of species) and evenness (distribution uniformity of individual number of each species). Pielou's evenness index is used to measure the uniformity of individual number distribution of each species in the community. The value of this index is between 0 and 1. The higher the value is, the more uniform the distribution of each species in the community is. The Simpson dominance index reflects the uneven distribution of a species or the concentration of dominant species in the community. The value of the index is between 0 and 1. The smaller the value, the more uniform the distribution of species in the community, and there is no obvious dominant species. The respective calculation formulas are presented below [37]:

$$\text{Shannon diversity index}: H = -\Sigma P_i lnP_i$$

$$\text{Pielou's evenness index}: E = \frac{H}{lnS}$$

$$\text{Simpson dominance index}: D = -\Sigma P_i^2$$

where P_i is the importance value of the *i*-th species, *S* represents the number of distinct species that are found within the quadrat, and *N* signifies the total number of species that are present within the quadrat.

2.5.3. Niche Breadth

Niche breadth, following Levin's definition, was quantified using Colwell's adjusted formula [36].

$$B_i = \frac{1}{r \times \sum_{h=1}^{r} (P_{ih})^2}$$

where B_i is the niche breadth of the *i*-th species, P_{ih} is the ratio of the importance value of the *i*-th species at the *h*-th resource level to the sum of the importance values of the species at all resource levels, *r* is the number of resource levels, and the range of the formula is [0, 1].

2.5.4. Niche Overlap

Pianka's formula was employed to calculate niche overlap [36].

$$O_{ih} = \frac{\int_{h=1}^{r} P_{ih}P_{jh}}{\sqrt{\int_{h=1}^{r} P_{ih}^2 \int_{h=1}^{r} P_{jh}^2}}$$

where O_{ij} is the niche overlap of populations *i* and *j*, p_{ih} and p_{ij} are the proportion of the importance values of the ith and the *j*-th species at the *h*-th resource level in the synthesis of the importance values of the species at all resource levels, *r* is the number of resource levels, and the range of the formula is [0, 1].

2.6. Redundancy Analysis (RDA) and Detrended Correspondence Analysis (DCA)

DCA is a method for sorting and visualizing ecological data distribution. It is usually applied to species abundance data to explore differences between samples or communities. The advantage of DCA is that it can better deal with the nonlinear trends common in ecological data. By removing trends, DCA provides more realistic and interpretable information about the structure of ecological communities. RDA is a multivariate regression technique commonly used to analyze the influence of environmental variables on species composition. It combines the concepts of multiple linear regression and principal component analysis. RDA visualizes the relationship between environmental variables and biological communities by maximizing a linear combination that explains differences in species composition [38].

RDA and DCA were used in this study to systematically analyze the species composition, community types and distribution characteristics of vegetation in the two different types of the WLFZs in the Xiangxi River of the TGR, as well as their relationships with environmental factors. We conducted DCA and RDA using the CANOCO package v.5.0 (Ithaca, NY, USA). A Monte Carlo test using 999 permutations ($p < 0.05$) was performed to test the significance of the analysis.

3. Results

3.1. Plant Community Characteristics of the RWLFZ and NRZ

A total of 78 vascular plant species, belonging to 62 genera of 24 families, were identified in the Xiangxi River Basin of the TGR (specific species are shown in Table A1 (Appendix A)). Of these, 49 vascular plant species of 44 genera and 23 families with 26 endemic species in 22 endemic genera and 8 endemic families were found in the RWLFZs. In the case of the families, few-species and single-species families were dominant, accounting for 61.22% of the total family, and the most abundant species were members of Compositae, followed by Gramineae, which accounted for 24.49% and 14.29% of the total species, respectively. These families contained 33 annual herbs, 13 perennial herbs and 3 trees, accounting for 67.35%, 26.53% and 6.12% of the total species, respectively.

A total of 52 vascular plant species were found in the NRZs, including 13 families and 34 genera (20 endemic species in 18 endemic genera and 3 endemic families), of which 25 annual herbs accounted for 65.79% of all plant species. Both Compositae and Gramineae were the dominant families, both accounting for 23.68% of the plant communities.

3.2. Plant Community Diversity of the RWLFZs and NRZs

There was no significant difference in the Shannon diversity index between NRZs and RWLFZs of the Xiangxi River of the TGR (Figure 3A), but there was a significant difference between NRZs and RWLFZs in Pielou's evenness index and Simpson dominance index (Figure 3). The result revealed that Pielou's evenness index of NRZs was significantly higher than that of the RWLFZs (Figure 3B), but the Simpson dominance index of RWLFZs was significantly higher than that of NRZs (Figure 3C).

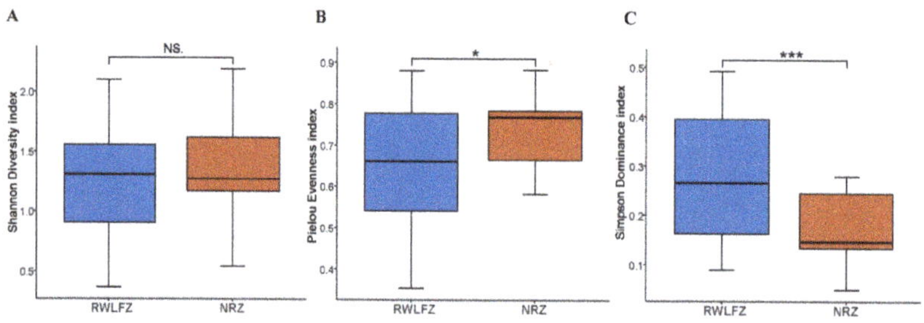

Figure 3. The plant community diversity indexes in the RWLFZ and NRZ. Significance levels: NS., $p > 0.05$; *, $p \leq 0.05$; ***, $p \leq 0.001$. (**A**), Shannon diversity index; (**B**), Pielou's evenness index; (**C**), Simpson dominance index.

3.3. Functional Traits of Dominant Plant Species of the RWLFZs and NRZs

Counter-seasonally hydrological alterations had significant effects on the aboveground biomass (AGB), net photosynthetic rate (Pn), WUE, specific leaf area (SLA) and vegetation coverage (VC) of dominant species in the RWLFZs. Compared with the NRZs, the SLA and VC of the RWLFZs increased by 74.0% and 27% ($p < 0.05$), respectively

(Figure 4C,E), while the WUE significantly decreased by 59.6% ($p < 0.05$) (Figure 4B) and the Pn and AGB between them showed no significant difference (Figure 4A,D). The results showed that counter-seasonally hydrological alterations had significant effects on the functional traits of dominant species in the RWFLZs and dominant species in the RWFLZs grew faster than those in the NRZs with higher SLA and VC, and still maintained high Pn.

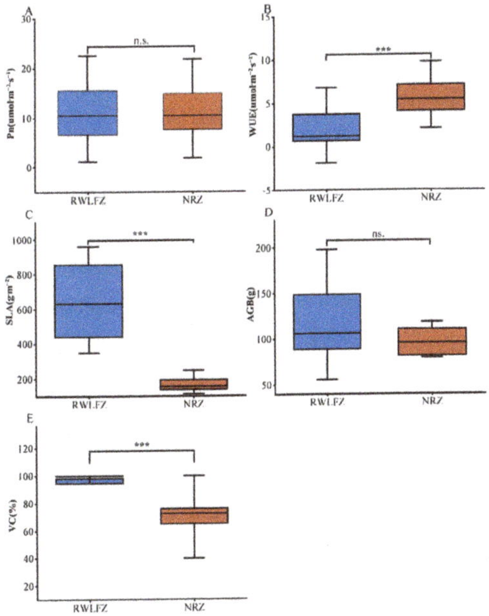

Figure 4. Functional traits of vegetation in the different drawdown zones. Significance levels: NS., $p > 0.05$; ***, $p \leq 0.001$. (**A**) Pn: net photosynthetic rate; (**B**) WUE: water-use efficiency; (**C**) SlA: specific leaf area; (**D**) AGB: above-ground biomass; (**E**) VC: vegetation coverage.

3.4. *Niche Overlap and Niche Breadth*

3.4.1. The Reservoir Water Level Fluctuation Zones (RWLFZs)

In total, 49 species of herbaceous plants were observed in the RWLFZs. The niche breadth of *C. dactylon*, *B. pilosa*, *Xanthium sibiricum*, *Echinochloa crusgalli* and *Polygonum hydropiper* was 0.789, 0.636, 0.505, 0.470 and 0.309, respectively, greater than those of other species. Niche breadth in the RWLFZs varied from 0.0111 to 0.7891. *C. dactylon*, *B. pilosa* and *X. sibiricum* had niche breadths greater than 0.5. Species with niche breadths less than 0.2, accounting for 79.59% of all the species found in the RWLFZs, were mostly distributed in the high elevation areas. And 14 species pairs with niche overlap greater than 0.5 accounted for 17.95% of all species pairs; 22 pairs with niche overlap less than 0.2, accounting for 28.21% of all the species pairs; and 3 pairs with niche overlap of 0, accounting for 3.85% of all the species pairs (Table A1), were recorded. Among them, the niche overlap value for *Eclipta prostrata* and *Rumex acetosa* was the largest (0.712), followed by *X. sibiricum* (0.662), *C. dactylon* (0.662), *B. pilosa* (0.622), *Abutilon theophrasti* (0.604); *P. hydropiper* (0.649), *E. crusgalli* (0.649), *Eleusine indica* (0.644), *E. prostrata* and *Arthraxon hispidus* (0.688). Species pairs with large niche overlap values were mostly those species with a large niche breadth and high importance values, but there were also a few species with large niche overlap, low importance values and low niche breadth, such as *R. acetosa* and *A. hispidus* (Figure 5A).

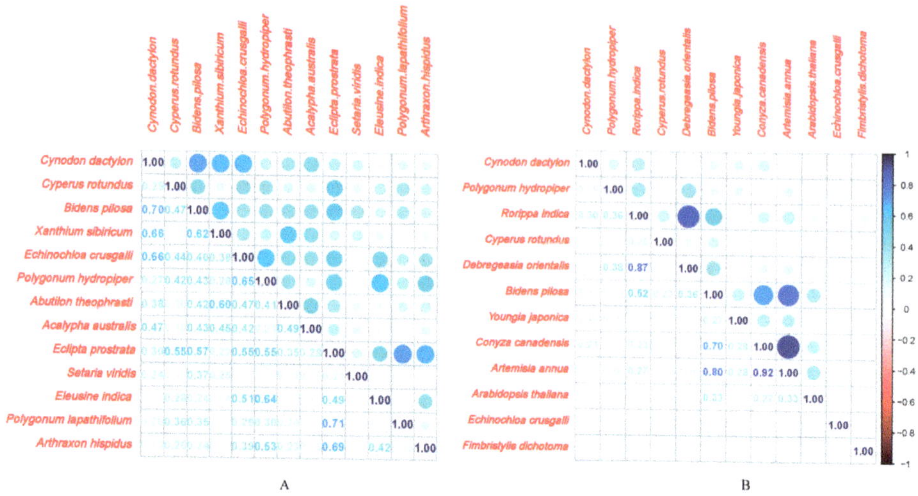

Figure 5. Niche overlap values for dominant herbaceous plants in the RWLFZ (**A**) and NRZ (**B**).

3.4.2. The Natural Riparian Zones (NRZs)

There were 38 herbaceous plants in the NRZs, and the niche overlap was calculated for the species pairs with importance values greater than 0.06 (Figure 5B). There were five pairs with niche overlap greater than 0.5, accounting for 7.58% of all species pairs; 46 pairs with an overlap of less than 0.2, accounting for 69.7% of all species pairs; and 29 pairs with niche overlap of 0, accounting for 43.9% of all species pairs. Among them, the highest values were for *Conyza canadensis-Artemisia annua* (0.922), followed by *R. indica-D. orientalis* (0.869), *B. pilosa-A. annua* (0.804), *B. pilosa-C. canadensis* (0.699), and *B. pilosa-R. indica* (0.521). Species with large niche overlap values were those species with a large niche breadth and high importance values, but *C. dactylon*, with the highest importance value, and *P. hydropiper* were not recorded. Among the five pairs of species with niche overlap values greater than 0.5, three were species pairs containing *B. pilosa*. The number of species with a niche overlap value of 0 in the NRZs was the highest in the two zones, accounting for 43.9% of all species pairs in the NRZs. The results showed that the degree of niche differentiation was high in the area. The niche breadth value of the NRZs varied from 0.0833 to 0.648 and was greater than 0.5 for *R. indica* and *B. Pilosa*. There were 21 species with a niche breadth of less than 0.2, accounting for 73.68% of all the species found in the NRZs.

3.5. DCA Ordination of Sampling Sites and Species

The DCA ordination results of 105 sampling sites in the Xiangxi River Basin were shown in Figure 6a. On the DCA ordination map, the distribution patterns for each community type were obvious; the vegetation community types in the source area of the NRZ were on the right side of Figure 6a. The distribution patterns of species in the DCA ordination map were coincidently similar to those of community types (Figure 6b). From the horizontal axis of the species DCA ordination map (Figure 5B), at the left end of the horizontal axis, the species were *X. sibiricum*, *C. dactylon*, *B. pilosa*, and *A. theophrasti* growing in the RWLFZs, while the right end was occupied by typical species such as *Debregeasia orientalis*, *Ricinus communis*, and *Rorippa indica* growing in the NRZs. The variety of species across the distribution reflects the changes in the water level of rhythm characteristics from the NRZs to the RWLFZs of the Xiangxi River Basin. The species at the left end of the horizontal axis were mainly flooding-resistant species such as *C. dactylon*, *X. sibiricum*, *B. pilosa*, and *A. theophrasti*, while at the right end of the horizontal axis, the species were mainly drought-resistant species such as *D. orientalis*, *R. communis*, and *R.*

indica. Differences in species composition in different regions indicated the effect of water level rhythms on vegetation.

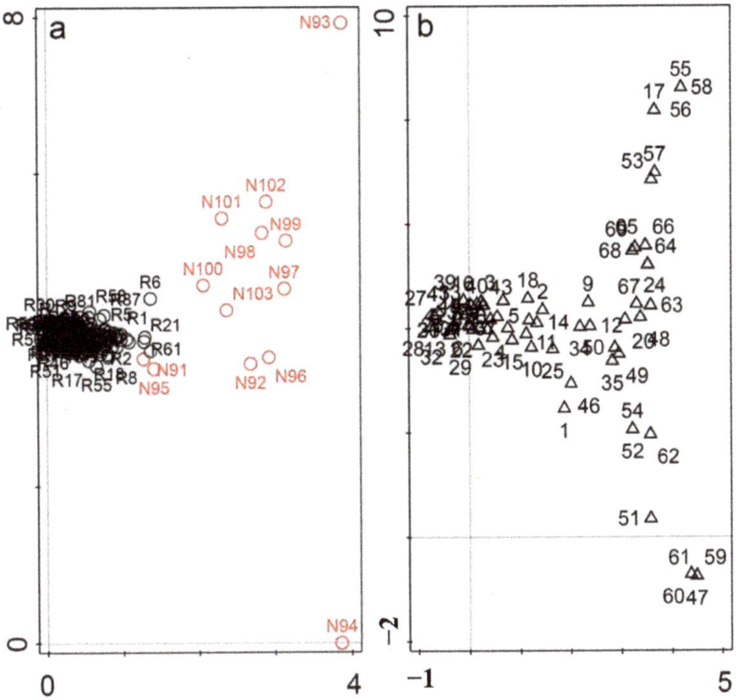

Figure 6. DCA ordination diagram of sampling sites in the Xiangxi River (**a**); DCA ordination diagram of 69 species in the Xiangxi River (**b**). Notes: (**a**) R1-90 sampling sites in the RWLFZ and N90-103; N91-104 sampling sites in the NRZ; (**b**) the species represented by each number are presented in Table A2 (Appendix A).

3.6. RDA of Species and Environmental Variables

Response data were compositional and had a gradient 2.8 SD units long, so the linear method was recommended. Therefore, the redundant analysis (RDA) method was used to rank the communities, and the relationship between the distribution pattern of the communities and environmental factors was studied. The ordination analysis revealed the environmental gradient in the area. The results of the RDA showed that elevation was the best explanatory variable (Figure 7). The conditional effects indicated the following:

(1) The contributions of elevation (F = 3.3, $p < 0.01$), TK (F = 3.2, $p < 0.01$) and flooding duration (F = 3.1, $p < 0.01$) were significant during the forward selection; (2) the contributions of the remaining variables were not significant. The RDA results also showed that there was a strong correlation between vegetation and environmental factors, with species–environment correlations of 0.9858 on the first axis and 0.9236 on the second axis. The eigenvalues for the first four RDA axes were 0.2388, 0.128, 0.1133 and 0.0771. The cumulative percentage variances of species–environment relations of the first axis and second axis were 30.42% and 46.73%, respectively. The sum of all the RDA canonical eigenvalues was 0.785. The angles between the vector segments of elevation, TK and flooding duration and the first order axis were small, indicating that these three environmental indicators mainly affected the distribution of species on the first order axis. The arrow of elevation and TK were located in the first quadrant, indicating that elevation and TK were positively related to the first sorting axis. *E. villosa*, *S. viridis* and *X. sibiricum* were distributed in the

higher elevation with a high TK content. Flooding duration was located in the second quadrant and had a negative correlation with the first sort axis. *E. indica*, *P. oleracea* and *L. procumbens* were distributed in the lower elevation. So, the main factors affecting the distribution of the plant community in TGR were elevation, TK and flooding duration.

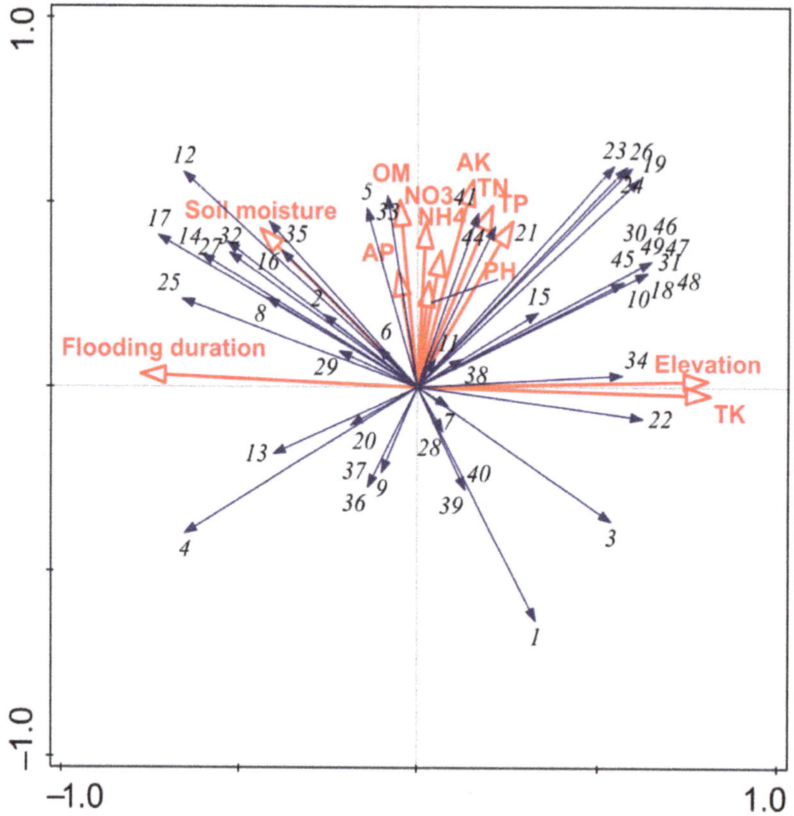

Figure 7. Redundancy analysis (RDA) ordination map of the important value index of species and environment variables. The red arrow represents the environmental factor (explanatory variable), and the length of the arrow represents the intensity of the influence of the environmental factor on community change. The longer the length, the greater the influence of the environmental factor (otherwise the opposite). The blue arrow represents the plant species (response variable). The angle between species (blue arrow) and environmental factors (red arrow) can represent the correlation between species and environmental factors. Environmental variables: elevation, flooding duration, soil moisture, OM (organic matter), TN (total nitrogen), TP (total phosphorus), TK (total potassium), AP (available phosphorus), AK (available potassium), NO_3 (nitrate); NH_4 (ammonium). The species sequence in the table was the same as the order of important value of species in RWLFZ. The numbers represent the species details in Table A3 (Appendix A).

4. Discussion

4.1. Plant Community Characteristics and Their Ecological Adaptation Strategies in Different Types of the WLFZs

With the filling of the TGR, original vegetation in the WLFZ between the elevations of 145 and 175 m disappeared and the species richness and diversity significantly decreased due to the great hydrological alterations by the TGD construction [6,39]. The analysis of vegetation responses to river regulation through this dam project is highly important to assess the effects of counter-seasonally hydrological fluctuations on newly formed riparian ecosystems. In the present study, to better understand the relationships between naturally recovered plant community characteristics and their responses to environmental factors for reconstructing floristically diverse riparian forests, we detected two different types of WLFZs in the Xiangxi River of the TGR after long-term tracking and monitoring [12,40]. The findings revealed a total of 78 species of vascular plants, belonging to 62 genera and 24 families, within the two distinct types of the WLFZs along the Xiangxi River. Compositae plants accounted for the highest proportion (20.25%), followed by Gramineae (15.38%) and Polygonaceae (7.70%) of all the species detected. There are seven common families, 15 common genus and 16 common species in the two types of the WLFZs. There were 15 species belonging to one genus of one family, accounting for 19.23% of all the total species. Most of the dominant species were annual herbs in the two different types of the WLFZs after 20 cycles of water level fluctuations and the proportion of the annual dominant species increased from 65.79% in the NRZs up to 67.34% in the RWLFZs.

However, differences in plant community composition among the different types of the WLFZs were found. Only 38 of vascular plant species were identified in the NRZs affected by summer floods, but 49 species were found in the RWLFZs with the opposite water level rhythm. The taxonomic composition in the RWLFZs was predominantly characterized by the family Compositae, which was the most frequently observed taxon. In contrast, the NRZs exhibited a more balanced distribution, with both Compositae and Gramineae families being equally represented and accounting for the same proportion of the total species count. This distinction in the taxonomic dominance between the two zones highlights the varying ecological preferences and adaptability of these plant families to the specific environmental conditions within the WLFZs. Only 15 species appeared at the same time in the RWLFZs and the NRZs, further highlighting the impact of the anti-seasonally hydrological rhythm of the TGR on the vegetation of Xiangxi River. There were 26 endemic species in 22 endemic genera and eight endemic families in the RWLFZs and 20 endemic species in 18 endemic genera and three endemic families in the NRZs. Thus, it can be seen that the hydrological rhythms of the TGR had significant effects on plant community composition among different water level fluctuation zones. The dominant species evolved from Cruciferae, Urticaceae, Compositae and Gramineae to Compositae and Gramineae from the NRZs to RWLFZs. The species with the highest importance value in the RWLFZs was *C. dactylon* (0.4445), followed by *C. rotundus* (0.1266) and *E. crusgalli* (0.0965). Perennial plant *C. dactylon* exhibited an absolute advantage in this type of RWLFZ. The species with the highest importance value in the NRZs was *C. dactylon* (0.3884), followed by *P. hydropiper* (0.1694) and *R. indica* (0.1585), with the dominant species being members of Gramineae, Polygonaceae and Compositae. In the two different types of the WLFZs, *C. dactylon* was obviously dominant, showing its stronger ecological adaptability to different environments than other plant species by rapid growth and lateral expansion with seed reproduction and clonal propagation [41,42]. Gramineae species in the RWLFZs demonstrated robust adaptation to the irregular, anti-seasonal water level fluctuations. In 2010, Wang et al. [40] found that 64 vascular plant species belonging to 59 genera of 29 families were identified in the same field vegetation survey of the RWLFZs of the Xiangxi River. After seven years following the impoundment of the TGR, Gramineae and Compositae families constituted 17.19% and 14.06% of the total identified plant species, respectively. Fu et al. [12] reported that 50 vascular plant species of 49 genera of 23 families occurred in the RWLFZs of the Xiangxi River through the same field study and fixed observation, with Compositae and Gramineae accounting for 16.68% and 13.73%, respectively, after 11 years of anti-seasonal flooding. In the present study, a total of 49 vascular plant species were identified, belonging to 44 genera of 23 families in the same studied area of the Xiangxi River with Compositae and Gramineae accounting for 20.25% and 15.38%, respectively, 20 years after the impoundment of the TGR. Thus, after experiencing 20 years of off-seasonal water level fluctuations, it was found that the number of plant species considerably decreased in the RWLFZs of the TGR. Compositae plants showed an increasing trend with the number of flood–dry–flood cycles and the number of Compositae has become higher than Gramineae. The reason may be that Compositae has a unique nitrogen–carbon balance system, resulting in enhanced N uptake and fatty acid biosynthesis, which may explain their rich biodiversity and excellent adaptive characteristics [43]. Under this type of water level rhythm with a high water level in winter and low water level in summer, Compositae and Gramineae plants showed obvious advantages in species number and population size. These results showed that if the water-level fluctuation rhythm changes, the composition of annual herb communities will change because of their rapid responses to the water level fluctuations. A major underlying mechanism that allows annual plants to persist even after years of water level fluctuations with no reproduction is the maintenance of a persistent seed bank [44]. Therefore, as environmental conditions vary, plant community composition would shift in response to hydrologic changes.

Additionally, in the process of plant community succession, the effects of stress and disturbance on the vegetation assembly result in a particular adaptation strategy of plant species in a particular stage of succession. The competitive (C) stress-tolerant (S) ruderal (R) (CSR) model of Grime [45] is an attempt to describe the three primary plant adaptation strategies that are linked to the different stages of succession and includes the secondary strategies (CR, SR, SC, and CSR) that correspond to the various trade-offs in the adaptations to competition, stress and disturbance [45–50]. In the present study, the Pn and WUE of dominant plant species from the RWLFZs and NRZs had no significance between them, but the SLA of the former was significantly higher than that of the latter. These indicated that the annual species in the RWLFZs adopt the R adaptation strategy of being fast-growing, able to effectively disperse with a short lifespan and quickly acquiring and investing resources by altering leaf morphology, including expanding the leaf area into the production of offspring in the primary stage of plant community succession. In habitats with long-term poor conditions and frequent and severe disturbances in the RWLFZs, the resources captured could be quickly reintroduced or redistributed for rapid settlement in the area; annual plants would allocate much more to reproduction in order to be able to maximize the fitness gain from reproductive output as soon as the environment allows them [51]. Plant individuals are relatively small and have a short life history and high fertility, with annuals germinating in spring and fructifying in autumn or annuals germinating in late winter and fructifying in summer, while in the high elevation areas of the RWLFZs, annual herbs, such as *X. sibiricum*, grow tall and allocats much more to vertical vegetative growth at the expense of lower reproductive output in order to be able to maximize this species fitness. The progressive replacement of the R strategists by the SC or the CSR strategists occurs in the long-term succession stages of naturally recovered plant communities and perennial plants such as *C. dactylon* adopt the rapid capture and monopolization of resources with the spatially dynamic foraging of roots and shoots according to the C adaptation strategy of the competitive CSR model of Grime. Thus, plant community species at different hydrological regimes adopt different adaptation strategies for different types of water level fluctuations.

4.2. Dominant Plant Niches and Biodiversity Coexistence and Resource Utilization Mechanism in the Different Water Level Fluctuation Zones

It is universally acknowledged that the number of related species capable of coexisting in a particular community is contingent upon the niche breadths of individual species as well as the extent to which their niches overlap [52]. The niche is a valid indicator for better understanding the relationship between species and habitats. It not only reflects the ecological adaptability and distribution range of species but also provides a description of the scarce resources for which species compete [16]. Niche breadth reflects the adaptation of species to the environment and the degree of resource utilization. In general, the greater the niche breadth of the species, the stronger its ability to use resources and adapt to the environment [53]. In the present study, we found the dominant plant species with a niche breadth value of more than 0.5 in the RWLFZs to be *C. dactylon*, *B. pilosa* and *X. sibiricum*. *C. dactylon* had both the highest importance value and the highest niche breadth value. *B. pilosa* and *X. sibiricum* also had relatively high important values and niche breadth values. Species with a niche breadth of less than 0.2 accounted for 79.59% of all the species found in the RWLFZs. However, Wang et al. [54] found that the initial flooding changed the vegetation composition of the original RWLFZs, and many specialized species started to appear and gradually occupied a dominant position (the number of species with a niche breadth less than 0.2 accounting for 90% of all the species pairs) 5 years after the impoundment in 2008. After 20 years, the proportion of the species with a niche breadth of less than 0.2 (79.59%) in the RWLFZs decreased by approximately 10%, and the dominant position of the specialized species appeared. However, there were only three species (*C. dactylon*, *B. pilosa* and *X. sibiricum*) in the region with a niche breadth of more than 0.5, which reflects their stronger ecological adaptability and ability to use resources, thus quickly expanding their own niche space and the wider distribution range of species in the RWLFZs of the TGR. The NRZ is the zone that was closest to the environment before the construction of the TGR, and the study of the plant communities in this area can highlight changes in vegetation in the Xiangxi River during off-seasonal flooding. A broader niche occupancy, characterized by a wider utilization of environmental resources and spatial domains, typically confers a competitive advantage to certain species. This competitive edge can lead to the exclusion of species with narrower niche widths, thereby facilitating the overall growth and development of the dominant species' guilds within the ecosystem [55]. Therefore, different species have different ecological adaptation strategies under periodic or aperiodic flooding stress in the different types of WLFZs. In this study, the anti-seasonal flooding rhythm expanded the niche breadth of *C. dactylon*, *B. pilosa*, *X. sibiricum*, *E. crusgalli*, *P. hydropiper* and *A. theophrasti*, showing the ability of these plant species to utilize resources and exhibit high adaptability to dry–wet alternating environments in the RWLFZs. The plant community in the RWLFZs was dominated by the species with the highest or/and second highest niche breadths, revealing that these species had broader environmental tolerances and were able to use a wider range of resources. *C. dactylon* with the highest important values and niche breadth formed a single-species community in the mid-elevation region of the RWLFZs, and a single species community poses a great threat to the stability of the communities.

Niche overlap is the similarity between two species in relation to ecological factors; when two species use the same resource or share a certain resource, niche overlap occurs [56]. Moreover, niche overlap can to a certain extent reflect the utilization of common resources among populations and also reflect the degree of interlaced distribution between species [53]. Under the effect of long-term and anti-seasonal water level fluctuations of the TGR, the dry–wet alternating process leads to the surviving resource scarcity of species. Li et al. [57] found that the soil nutrient content decreased significantly after dry–wet alternation. Plants need to expand their resource utilization scope to increase their survival opportunities in unstable habitats to promote niche differentiation and increase niche breadth and niche overlap values. Although the niche breadth of species in the RWLFZs was generally small, the overall niche overlap was relatively high, that is, species competition in the habitat was intense, which was not conducive to the coexistence of species, and the community structure was unstable. Based on the comprehensive analysis of the overlap of plant niches in the two different WLFZs, it can be inferred that after experiencing anti-seasonal flooding, the overlap of niches between all species in the RWLFZs was relatively high, which showed that the competition in the community was much more intraspecific than interspecific and would facilitate the expansion of species niches. This is consistent with the conclusions drawn by Wang et al. [58] who found that "as the number of reservoir water level fluctuation cycles increases, the niche width and niche overlap of dominant plant populations in the drawdown zone will increase". The reason for this may be the impact of habitat stability. The habitat of the drawdown zone is constantly changing, and in order to gain an advantage under habitat pressure and competition, plant populations tend to expand their range of resource utilization, thus increasing the niche width. Consequently, the likelihood of plant populations utilizing similar resources increases, and the proportion of niche similarity and the value of niche overlap are correspondingly enhanced [59].

However, the competition in the community at the NRZs was much more interspecific than intraspecific. The intensive interspecies competition not only excluded some species with poorer adaptive abilities but also segregated and optimized niches of the remaining species, leading to a decline in the total niche overlap of all paired species in the community. The extent of interspecific competition in the community was also weakened and most species could utilize the environmental resources more effectively and adequately, which showed that the community structure in the NRZs was extremely stable.

The species of coexistence not only compete for light, heat, water and nutrient resources, but there is also an obvious facilitation effect, especially in conjunction with high environmental stress [60]. The most significant problem under flooding stress is the reduction in oxygen supply, so plants have to switch from aerobic respiration to anaerobic respiration and increase the consumption of nutrients. Some flood-tolerant wetland plants can alleviate soil anoxia by releasing oxygen into the soil in flooded environments; simultaneously, neighboring flood-sensitive species should quickly grow under high water levels because flood-tolerant plants reduce soil anoxia stress [61]. In fact, the improvement of anoxic environments depends mainly on the roots of the plant. According to the survey, *C. dactylon* is a dominant plant species at different altitude gradients of the RWFLZs and NRZs; this is mainly due to the larger root porosity and the effective scavenging mechanism of reactive oxygen from *C. dactylon* under inundation conditions. Different species adopt different strategies to cope with the flooding stress. Some species with high flood tolerance can promote the growth of stems and leaves to achieve higher biomass through the elongation of stems and leaf area expansion. It facilitates the formation of a better living environment to achieve coexistence [62].

4.3. Distribution Patterns of Species Diversity in the RWLFZ

Environmental gradients provide a useful framework for evaluating the relative importance of differential selection for adaptive traits that maximize species fitness [63,64]. In the present study, an intriguing consequence of our study was that there was no significant difference between NRZs and RWLFZs for the Shannon diversity index (Figure 3A), but there was a significant difference between NRZs and RWLFZs for Pielou's evenness index and the Simpson dominance index (Figure 3). Pielou's evenness index of NRZs was significantly higher than that of RWLFZs (Figure 3B), while the Simpson dominance index of RWLFZs was significantly higher than that of NRZs (Figure 3C), which showed that species patterns of plant communities in the RWLFZs were more uneven than in the NRZs and there are more specialized species in the RWLFZs than in the NRZs. The diversity evidence further confirmed that the plant community structure in the RWLFZs was still unstable.

The distribution patterns of plant community species diversity along an elevation gradient and in different habitats may be influenced by environmental factors (e.g., latitude and longitude, elevation, flooding time, flooding depth, slope, aspect, human activity) [65]. Chen et al. [11] reported that the steep slope area above 25° in the Xiangxi River of TGR accounted for 41.19% of the RWLFZs, which belongs to the semi-submerged and semi-exposed type of the RWLFZs, especially often under high temperatures and drought stress in summer, and the soil organic matter is not readily retained (Figure 2), resulting in a slight decrease in species diversity in the RWLFZs. All the soil nutrients measured also showed an inverted hump-shaped pattern, except TK in the mid-elevation area. Species with a short life history like *E. indica*, *P. oleracea* and *L. procumbens* were associated with flooding duration in the RDA, whereas tall and loosely branched species, like *E. villosa*, *S. viridis* and *X. sibiricum*, indeed showed a negative relationship with flooding duration (Figure 7) in the RWLFZs. The left to the right ordination diagram of the horizontal axis of RDA showed that from low elevation to high elevation, plant community architecture changed from small, compact plants in the low-elevation areas to tall, loosely branched plants in the high-elevation areas, where biomass, height and density of neighbor plants were relatively high. The distribution trend was basically consistent with its dominant species community. The results of the Monte Carlo random displacement test showed that the simple term effect of interpretation of species distribution on elevation, TK and flooding duration reached a significant level. The concentration of total K increased from 2008 to 2016, while the concentrations of OM and available K declined over the same period in the TGR [66]. K can regulate water and energy metabolism and improve plant resistance [67,68]. The close correlation between TK and elevation explained the trend of community diversity. Therefore, the main environmental constraint is most likely elevation, TK and flooding duration. Environmental factors drove the succession of vegetation. Vegetation also plays an important role in nutrient cycling directly via plant uptake, decomposition, and deposition of alluvial materials, and indirectly through stimulation of microbial processes in the rhizosphere [66]. The synergy between environmental factors and vegetation played an important role in the natural vegetation restoration in the TGR.

In the study of long-term responses of the WLFZ vegetation to water level regulation, Nilsson et al. [10,69–71] assessed the long-term effect of water level regulation on riparian plant communities for storage reservoirs and run-of-river impoundments in central and northern Sweden after regulation. The results showed that transformed species richness in pre-upland sites along storage reservoirs gradually increased during the first 34 years but decreased thereafter between 1 and 70 years after dam closure. Along run-of-river impoundments, pre-upland sites had maximum species richness at about 13 years, while pre-riparian sites had maximum species richness at about 18 years after regulation. In the present study, after the first impoundment in June 2003, the TGR experienced 20 times the amount of water level fluctuation events, demonstrated simple species composition and community structure, lower niche differentiation between different dominant plants, and more intense inter-specific competition.

4.4. Restoration Implications

The new insight into the vegetation distribution patterns of NRZs and RWLFZs through plant niche analyses showed that the inter-specific competition of vegetation distribution in the RWLFZs was more intense due to repeated anti-seasonal flooding stress compared with the NRZs. Plant community architecture changed from small, compact plants in the low-elevation areas to tall, loosely branched plants in the high-elevation areas, where the biomass, height and density of neighbor plants were also relatively high, and *C. dactylon*, *B. pilosa*, *X. sibiricum*, *P. hydropiper*, *S. viridis*, *E. crusgalli* had larger niche breadths, their resource utilization and environmental adaptability were strong, and they had both ecological and economic values. Accordingly, these species should be prioritized for optimal selection and scientific arrangement in the ecological restoration and reconstruction of vegetation in the RWLFZs of the TGR. Moreover, annual herbaceous plants are better suited to the fragmented conditions of the RWLFZs compared to perennial herbaceous species. It is therefore recommended that the general policy for restoring and establishing artificial plant communities in the reservoir area should primarily consist of herbaceous species, complemented by flooding-tolerant shrubs or small trees such as *Distylium chinense* [72].

In order to establish a complete reference system for vegetation restoration, restoration monitoring plots in the different succession stages should be established in the different water level fluctuation zones of the TGR, and their environmental conditions, community structures and inter-specific relationships should also be analyzed. Therefore, our findings are useful for anticipating the highly adaptable and widely distributed species with large niche breadths and high important values selected as the candidate species in ecological restoration and revegetation in the newly formed hydro-fluctuation zone and the other similar degenerative riparian area and stream ecosystems [73].

5. Conclusions

A total of 76 species of vascular plants were found in the WLFZs of the Xiangxi River Basin. There were 49 and 38 species in the RWLFZs and NRZs, respectively. There were 26 endemic species in 22 endemic genera and eight endemic families in the RWLFZs and 20 endemic species in 18 endemic genera and three endemic families in the NRZs. The dominant species belonged to *Cruciferae, Urticaceae, Compositae* and *Gramineae* to *Compositae* and *Gramineae* in the NRZs to the RWLFZs. The analysis of niche overlap revealed the instability of the plant community in the RWLFZs, which was attributed to intense intraspecific competition and incomplete niche differentiation. Specifically, species with larger niche widths in the RWLFZs exhibited a high degree of niche overlap, indicating the intensity of resource competition within the plant community in this area. In contrast, the plant communities in the NRZs show lower levels of niche overlap, reflecting greater niche differentiation and a more stable community structure. The results of the RDA indicated that the main factors influencing the vegetation characteristics were the elevation (flooding depth), TK and flooding duration. Our results had some implications for the vegetation restoration efforts of the RWLFZs. The highly adaptable and widely distributed species with large niche breadths and high important values could be selected as candidate species in ecological restoration and revegetation in the newly formed hydro-fluctuation zones and the other similar degenerative riparian area and stream ecosystems.

Author Contributions: Conceptualization, X.L. and Z.Y.; methodology, D.D.; software, S.X. and W.Y.; validation, D.H., Q.M. and G.C.; formal analysis, W.Y.; investigation, J.Y.; resources, G.H.; data curation, M.H.; writing—original draft preparation, D.H. and S.X.; writing—review and editing, W.Y. and C.Y.; visualization, X.L. and Z.Y.; supervision, X.L. and C.Y.; project administration, X.L.; funding acquisition, X.L. All authors have read and agreed to the published version of the manuscript.

Funding: This work was supported by the National Natural Science Foundation of China (No. 51779127), partially funded by the National Key R&D Program of China (2023YFF1305204), Research Project on Seed Preservation Technology and Facilities of Rare Plants in the Three Gorges Reservoir Area—Investigation and Collection of Flooding-Tolerant Germplasm Resources (SDHZ2021346) and the 111 Project of China (No. D20015).

Data Availability Statement: The datasets used in this study can be obtained by contacting the corresponding author upon request.

Acknowledgments: The authors would like to acknowledge Lei Sun, Lei Yue, Ling Xiang, Jie Huang and Cheng-Rui Yu for their assistance during field sampling and laboratory analysis. The authors would also like to express their gratitude to the anonymous reviewers for their insightful feedback and constructive suggestions, which significantly enhanced the quality of this manuscript.

Conflicts of Interest: Guiyun Huang and Meixiang Hu are employed by Three Gorges Corporation Co., Ltd. The remaining authors declare that the research was conducted in the absence of any commercial or financial relationships that could be construed as a potential conflict of interest.

Appendix A

Table A1. Changes in composition and niche breadth of dominant herbaceous plants.

Species	Life Form	Family	Genera	WLFZ		NRZ	
				L_v	B_i	L_v	B_i
Cynodon dactylon	Perennial herb	*Gramineae*	*Cynodon*	0.789	0.444	0.478	0.289
Bidens pilosa	Annual herbaceous	*Compositae*	*Bidens*	0.636	0.059	0.461	0.076
Xanthium sibiricum	Annual herbaceous	*Compositae*	*Xanthium*	0.505	0.068	0.228	0.101
Echinochloa crusgalli	Annual herbaceous	*Gramineae*	*Echinochloa*	0.47	0.097	0.371	0.193
Polygonum hydropiper	Annual herbaceous	*Polygonaceae*	*Polygonum*	0.309	0.065	0.197	0.088

Table A1. *Cont.*

Species	Life Form	Family	Genera	WLFZ		NRZ	
				L_v	B_i	L_v	B_i
Abutilon theophrasti	Annual herbaceous	*Malvaceae*	*Abutilon*	0.305	0.061	0.48	0.047
Acalypha australis	Annual herbaceous	*Euphorbiaceae*	*Acalypha*	0.293	0.028	0.056	0.008
Eclipta prostrata	Annual herbaceous	*Compositae*	*Eclipta*	0.28	0.056	0.465	0.065
Cyperus rotundus	Perennial herb	*Cyperaceae*	*Cyperus*	0.249	0.127	0.1	0.073
Setaria viridis	Annual herbaceous	*Gramineae*	*Setaria*	0.24	0.078	0.376	0.112
Digitaria sanguinalis	Annual herbaceous	*Gramineae*	*Digitaria*	0.131	0.041	0.192	0.021
Eleusine indica	Annual herbaceous	*Gramineae*	*Eleusine*	0.129	0.085	0.289	0.057
Polygonum lapathifolium	Annual herbaceous	*Polygonaceae*	*Polygonum*	0.125	0.023	-	-
Arthraxon hispidus	Annual herbaceous	*Gramineae*	*Arthraxon*	0.108	0.023	-	-
Phyllanthus urinaria	Annual herbaceous	*Euphorbiaceae*	*Phyllanthus*	0.089	0.02	-	-
Amaranthus retroflexus	Annual herbaceous	*Amaranthaceae*	*Amaranthus*	0.085	0.042	0.13	0.038
Portulaca oleracea	Annual herbaceous	*Portulacacea*	*Portulaca*	0.085	0.037	0.111	0.01
Eriochloa villosa	Annual herbaceous	*Gramineae*	*Eriochloa*	0.08	0	-	-
Conyza canadensis	Annual herbaceous	*Compositae*	*Conyza*	0.077	0.014	0.212	0.053
Solanum nigrum	Annual herbaceous	*Solanaceae*	*Solanum*	0.076	0.017	0.143	0.072
Alternanthera philoxeroides	Perennial herb	*Amaranthaceae*	*Alternanthera*	0.071	0.02	-	-
Pharbitis purpurea	Annual herbaceous	*Convolvulaceae*	*Pharbitis*	0.049	0.121	-	-
Artemisia argyi	Perennial herb	*Compositae*	*Artemisia*	0.044	0.013	0.124	0.235
Humulus scandens	Annual twining herb	*Moraceae*	*Humulus*	0.042	0.045	0.148	0.021
Lindernia procumbens	Annual herbaceous	*Scrophulariaceae*	*Lindernia*	0.041	0.026	0.102	0.02
Artemisia capillaris	Perennial herb	*Compositae*	*Artemisia*	0.041	0.017	-	-
Mazus japonicus	Annual herbaceous	*Scrophulariaceae*	*Mazus*	0.035	0.031	-	-
Gnaphalium affine	Annual herbaceous	*Compositae*	*Gnaphalium*	0.033	0.005	-	-

Table A1. *Cont.*

Species	Life Form	Family	Genera	WLFZ		NRZ	
				L_v	B_i	L_v	B_i
Crassocephalum crepidioides	Annual herbaceous	*Compositae*	*Crassocephalum*	0.033	0.01	0.083	0.034
Kummerowia striata	Annual herbaceous	*Leguminosae*	*Kummerowia*	0.033	1.887	0.138	0.015
Daucus carota	Perennial herb	*Umbelliferae*	*Daucus*	0.025	0.012	-	-
Phytolacca acinosa	Perennial herb	*Phytolaccaceae*	*Phytolacca*	0.022	0.046	-	-
Chenopodium album	Annual herbaceous	*Chenopodiaceae*	*Chenopodium*	0.022	0.053	-	-
Siegesbeckia orientalis	Annual herbaceous	*Compositae*	*Siegesbeckia*	0.021	0.038	-	-
Polygonum plebeium	Annual herbaceous	*Polygonaceae*	*Polygonum*	0.019	0.005	0.037	0.006
Stellaria media	Biennial herbs	*Caryophyllaceae*	*Stellaria*	0.011	0.002	0.083	0.041
Gynura formosana	Perennial herb	*Compositae*	*Gynura*	0.011	0.004	-	-
Clinopodium chinense	Perennial herb	*Labiatae*	*Clinopodium*	0.011	0.002	-	-
Sapium sebiferum	Tree	*Euphorbiaceae*	*Sapium*	0.011	0.004	-	-
Convolvulus arvensis	Perennial herb	*Convolvulaceae*	*Convolvulus*	0.011	0.003	-	-
Paederia scandens	Perennial twining herb	*Rubiaceae*	*Paederia*	0.011	0.005	-	-
Lindernia procumbens	Annual herbaceous	*Scrophulariaceae*	*Lindernia*	0.011	0.011	-	-
Cucumis sativus	Annual Vines herb	*Cucurbitaceae*	*Cucumis*	0.011	0.003	-	-
Melilotus officinalis	Biennial herbs	*Leguminosae*	*Melilotus*	0.011	0.039	-	-
Dendranthema indicum	Perennial herb	*Compositae*	*Dendranthema*	0.011	0.009	-	-
Broussonetia papyrifera	Tree	*Moraceae*	*Broussonetia*	0.011	0.024	-	-
Corydalis pallida	Annual herbaceous	*Papaveraceae*	*Corydalis*	0.011	0.019	-	-
Ixeridium chinense	Perennial herb	*Compositae*	*Ixeridium*	0.011	0.031	-	-
Rhus chinensis	Deciduous small trees or shrubs	*Anacardiaceae*	*Rhus*	0.011	0.071	-	-
Pennisetum alopecuroides	Perennial herb	*Gramineae*	*Pennisetum*	-	-	0.184	0.035
Chenopodium ambrosioides	Annual or perennial herbs	*Chenopodiaceae*	*Chenopodium*	-	-	0.101	0.008

Table A1. *Cont.*

Species	Life Form	Family	Genera	WLFZ		NRZ	
				L_v	B_i	L_v	B_i
Salix variegata	shrub	*Salicaceae*	*Salix*	-	-	0.093	0.014
Ludwigia prostrata	Annual herbaceous	*Onagraceae*	*Ludwigia*	-	-	0.093	0.017
Cyperus difformis	Annual herbaceous	*Cyperaceae*	*Cyperus*	-	-	0.12	0.016
Artemisia annua	Annual herbaceous	*Compositae*	*Artemisia*	-	-	0.177	0.04
Ranunculus sieboldii	Perennial herb	*Ranunculaceae*	*Ranunculus*	-	-	0.056	0.005
Oxalis corniculata	Perennial herb	*Oxalidaceae*	*Oxalis*	-	-	0.037	0.008
Youngia japonica	Annual herbaceous	*Compositae*	*Youngia*	-	-	0.098	0.047
Rorippa indica	Biennial herbs	*Cruciferae*	*Rorippa*	-	-	0.648	0.158
Eragrostis minor	Annual herbaceous	*Poaceae*	*Eragrostis*	-	-	0.246	0.024
Oxalis corniculata	Annual herbaceous	*Oxalidaceae*	*Oxalis*	-	-	0.167	0.025
Debregeasia orientalis	Deciduous small trees or shrubs	*Urticaceae*	*Debregeasia Gaudich.*	-	-	0.496	0.098
Veronica didyma	Biennial herbs	*Scrophulariaceae*	*Veronica L.*	-	-	0.083	0.012
Cerastium arvense	Perennial herb	*Caryophyllaceae*	*Cerastium*	-	-	0.162	0.016
Arabidopsis thaliana	Annual or perennial herbs	*Cruciferae*	*Arabidopsis*	-	-	0.083	0.066
Cardamine hirsuta	Annual herbaceous	*Cruciferae*	*Cardamine*	-	-	0.166	0.039
Fimbristylis dichotoma	Annual herbaceous	*Cyperaceae*	*Fimbristylis Vahl*	-	-	0.083	0.06
Buddleja lindleyana	shrub	*Loganiaceae*	*Buddleja*	-	-	0.167	0.036
Rumex japonicus	Perennial herb	*Polygonaceae*	*Rumex*	-	-	0.083	0.016
Rumex acetosa	Perennial herb	*Polygonaceae*	*Rumex*	-	-	0.083	0.01
Polygonum capitatum	Perennial herb	*Polygonaceae*	*Polygonum*	-	-	0.083	0.011
Ambrosia artemisiifolia	Annual or perennial herbs	*Compositae*	*Ambrosia*	-	-	0.083	0.054
Galinsoga parviflora	Annual herbaceous	*Compositae*	*Galinsoga*	-	-	0.25	0.051
Polypogon fugax	Annual herbaceous	*Gramineae*	*Polypogon*	-	-	0.083	0.04

Table A1. *Cont.*

Species	Life Form	Family	Genera	WLFZ		NRZ	
				L_v	B_i	L_v	B_i
Cenchrus echinatus	Annual herbaceous	*Gramineae*	*Cenchrus*	-	-	0.152	0.048
Brachypodium sylvaticum	Perennial herb	*Gramineae*	*Brachypodium*	-	-	0.083	0.017
Vicia sepium	Perennial herb	*Leguminosae*	*Vicia*	-	-	0.083	0.009
Ricinus communis	Annual	*Euphorbiaceae*	*Ricinus*	-	-	0.083	0.033

Table A2. Each number in Figure 6b represents the species.

Number	Plant Species	Number	Plant Species
1	*Echinochloa crusgalli*	36	*Chenopodium album*
2	*Polygonum hydropiper*	37	*Humulus scandens*
3	*Portulaca oleracea*	38	*Melilotus officinalis*
4	*Cynodon dactylon*	39	*Artemisia capillaris*
5	*Eleusine indica*	40	*Cucumis sativus*
6	*Arthraxon hispidus*	41	*Eriochloa villosa*
7	*Cyperus rotundus*	42	*Kummerowia striata*
8	*Eclipta prostrata*	43	*Daucus carota*
9	*Amaranthus retroflexus*	44	*Dendranthema indicum*
10	*Polygonum plebeium*	45	*Corydalis pallida*
11	*Rumex acetosa*	46	*Broussonetia papyrifera*
12	*Lindernia procumbens*	47	*Pennisetum alopecuroides*
13	*Abutilon theophrasti*	48	*Artemisia annua*
14	*Bidens pilosa*	49	*Oxalis corniculata*
15	*Mazus japonicus*	50	*Youngia japonica*
16	*Phyllanthus urinaria*	51	*Rumex japonicus*
17	*Phytolacca acinosa*	52	*Polygonum lapathifolium*
18	*Xanthium sibiricum*	53	*Rorippa indica*
19	*Crassocephalum crepidioides*	54	*Arabidopsis thaliana*
20	*Acalypha australis*	55	*Veronica didyma*
21	*Gnaphalium affine*	56	*Fimbristylis dichotoma*
22	*Solanum nigrum*	57	*Debregeasia orientalis*
23	*Stellaria media*	58	*Vicia sepium*
24	*Gynura formosana*	59	*Ricinus communis*
25	*Setaria viridis*	60	*Buddleja lindleyana*
26	*Alternanthera philoxeroides*	61	*Ambrosia artemisiifolia*
27	*Clinopodium chinense*	62	*Galinsoga parviflora*
28	*Digitaria sanguinalis*	63	*Polypogon fugax*
29	*Sapium sebiferum*	64	*Cenchrus echinatus*
30	*Convolvulus arvensis*	65	*Polygonum capitatum*
31	*Pharbitis purpurea*	66	*Brachypodium sylvaticum*
32	*Siegesbeckia orientalis*	67	*Cerastium arvense*
33	*Artemisia argyi*	68	*Eragrostis minor*
34	*Conyza canadensis*	69	*Cardamine hirsuta*
35	*Paederia scandens*		

Table A3. Each number in Figure 7 represents the species.

Number	Plant Species	Number	Plant Species
1	*Cynodon dactylon*	26	*Artemisia capillaris*
2	*Bidens pilosa*	27	*Mazus japonicus*
3	*Xanthium sibiricum*	28	*Gnaphalium affine*

Table A3. *Cont.*

Number	Plant Species	Number	Plant Species
4	*Echinochloa crusgalli*	29	*Crassocephalum crepidioides*
5	*Polygonum hydropiper*	30	*Kummerowia striata*
6	*Abutilon theophrasti*	31	*Daucus carota*
7	*Acalypha australis*	32	*Phytolacca acinosa*
8	*Eclipta prostrata*	33	*Chenopodium album*
9	*Cyperus rotundus*	34	*Siegesbeckia orientalis*
10	*Setaria viridis*	35	*Polygonum plebeium*
11	*Digitaria sanguinalis*	36	*Stellaria media*
12	*Eleusine indica*	37	*Gynura formosana*
13	*Polygonum lapathifolium*	38	*Clinopodium chinense*
14	*Arthraxon hispidus*	39	*Sapium sebiferum*
15	*Phyllanthus urinaria*	40	*Convolvulus arvensis*
16	*Amaranthus retroflexus*	41	*Paederia scandens*
17	*Portulaca oleracea*	42	*Lindernia procumbens*
18	*Eriochloa villosa*	43	*Cucumis sativus*
19	*Conyza canadensis*	44	*Melilotus officinalis*
20	*Solanum nigrum*	45	*Dendranthema indicum*
21	*Alternanthera philoxeroides*	46	*Broussonetia papyrifera*
22	*Pharbitis purpurea*	47	*Corydalis pallida*
23	*Artemisia argyi*	48	*Ixeridium chinense*
24	*Humulus scandens*	49	*Rhus chinensis*
25	*Cenchrus echinatus*		

References

1. Benchimol, M.; Peres, C.A. Predicting local extinctions of Amazonian vertebrates in forest islands created by a mega dam. *Biol. Conserv.* **2015**, *187*, 61–72. [CrossRef]
2. Sun, L.; Ding, W.; Zhou, Y.; Wang, J.; Ouyang, X.; Fan, Z.; Yao, Y.; Zhang, C. Tree Radial Growth Responses to Climate and Reservoir Impoundment in Valleys in Southwestern China. *Forests* **2024**, *15*, 749. [CrossRef]
3. Barnes, B.; Sidhu, H.; Roxburgh, S. A model integrating patch dynamics, competing species and the intermediate disturbance hypothesis. *Ecol. Modell.* **2006**, *194*, 414–420. [CrossRef]
4. Ali, M.M. Shoreline vegetation of Lake Nubia, Sudan. In *Macrophytes in Aquatic Ecosystems: From Biology to Management: Proceedings of the 11th Internatonal Symposium on Aquatic Weeds, European Weed Research Society*; Springer: Berlin/Heidelberg, Germany, 2006; pp. 101–105. [CrossRef]
5. Zhang, Z.; Wan, C.; Zheng, Z.; Hu, L.; Feng, K.; Chang, J.; Xie, P. Plant community characteristics and their responses to environmental factors in the water level fluctuation zone of the three gorges reservoir in China. *Environ. Sci. Pollut. Res.* **2013**, *20*, 7080–7091. [CrossRef] [PubMed]
6. Ye, C.; Zhang, K.; Deng, Q.; Zhang, Q. Plant communities in relation toflooding and soilcharacteristics in the water level fluctuation zone of the Three Gorges Reservoir, China. *Environ. Sci. Pollut. Res.* **2013**, *20*, 1794–1802. [CrossRef] [PubMed]
7. Bao, Y.; Gao, P.; He, X. The water-level fluctuation zone of Three Gorges Reservoir—A unique geomorphological unit. *Earth-Sci. Rev.* **2015**, *150*, 14–24. [CrossRef]
8. Naiman, R.J.; Melillo, J.M.; Lock, M.A.; Ford, T.E.; Reice, S.R. Longitudinal patterns of ecosystem processes and community structure in a subarctic river continuum. *Ecology* **1987**, *68*, 1139–1156. [CrossRef]
9. Jansson, R.; Nilsson, C.; Dynesius, M.; Andersson, E. Effects of river regulation on river-margin vegetation: A comparison of eight boreal rivers. *Ecol. Appl.* **2000**, *10*, 203–224. [CrossRef]
10. Nilsson, C.; Jansson, R. Floristic differences between riparian corridors of regulated and free-flowing boreal rivers. *Regul. Rivers Res. Manag.* **1995**, *11*, 55–66. [CrossRef]
11. Chen, Z.L.; Yuan, X.Z.; Liu, H.; Li, B. Characteristics of plant communities in the hydrofluctuation zones of the TGR area under water level fluctuation. *Resour. Environ. Yangtze Basin* **2012**, *21*, 672–677.
12. Fu, J.; Li, X.L.; Dai, Z.; Zhang, H.F.; Luo, Y.H.; Xu, T.; Huang, Y.P. Effect of water-level fluctuation discrepancy on the composition of different annuals in TGR drawdown zone. *J. Wuhan Univ. (Nat. Sci. Ed.)* **2015**, *61*, 285–290. (In Chinese)
13. Su, X.; Bejarano, M.D.; Yi, X.; Lin, F.; Ayi, Q.; Zeng, B. Unnatural flooding alters the functional diversity of riparian vegetation of the Three Gorges Reservoir. *Freshw. Biol.* **2020**, *65*, 1585–1595. [CrossRef]
14. Noble, R.E.; Murphy, P.K. Short term effects of prolonged backwater flooding on understory vegetation. *Castanea* **1975**, *40*, 228–238.
15. Yuan, S.; Zeng, B.; Su, X.; Xu, J. Effect of water-level fluctuation discrepancy on the composition of different annuals in Three Gorges reservoir drawdown zone. *Acta Ecol. Sin.* **2014**, *34*, 6481–6488. (In Chinese)

16. Guo, Y.; Yang, S.; Shen, Y.F.; Xiao, W.F.; Rui, R.M. Composition and niche of the existing herbaceous plants in the water-level-fluctuating zone of the TGR Area, China. *China J. Appl. Ecol.* **2018**, *29*, 3559–3568. [CrossRef]
17. Guo, Q.; Kang, Y.; Hong, M.; Jin, J.; Zhu, N.; Nie, B.; Wang, Z. Responses of terrestrial plants in hydro-fluctuation belt of the Three Gorges Reservoir Area to the first time flooding-drying habitat change. *Sci. Silvae Sin.* **2013**, *49*, 1–9. [CrossRef]
18. Zhang, A.; Fan, D.; Li, Z.; Xiong, G.; Xie, Z. Enhanced photosynthetic capacity by perennials in the riparian zone of the Three Gorges Reservoir Area, China. *Ecol. Eng.* **2016**, *90*, 6–11. [CrossRef]
19. Harrison, M.T.; Kelman, W.M.; Moore, A.D.; Evans, J.R. Grazing winter wheat relieves plant water stress and transiently enhances photosynthesis. *Funct. Plant Biol.* **2010**, *37*, 726–736. [CrossRef]
20. Whittaker, R.H.; Levin, S.A.; Root, R.B. Niche, habitat, and ecotope. *Am. Nat.* **1973**, *107*, 321–338. [CrossRef]
21. Petraitis, P.S. Likelihood measures of niche breadth and overlap. *Ecology* **1979**, *60*, 703–710. [CrossRef]
22. Ganis, P. NICHE—Programs for Niche Breadth, Overlap and Hypervolumes. In *Computer Ssisted Egetation Analysis*; Springer: Berlin/Heidelberg, Germany, 1991; pp. 469–487. [CrossRef]
23. Slatyer, R.A.; Hirst, M.; Sexton, J.P. Niche breadth predicts geographical range size: A general ecological pattern. *Ecol. Lett.* **2013**, *16*, 1104–1114. [CrossRef]
24. Malkinson, D.; Tielbörger, K. What does the stress-gradient hypothesis predict? Resolving the discrepancies. *Oikos* **2010**, *119*, 1546–1552. [CrossRef]
25. Sun, L.; Li, X.L.; Wang, X.S.; Xiang, L.; Yang, J.; Min, Q.F.; Chen, G.H.; Chen, F.Q.; Huang, C.M.; Wang, G.X. Growth and respiratory metabolic adaptation strategies of riparian plant *Distylium chinense* to submergence by the field study and controlled experiments. *Plant Physiol. Biochem.* **2020**, *157*, 1–12. [CrossRef] [PubMed]
26. Yang, F.; Ji, D.B.; Wang, L.J.; Li, H.; Li, Y.J. Vertical distribution characteristics of dissolved oxygen and chlorophyll a in typical tributaries during the impoundment period of the Three Gorges Reservoir. *Environ. Sci.* **2020**, *41*, 2107–2115.
27. Tang, T.; Cai, Q.; Liu, J. Using epilithic diatom communities to assess ecological condition of Xiangxi River system. *Environ. Monit. Assess.* **2006**, *112*, 347–361. [CrossRef] [PubMed]
28. Yan, H.; Huang, Y.; Wang, G.; Zhang, X.; Shang, M.; Feng, L.; Dong, J.; Shan, K.; Wu, D.; Zhou, B. Water eutrophication evaluation based on rough set and petri nets: A case study in Xiangxi-River, Three Gorges Reservoir. *Ecol. Indic.* **2016**, *69*, 463–472. [CrossRef]
29. Maloof, J.N.; Nozue, K.; Mumbach, M.R.; Palmer, C.M. LeafJ: An ImageJ plugin for semi-automated leaf shape measurement. *J. Vis. Exp.* **2013**, *71*, e50028. [CrossRef]
30. Wu, X.; Fan, W.; Du, H.; Ge, H.; Huang, F.; Xu, X. Estimating crown structure parameters of moso bamboo: Leaf area and leaf angle distribution. *Forests* **2019**, *10*, 686. [CrossRef]
31. Kabała, C.; Musztyfaga, E.; Gałka, B.; Łabuńska, D.; Mańczyńska, P. Conversion of Soil pH 1:2.5 KCl and 1:2.5 H_2O to 1:5 H_2O: Conclusions for Soil Management, Environmental Monitoring, and International Soil Databases. *Pol. J. Environ. Stud.* **2016**, *2*, 25. [CrossRef] [PubMed]
32. Forch, J. A simple method for the isolation and purification of total lipids from animal tissues. *J. Biol. Chem.* **1957**, *226*, 498–509. [CrossRef]
33. Beers, R.F.; Sizer, I.W. A spectrophotometric method for measuring the breakdown of hydrogen peroxide by catalase. *J. Biol. Chem.* **1952**, *195*, 133–140. [CrossRef]
34. Williams, V.; Twine, S. Flame photometric method for sodium, potassium and calcium. *Mod. Methods Plant Anal.* **1960**, *5*, 3–5.
35. Ye, C.; Li, S.; Zhang, Y.; Zhang, Q. Assessing soil heavy metal pollution in the water-level-fluctuation zone of the Three Gorges Reservoir, China. *J. Hazard. Mater.* **2011**, *191*, 366–372. [CrossRef] [PubMed]
36. Tan, L. *Statistical Ecology*; China Forestry Press: Beijing, China, 2009.
37. Curtis, J.T.; McIntosh, R.P. An upland forest continuum in the prairie-forest border region of Wisconsin. *Ecology* **1951**, *32*, 476–496. [CrossRef]
38. Takane, Y.; Hwang, H. Regularized linear and kernel redundancy analysis. *Comput. Stat. Data Anal.* **2007**, *52*, 394–405. [CrossRef]
39. Zhu, K.W.; Chen, Y.C.; Zhang, S.; Lei, B.; Yang, Z.M.; Huang, L. Vegetation of the water-level fluctuation zone in the Three Gorges Reservoir at the initial impoundment stage. *Glob. Ecol. Conserv.* **2020**, *21*, e00866. [CrossRef]
40. Wang, F.; Xu, T.; Huang, Y.P. Investigation on plant community and distribution characteristics of Xiangxi River bank. *Int. J. Green Technol.* **2014**, *1*, 88–91.
41. Xu, Y.; Chena, F. Effects of concrete content in vegetation concrete matrix on seed germination and seeding establishment of cynodon dactylon. *Procedia Eng.* **2012**, *28*, 105–109. [CrossRef]
42. Singh, K.; Pandey, V.C.; Singh, R.P. Cynodon dactylon: An efficient perennial grass to revegetate sodic lands. *Ecol. Eng.* **2013**, *54*, 32–38. [CrossRef]
43. Shen, F.; Qin, Y.; Wang, R.; Huang, X.; Wang, Y.; Gao, T.; Yang, X. Comparative genomics reveals a unique nitrogen-carbon balance system in Asteraceae. *Nat. Commun.* **2023**, *14*, 4334. [CrossRef]
44. Zhang, A.Y.; Xiong, G.M.; Fan, D.Y.; Xie, Z.Q. Effects of damming on plant diversity in the inundated and riparian zones of the TGR Area, China. *China J. Ecol.* **2016**, *35*, 2505–2518.
45. Grime, J.P. *Plant Strategies, Vegetation Processes, and Ecosystem Properties*; John Wiley & Sons Ltd.: Hoboken, NJ, USA, 2001.
46. Grime, J.P. Evidence for the existence of three primary strategies in plants and its relevance to ecological and evolutionary theory. *Am. Nat.* **1977**, *111*, 1169–1194. [CrossRef]

47. Grime, J. Dominant and subordinate components of plant communities: Implications for succession, stability and diversity. In Proceedings of the Symposium of the British Ecological Society, London, UK, 3 December 1987.
48. Liao, J.; Jiang, M.; Li, L. Effects of simulated submergence on survival and recovery growth of three species in water fluctuation zone of the Three Gorges reservoir. *Acta Ecol. Sin.* **2010**, *30*, 216–220. [CrossRef]
49. Chai, Y.; Yue, M.; Wang, M.; Xu, J.; Liu, X.; Zhang, R.; Wan, P. Plant functional traits suggest a change in novel ecological strategies for dominant species in the stages of forest succession. *Oecologia* **2016**, *180*, 771–783. [CrossRef] [PubMed]
50. Yi, X.; Huang, Y.; Ma, M.; Wen, Z.; Chen, J.; Chen, C.; Wu, S. Plant trait-based analysis reveals greater focus needed for mid-channel bar downstream from the Three Gorges Dam of the Yangtze River. *Ecol. Indic.* **2020**, *111*, 105950. [CrossRef]
51. Pisula, W.; Łukasik, A.; Kawa, R.; Pisula, E. Exploratory analysis of the links among life history, reproductive strategy, autism-spectrum quotient, and quality of life. *Psychology* **2018**, *9*, 2312. [CrossRef]
52. Pielou, E. Niche width and niche overlap: A method for measuring them. *Ecology* **1972**, *53*, 687–692. [CrossRef]
53. Xu, Z.G.; He, Y.; Yan, B.X.; Song, C.C. Niche characteristics of typical marsh wetland plant populations in Sanjiang Plain. *J. Appl. Ecol.* **2007**, *18*, 783–787. (In Chinese)
54. Wang, X.R.; Cheng, R.M.; Xiao, W.F.; Pan, L.; Zeng, L.X. Characteristics of niche changes of main dominant herbaceous plants in the initial stage of flooding in the hydro-fluctuation zones of the TGR area. *Resour. Environ. Yangtze Basin* **2016**, *25*, 8. (In Chinese)
55. Li, D.Z.; Shi, Q.; Zang, R.G.; Wang, X.P.; Sheng, L.J.; Zhu, Z.L.; Wang, C.A. Models for niche breadth and niche overlap of species or populations. *Sci. Silvae Sin.* **2006**, *42*, 95–103.
56. Escobedo, V.M.; Rios, R.S.; Salgado-Luarte, C.; Gianoli, E. Correlation of plasticities to drought and shade: Implications for environmental niche overlap in drylands. *Oikos* **2023**, *2024*, e09766. [CrossRef]
57. Li, M.X.; Wang, D.M.; Ren, Y.; Wang, X.L. Influence of different drying-rewetting frequencies on available soil nutrients and DOC. *Acta Ecol. Sin.* **2018**, *38*, 1542–1549.
58. Wang, X.R.; Cheng, R.M.; Xiao, W.F.; Pan, L.; Zeng, L.X. Niche variation of dominant herbaceous plants in waterlevel-fluctuating zone of Three Gorges Reservoir at the beginning after charging water. *Resour. Environ. Yangtze Basin* **2016**, *25*, 404–411.
59. Jian, Z.J.; Ma, F.Q.; Guo, Q.S.; Qin, A.L.; Xiao, W.F. Niche of dominant plant populations in the water level fluctuation zone of canyon landform area of the Three Gorges Reservoir. *Chin. J. Ecol.* **2017**, *36*, 328–334.
60. Crain, C.M. Interactions between marsh plant species vary in direction and strength depending on environmental and consumer context. *J. Ecol.* **2008**, *96*, 166–173. [CrossRef]
61. Graffmann, K.; Grosse, W.; Junk, W.J.; Parolin, P. Pressurized gas transport in Amazonian floodplain trees. *Environ. Exp. Bot.* **2008**, *62*, 371–375. [CrossRef]
62. Chen, J.; Zeng, C.; Wei, H.; Liu, Y.; Wang, Z.; Jia, Z. Effects of mixed intercropping of Cynodon dactylon and Hemarthria altissima on their biomass under different flooding conditions. *Acta Ecol. Sin.* **2017**, *37*, 54–61. [CrossRef]
63. Petrů, M.; Tielbörger, K.; Belkin, R.; Sternberg, M.; Jeltsch, F. Life history variation in an annual plant under two opposing environmental constraints along an aridity gradient. *Ecography* **2006**, *29*, 66–74. [CrossRef]
64. Herben, T.; Klimešová, J.; Chytrý, M. Effects of disturbance frequency and severity on plant traits: An assessment across a temperate flora. *Funct. Ecol.* **2018**, *32*, 799–808. [CrossRef]
65. Chesson, P. Mechanisms of maintenance of species diversity. *Annu. Rev. Ecol. Syst.* **2000**, *31*, 343–366. [CrossRef]
66. Ye, C.; Chen, C.; Butler, O.M.; Rashti, M.R., Esfandbod, M.; Du, M.; Zhang, Q. Spatial and temporal dynamics of nutrients in riparian soils after nine years of operation of the Three Gorges Reservoir, China. *Sci. Total Environ.* **2019**, *664*, 841–850. [CrossRef]
67. Wilson, P.J.; Thompson, K.; Hodgson, J.G. Specific leaf area and leaf dry matter content as alternative predictors of plant strategies. *New Phytol.* **1999**, *143*, 155–162. [CrossRef]
68. Larbi, A.; Kchaou, H.; Gaaliche, B.; Gargouri, K.; Boulal, H.; Morales, F. Supplementary potassium and calcium improves salt tolerance in olive plants. *Sci. Hortic.* **2020**, *260*, 108912. [CrossRef]
69. Nilsson, C.; Aradóttir, Á.L. Ecological and social aspects of ecological restoration: New challenges and opportunities for northern regions. *Ecol. Soc.* **2013**, *18*, 35. [CrossRef]
70. Nilsson, C.; Polvi, L.E.; Gardeström, J.; Hasselquist, E.M.; Lind, L.; Sarneel, J.M. Riparian and in-stream restoration of boreal streams and rivers: Success or failure? *Ecohydrology* **2015**, *8*, 753–764. [CrossRef]
71. Nilsson, C.; Svedmark, M. Basic principles and ecological consequences of changing waterregimes: Riparian plant communities. *Environ. Manag.* **2002**, *30*, 468–480. [CrossRef] [PubMed]
72. Yue, L.; Yu, C.; Abdoussalami, A.; Li, X.; Lv, K.; Huang, G.; Hu, M.; Yang, Z. Growth, Morphological Alterations, and Enhanced Photosynthetic Performance Promote Tolerance of *Distylium chinense* to Alternate Drought–Flooding Stresses. *Forests* **2024**, *15*, 125. [CrossRef]
73. Chi, X.; Liu, J.M.; Yan, Q.; Li, P.; Ai, H.; Gao, P. Niche Characteristics of Dominant Populations of Drepanostachyum luodianense. *Guizhou Agric. Sci.* **2014**, *42*, 180–183.

water

MDPI

Article

Excessive Eutrophication as a Chemical Barrier for Fish Fauna Dispersion: A Case Study in the Emblematic Tietê River (São Paulo, Brazil)

Bruna Urbanski and Marcos Nogueira *

Department of Biodiversity and Biostatistics, Institute of Biosciences, State University of São Paulo, Campus of Botucatu, 250, Rubião Junior, Botucatu 18618-689, SP, Brazil; brurbanski@outlook.com

* Correspondence: marcos.nogueira@unesp.br; Tel.: +55-14-99772-5668

Abstract: The Tietê River receives most of the effluents and diffuse wastes from the São Paulo metropolis (21.9 million inhabitants). The study aimed to assess the extent to which environmental changes affected the fish fauna. We compared, in rainy and dry seasons, three sites in Tietê and three in tributaries with much better water quality conditions. No physical barriers exist between the sites. Fish were sampled with gillnets (exposed overnight) and the assemblage's ecological attributes calculated. Water depth, transparency, temperature, electrical conductivity, dissolved oxygen, turbidity, total dissolved solids, pH, redox potential, total phosphorus, total nitrogen, and chlorophyll were simultaneously determined. Low oxygen concentrations (near zero in the rainy period), and the opposite for nutrients and total solids, separated the predominantly hypereutrophic Tietê from the tributaries (PCA). Differences in the fish assemblages were also evidenced (nMDS), including a higher richness per sample in tributaries (11 to 14 spp.) compared to Tietê (3 to 4 spp.). Siluriformes with accessory breathing dominated in Tietê and the highly tolerant detritivorous *Prochilodus lineatus* (the main commercial fish) was the only species found in all sites. The species correlated positively with oxygen in the tributaries and with turbidity, redox potential, and nutrients in Tietê (DistLM) (rainy season). Recovery measures are urgently required.

Keywords: ecological barrier; environmental degradation; fish tolerance; pollution

1. Introduction

Dispersal plays a fundamental role in the maintenance of the structure and functioning of populations and, consequently, of the species they represent, assuring a continuous genetic flow, over space and time [1–3]. This ecological process is related to the capability of biota to transpose relative barriers and move among distinct habitats in response to internal (e.g., genetic and physiological) and external signals (e.g., light, temperature, water quality, and predation pressure) [4], this being one of the main explanatory models of the distribution patterns of organisms on planet [5].

In the current scenario of massive habitat degradation and climate uncertainties, understanding how, where, and why organisms disperse is extremely important [2,3]. Numerous artificial barriers to dispersion have emerged because of significant population growth and economic development. In the case of Brazilian fluvial environments, the construction of hydroelectric dams, for instance, is a major interference in the free movement of the biota, especially fish [6–11]. However, other types of environmental disturbances, such as pollution/eutrophication, can also affect the distribution of organisms and other ecological processes related to the different components of the ecosystems [12].

The Tietê River is a notorious example of severe environmental degradation, particularly in its upper section that crosses the metropolitan region of São Paulo, the most populous in Brazil (around 21.9 million inhabitants) [13]. The history of water quality deterioration in this river began with organic contamination in the mid-nineteenth century. Nevertheless, across the twentieth century, in addition to high loads of nutrients, degradation became a more complex scenario. There was a considerable increase in inorganic contamination due to the introduction of a myriad dissolved substances, such as pesticides, herbicides, medicines, and cosmetics. The Tietê River represents a significant threat to society, in terms of to water security, due to the potential adverse effects on human health [14].

After a critical water crisis due to a deficit in precipitation, with an enormous impact in São Paulo State in the year 2014, technical and scientific studies tried to explain the complex degradation process of the Tietê River [15]. However, the discussion, especially in terms of governance, remained restricted to the São Paulo metropolitan area. The fact that this huge impact transcends to a much wider regional spatial scale, i.e., dramatically spreads towards the São Paulo State countryside, seems to be neglected.

In a long-term study, from the end of the 1970s to the beginning of the 2000s [16], a clear trend of water quality deterioration in the middle Tietê River was identified, more than 300 km downstream of São Paulo city. The electrical conductivity of water, for instance, increased from 103 to 370 µS cm^{-1}; phosphorus concentration from 15 to 40 µg L^{-1}; nitrate from 200 to 3480 µg L^{-1}; and ammonium from 16 to 232 µg L^{-1}. The authors also highlighted the increased potential for toxicity associated with the recurrent cyanobacterial blooms. After that, during the last two decades, there has been no signal of consistent improvement in the river water quality; maybe the contrary [17,18].

The extraordinary increase in nutrients introduced in the Tietê River resulted in a remarkable eutrophication, immediately changing the basic levels of the aquatic food chains (magnification in the biomass of phyto- and zooplankton-tolerant species). Through direct and indirect transference processes, such changes affected the upper trophic levels represented by the different species of fish. When trophic conditions become extreme, recurrent decreases in dissolved oxygen concentration occur, directly influencing the ichthyofauna [19]. The consequences of this type of interference are reflected in the composition and other structural characteristics of fish. Previous studies in the Tietê River have already pointed to a reduction in the size of native populations and their reproductive performance, and a general decrease in species richness and diversity [11,20].

In this context, the objective of this study was to assess the extent to which changes in environmental factors have affected the regional fish fauna in the middle Tietê River basin (SE Brazil). The selected study sites, in addition to the main river, include the lower stretches of the tributaries Capivari (right bank), Sorocaba, and Peixe (both on the left bank). These environments are widely connected, without any natural (e.g., falls and rapids) or artificial (e.g., dams and reservoirs) physical barriers that could interfere with the free movement of the organisms. However, they exhibit contrasting differences in terms of water quality, with the tributaries in a much better condition. Therefore, we hypothesized that water quality degradation acts as a chemical barrier to the dispersion of species, on a regional scale.

2. Material and Methods

The Tietê River is 1100 km long (~1300 m^3 s^{-1} in its final stretch) [20] and flows from east to west along almost the entire state of São Paulo, southeast Brazil. The study sites are located approximately 300 km downstream of São Paulo city.

Samplings occurred at the end of the rainy period (April 2019), early autumn, and at the end of the dry period (October 2019), spring, at six selected sites (Figure 1; Table 1). The study considered six sampling sites, without replicates, due to the difficulties found during the fieldwork—excessive amount of solid waste tapped into the nets and boat engine and health risks due to the inevitable direct contact with the Tietê contaminated water. Three sites were distributed in the Tietê River along a stretch of 40 km, one further upstream (TI), above the mouth of the Sorocaba (Sor) and Capivari (Cap) rivers; one site in the middle section (TII); and one further downstream (TIII), below the mouth of the Peixe (Pei) River (upstream of the Barra Bonita Reservoir). Three additional sites corresponded to the lower reaches of the mentioned tributaries.

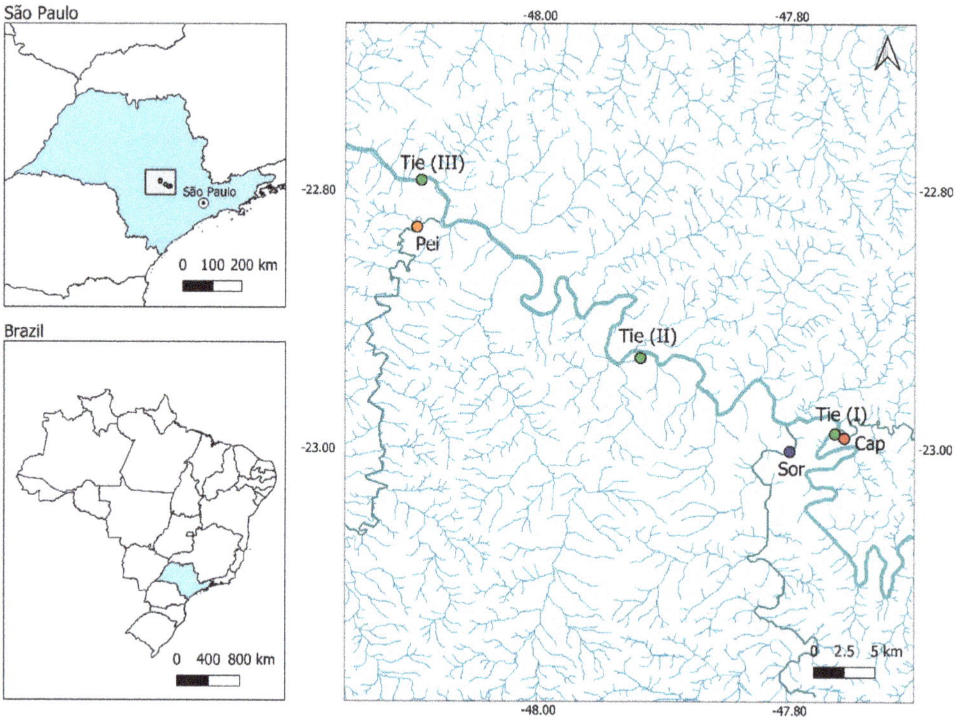

Figure 1. Distribution of sampling sites (**on the right**) in the Tietê River (green circles), Sorocaba River (blue circle), Capivari River (red circle), and Peixe River (orange circle). The highlighted region (**upper left**), in gray, corresponds to the study area, in the center of São Paulo State, SE Brazil.

Table 1. Sampling site denomination and respective locations.

Site	River	Municipality	Latitude	Longitude
Tie (I)	Tietê River	Laranjal Paulista	22° 59′ 17.6″ S	47° 46′ 00.1″ W
Tie (II)	Tietê River	Laranjal Paulista	22° 55′ 47.6″ S	47° 55′ 22.8″ W
Tie (III)	Tietê River	Anhembi	22° 47′ 31.0″ S	48° 05′ 48.8″ W
Cap	Capivari River	Laranjal Paulista	22° 59′ 29.9″ S	47° 45′ 34.9″ W
Sor	Sorocaba River	Laranjal Paulista	23° 00′ 07.6″ S	47° 48′ 12.2″ W
Pei	Peixe River	Anhembi	22° 49′ 42.8″ S	48° 06′ 01.5″ W

It is worth observing, through Figure 1 (detailed scale of 1:10,000), the extensive and complex net of small and middle-size tributaries distributed along the Tietê basin.

In each sampling site, the following environmental variables were measured in situ: local depth (m) (portable Speedtech sonar) (Great Falls, MT, USA); water transparency (m) (Secchi disc); and vertical profiles (measurements at 0.5 to 1 m intervals) (multi-parameter probe Horiba U-5000, previously calibrated) (Kyoto, Japan) of water temperature (°C); electrical conductivity (μS cm^{-1}); dissolved oxygen (DO) (mg L^{-1}); turbidity (NTU); total dissolved solids (TDS) (g L^{-1}); pH; and redox potential (ORP) (mV). Surface water samples were collected to analyze total phosphorus, total nitrogen, and chlorophyll a [21].

The trophic status index (modified TSI) [22] was calculated using the total phosphorus index and chlorophyll index, following the proposal of the Environmental Company of São

Paulo State (CETESB, in Portuguese) for river systems (https://cetesb.sp.gov.br/aguas-interiores/wp-content/uploads/sites/12/2019/10/Ap%C3%AAndice-D_-%C3%8Dndices-de-Qualidade-das-%C3%81guas.pdf) (accessed on 15 April 2024):

TSI (Chlorophyll) = $10 \times (6 - ((-0.7 - 0.6 \times (\ln CL))/\ln 2)) - 20$;

TSI (Total Phosphorus) = $10 \times (6 - ((0.42 - 0.36 \times (\ln TP))/\ln 2)) - 20$;

TSI = [TSI (TP) + IET (CHL)]/2.

Values of the TSI include six categories: ultraoligotrophic (TSI $\leq$ 47), oligotrophic (47 < TSI $\leq$ 52), mesotrophic (52 < TSI $\leq$ 59), eutrophic (59 < TSI $\leq$ 63), supereutrophic (63 < TSI $\leq$ 67), and hypereutrophic (TSI > 67).

The studied parameters were compared with the reference standards established by national legislation (Table 2), following CONAMA (National Environment Council, in Portuguese) Resolution n° 357/2005 [23] (conditions for water quality categories for Brazilian aquatic ecosystems) for Class 2 waters–the present classification of the considered Tietê River stretch and its tributaries. Class 2 waters are suitable (presently or intended in the future after recovery actions) for: (a) supply for human consumption, after conventional treatment; (b) the protection of aquatic communities; (c) primary contact recreation, such as swimming, skiing, and diving; (d) irrigation of vegetables, fruit plants, parks, gardens, and sports and leisure fields, with which the public may have direct contact; and (e) aquaculture and fishing activities.

Table 2. Standard references established by CONAMA (National Environment Council, in Portuguese) Resolution 357/2005 (conditions for water quality categories for Brazilian aquatic ecosystems) for Class 2 waters (see Material and Methods).

Parameter	Reference
pH	6 to 9
Turbidity (NTU)	<100 NTU
DO (mg·L^{-1})	>5 mgL^{-1}
TDS (g·L^{-1})	<0.5 gL^{-1}
Total phosphorous (mg·L^{-1}) *	<0.10 mgL^{-1}
Total nitrogen (mg·L^{-1})	<2.18 mgL^{-1}
Chlorophyll-a (μg·L^{-1})	<30 μgL^{-1}

Note: * Values for lotic environment.

The ichthyofauna was collected (sampling license 13794-1 conceived to MGN by the National Institute of Biodiversity Conservation) (ICMBio, in Portuguese) with gillnets, exposed for approximately 16 h (overnight). Meshes of 30, 40, 50, 60, 70, 80, 100, 120, 140, 160, and 180 mm (between knots) were installed from the marginal habitats (smaller apertures) towards the river's central channel (larger apertures). Each net has a standard extension of 10 m and the heights vary between 1.5 m (smaller apertures) and 2.5 m (larger apertures).

Captured specimens were kept in buckets with water to be identified, photographed, counted, and measured (standard length and total weight) and then returned to the original water body. Morphological identification was performed at the lowest taxonomic level based on the specialized literature [24–27].

The fish abundance was standardized in terms of CPUE (catch per unit effort) in numbers of individuals (CPUEn) and biomass (CPUEb), considering 1000 m^2 of gillnets and 16 h of exposure. The assemblage's alpha (α) diversity was determined with the Shannon–Wiener index and Pielou's equability. Beta diversity (β) was calculated based on the Whittaker index. Analyses were performed with the software Primer v.6.1.12 and Permanova + v.1.0.2 [28] and PAST 1.48 [29].

We also determined the trophic guilds of the fish (indicative of functional diversity) based on "Diet" or "Food items" reported by Froese and Pauly [30] and/or from published studies using the search tools Google Scholar (https://scholar.google.com.br/) (accessed on 5 April 2024) and SciELO (http://www.scielo.org/php/index.php) (accessed on 5 April 2024).

For the ordination of the studied rivers, based on the environmental variables, we used principal component analysis (PCA). A non-metric multidimensional scaling (nMDS) analysis was performed to separate the species groups per similarity among sampling sites. Finally, a distance-based linear model (DistLM) [31] was used to correlate the ichthyofauna structure (abundances per species per site) with the environmental variables. All parameters (except pH) were log (x + 1) transformed prior the analyses and, posteriorly, normalized for PCA. For nMDS, the Bray–Curtis dissimilarity coefficient was used and the matrix was later transformed into the square root [28]. The DistLM analysis used the Best method with the Akaike Information Criterion (AIC) derived from the Bray–Curtis similarity matrix with a dummy effect of 0.01 [31]. These tests were performed using Primer v.6.1.12 and Permanova + v.1.0.2 [28] and R [32].

3. Results

3.1. Environmental Conditions

Means and standard deviations (vertical profiles) of the environmental variables for each site and sampling period are shown in Table 3. Figure 2 represents the vertical distribution of selected variables, dissolved oxygen and temperature.

Table 3. Mean values and standard deviation (water column) of the environmental variables measured at the sampling sites (Tietê River and tributaries), in the rainy (R) and dry (D) periods. Values in bold are not in accordance with the limits established by CONAMA (National Environment Council, in Portuguese) Resolution 357/2005 for Class 2 waters (see Material and Methods).

Variables	**Tie (I)**		**Tie (II)**		**Tie (III)**		**Cap**		**Sor**		**Pei**	
	R	**D**	**R**	**D**	**R**	**D**	**R**	**D**	**R**	**D**	**R**	**D**
Depth (m) *	5.00	3.40	6.00	5.40	7.80	5.00	4.50	0.90	5.70	6.40	5.50	2.00
Transparency (m) *	0.20	0.45	0.15	0.60	0.40	0.80	0.20	0.45	0.30	1.20	0.30	0.80
Temperature (°C)	24.30 (±0.00)	26.11 (±0.00)	24.36 (±0.01)	26.13 (±0.00)	26.09 (±0.01)	25.34 (±0.12)	23.61 (±0.00)	26.84 (±0.00)	23.54 (±0.02)	25.71 (±0.13)	23.60 (±0.02)	24.35 (±0.15)
pH	6.68 (±0.07)	7.45 (±0.07)	6.54 (±0.12)	7.28 (±0.19)	6.45 (±0.16)	7.05 (±0.20)	6.80 (±0.06)	7.76 (±0.01)	6.49 (±0.24)	6.57 (±0.23)	6.21 (±0.11)	6.61 (±0.23)
ORP (mV)	242.38 (±4.06)	224 (±3.54)	415.80 (±8.42)	254.67 (±16.01)	228.40 (±4.10)	283.86 (±16.18)	257.17 (±7.06)	211.67 (±2.62)	268.29 (±20.73)	260.14 (±15.49)	228.86 (±17.98)	344.67 (±14.08)
Conductivity (μScm^{-1})	274.75 (±0.43)	690 (±0.00)	249 (±0.00)	569.83 (±0.37)	409.80 (±0.40)	557.71 (±1.28)	149 (±0.00)	270.67 (±0.47)	121.86 (±0.35)	241.43 (±0.73)	92.29 (±0.45)	124 (±0.00)
Turbidity (NTU)	**171.38** (±20.35)	25.98 (±1.69)	**237.20** (±3.97)	26.47 (±1.07)	**120.40** (±8.01)	16.33 (±0.72)	**334.50** (±13.99)	27.13 (±0.40)	**122.83** (±12.72)	8.97 (±0.64)	**161.43** (±2.61)	15.83 (±0.96)
DO (mgL^{-1})	**0.18** (±0.15)	**2.92** (±0.08)	**1.50** (±0.04)	**4.94** (±0.11)	**0.53** (±0.21)	**0.59** (±0.76)	6.82 (±0.15)	10.61 (±0.22)	7.73 (±0.11)	**4.72** (±0.17)	5.47 (±0.10)	**3.25** (±0.30)
TDS (gL^{-1})	0.18 (±0.00)	0.44 (±0.00)	0.16 (±0.00)	0.36 (±0.00)	0.27 (±0.00)	0.36 (±0.00)	0.10 (±0.00)	0.18 (±0.00)	0.08 (±0.00)	0.16 (±0.00)	0.06 (±0.00)	0.08 (±0.00)
Total phosphorous (mgL^{-1})	**0.16** (±0.01)	**1.29** (±0.03)	**0.16** (±0.01)	**1.17** (±0.02)	**0.26** (±0.01)	**0.58** (±0.04)	0.08 (±0.01)	**0.16** (±0.01)	0.06 (±0.00)	**0.17** (±0.00)	0.04 (±0.00)	0.03 (±0.00)
Total nitrogen (mgL^{-1})	**3.54** (±0.18)	**15.27** (±0.33)	**3.97** (±0.01)	**13.40** (±0.18)	**7.02** (±0.17)	**14.49** (±0.53)	1.48 (±0.03)	**4.90** (±0.15)	1.59 (±0.02)	**3.57** (±0.10)	0.85 (±0.02)	0.67 (±0.04)
Chlorophyll-a (μgL^{-1})	17.95 (±4.66)	**86.73** (±1.43)	14.29 (±1.55)	**139.19** (±12.09)	**50.27** (±0.00)	**96.37** (±3.83)	12.09 (±3.89)	**92.03** (±1.42)	6.78 (±0.26)	8.24 (±0.16)	10.26 (±4.14)	1.81 (±0.08)

Note: * Single measurements.

All sampling sites exhibited shallow depths. The minimum was 0.90 m in Capivari River, during the dry period, and the maximum was 7.80 m at Tie (III), during the rainy period. The lowest value of water transparency, 0.15 m, was found at site Tie (II) during the rainy period, and the highest, 1.20 m, at Sorocaba River during the dry period.

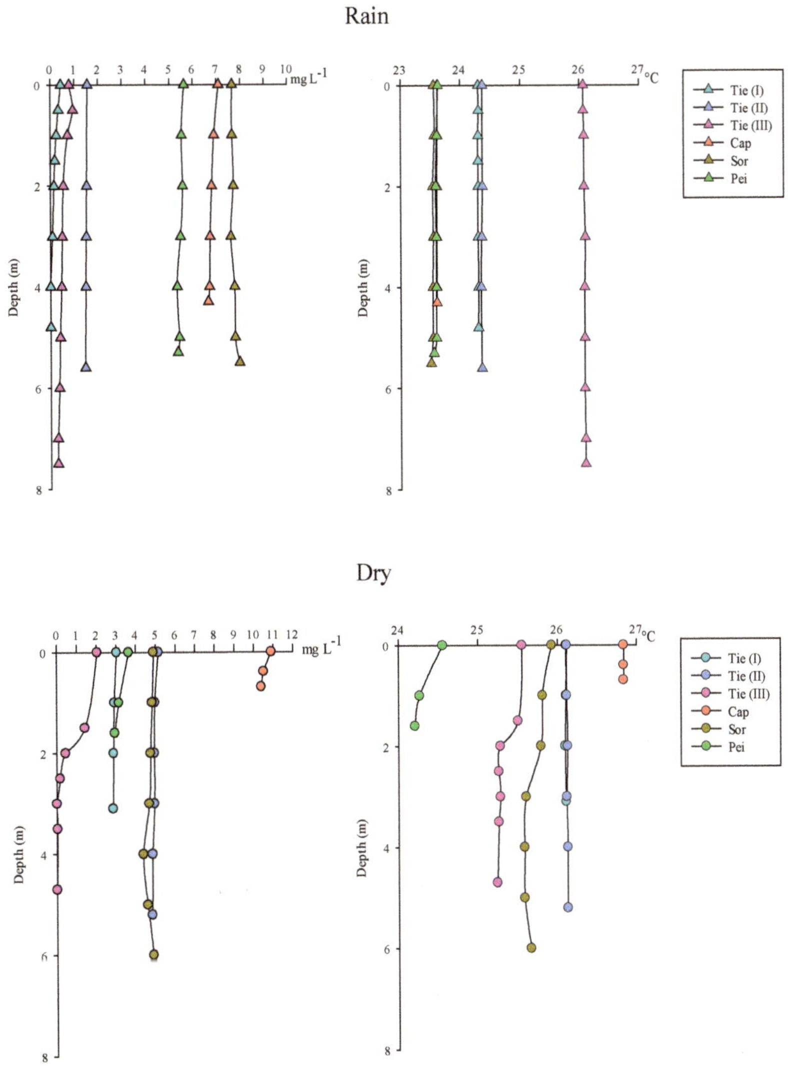

Figure 2. Vertical profiles of dissolved oxygen (**left**) and temperature (**right**) at sampling sites of the Tietê River and tributaries, during the rainy and dry periods.

For turbidity, higher values occurred in the rainy period, exceeding the national standards for Class 2 waters (see Material and Methods item) (maximum of 100 NTU) in all sites. The Capivari River had the highest averages for both periods analyzed, reaching 334.50 NTU and 27.13 NTU in the rainy and dry periods, respectively.

Lower mean pH values (water column) were observed in the rainy period, ranging from 6.21 to 6.80, compared to the dry period, from 6.57 to 7.76. Total dissolved solids varied from 0.06 to 0.27 g L^{-1} in the rainy period and from 0.08 to 0.44 g L^{-1} in the dry period. Both parameters were in conformity with the legislation.

The electrical conductivity values were high, or relatively high, in all sampling sites, reaching 690 µS cm^{-1} at Tie (I) during the dry period. The values in this sampling campaign were higher for all sites, compared with measurements taken during the rainy period, when the Peixe River had the lowest value, 92 µS cm^{-1}. For ORP, the maximum value of 416 mV was measured at Tie (II) during the rainy period, and the lowest, 212 mV, at the Capivari River during the dry period. The Tietê River had higher salt concentration than the tributaries, especially in the dry period.

Nutrients, total phosphorus and total nitrogen, exhibited high values in the Tietê River sites during both sampling periods. Maximum concentrations, 1.29 mg L^{-1} of phosphorus and 15.27 mg L^{-1} of nitrogen, occurred at Tie (I), the upstream Tietê River sampling site, in the dry period. During this period, higher concentrations of nutrients were observed, except for the Peixe River, which had the lowest values among the analyzed sites: 0.03 mg L^{-1} of phosphorus and 0.67 mg L^{-1} of nitrogen. The values for the Peixe River in both periods and for the other tributaries in the rainy period did not exceed the limits established by the legislation.

Higher chlorophyll *a* concentrations were observed in the dry period, except for the Peixe River (1.81 μg L^{-1}) where the value was considerably higher in the rainy season (10.26 μg L^{-1}). Capivari was the only tributary that showed a high chlorophyll concentration, above the reference limits, reaching 92.03 μg L^{-1} in the dry period. At Tie (III), the values exceeded the limits in both sampling periods, which was also verified at the Tie (I) and Tie (II) sites in the dry period.

The highest average temperatures occurred in the Tietê River sites, from 24.30 °C to 26.09 °C during the rainy period and from 25.34 °C to 26.13 °C during the dry period. The tributaries showed a small variation in temperature during the rainy period, between 23.54 °C and 23.61 °C, compared to the dry period, from 24.35 °C to 26.84 °C.

The Tietê River sites showed extremely low concentrations of dissolved oxygen, tending to zero in the rainy period. A slight increase was observed in the dry period, especially at Tie (II), with a mean value (water column) of 4.94 mg L^{-1}. For both periods, all values in the main river were lower than the minimum established for Class 2 waters (5 mg L^{-1}). For the tributaries, the values measured in the Sorocaba River and Peixe River during the dry period were also below the limits. As expected for shallow rivers, the water column profiles (vertical dimension) for temperature and dissolved oxygen (Figure 2) show homogeneous conditions.

The PCA, based on the environmental variables explained (considering the first two dimensions), 84.4% and 84.7% for the rainy and dry periods, respectively, with a clear difference between the main river and the tributaries (Figure 3). The score values for the first and second PCA dimensions are shown in Table 4.

Table 4. Coefficients of linear correlations between the environmental variables and the first two dimensions of the PCA, for rainy and dry periods, based on the environmental variables measured in the Tietê Rivers and tributaries.

Variable	Rainy		Variable	Dry	
	Dim1	Dim2		Dim1	Dim2
Temperature	0.99286298	0.01413478	Temperature	0.73241485	−0.43456149
pH	−0.09129626	0.79169072	pH	0.86493133	−0.47645186
ORP	−0.06841088	0.61142919	ORP	−0.74714234	0.32708385
Conductivity	0.96290487	0.25353582	Conductivity	0.83829755	0.53407154
Turbidity	−0.45106079	0.71165828	Turbidity	0.85218413	−0.38708075
DO	−0.78753136	−0.36912688	DO	0.15311384	−0.91894482
TDS	0.96267064	0.25462677	TDS	0.83998185	0.53152241
Chlorophyll a	0.95351054	−0.09248894	Chlorophyll a	0.90682814	0.06351206
TP	0.96125535	0.27431054	TP	0.81519443	0.47951963
TN	0.98131456	0.16997972	TN	0.81519397	0.55900840
Depth	0.85914491	−0.34279742	Depth	−0.05614826	0.78493412
Transparency	0.49713473	−0.81801113	Transparency	−0.78653772	0.39283687

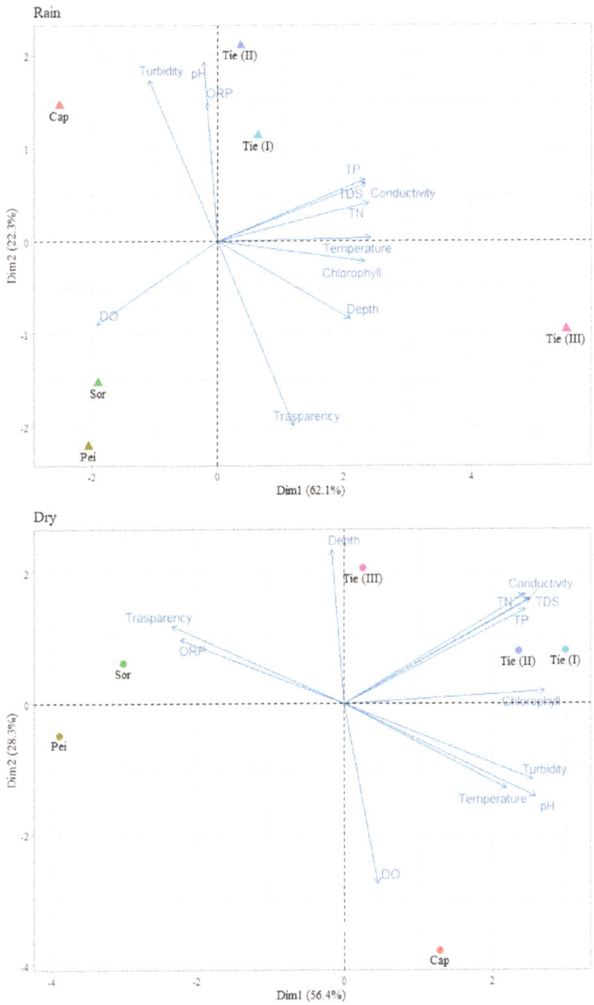

Figure 3. Biplot (Dimensions 1 and 2) representation of the PCA showing the distribution of the sampling sites of Tietê River and tributaries during the rainy and dry periods. DO: dissolved oxygen; TDS: total dissolved solids; ORP: redox potential; TP: total phosphorus; and TN: total nitrogen.

In the rainy season, the positioning of the tributaries, on the left side of Dimension 1, was determined by the higher concentrations of DO. In contrast, the sites of the Tietê River were placed on the right side of Dim1, determined by the lower values of oxygen and higher values of temperature, conductivity, total dissolved solids, chlorophyll, and nutrients (phosphorus and nitrogen). A positive correlation with turbidity explained the positioning of the Capivari River in the superior left quadrant.

This discrimination between the main river and tributaries also occurred in the dry period, with the Tietê sampling points ordinated in the superior right quadrant, positively correlated with conductivity, total dissolved solids, and nutrients, Dim1, in opposition to higher dissolved oxygen values, Dim2.

The trophic status index (TSI) (Table 5) showed a higher degree of eutrophication in the Tietê River compared to the tributaries. Tie (III) was classified as hypereutrophic in both periods. Tie (I) and Tie (II) were supereutrophic in the rainy season and hypereutrophic in the dry season. The tributaries Capivari and Sorocaba were considered mesotrophic in the rainy period and hypereutrophic and eutrophic, respectively, in the dry period. The Peixe River exhibited the best trophic condition, classified as mesotrophic and ultraoligotrophic in the rainy and dry periods, respectively.

Table 5. Trophic state classification of the Tietê River and tributary sampling points during the rainy and dry periods.

Sites	Rainy		Dry	
	TSI	Classification	TSI	Classification
Tie (I)	64.9	Supereutrophic	87.7	Hypereutrophic
Tie (II)	63.8	Supereutrophic	89.4	Hypereutrophic
Tie (III)	73.5	Hypereutrophic	82.4	Hypereutrophic
Cap	57.5	Mesotrophic	72.9	Hypereutrophic
Sor	52.5	Mesotrophic	61.5	Eutrophic
Pei	52.4	Mesotrophic	40.2	Ultraoligotrophic

All sites, except the Peixe River, showed higher eutrophication in the dry compared to the rainy period.

3.2. Ichthyofauna

Twenty-five fish species were collected, considering the six sampling sites, belonging to 12 families and four orders (Table 6). In the rainy period, only the orders Characiformes and Siluriformes were observed. Characiformes fish were relatively more abundant in the tributaries, corresponding to 100% of the captures for Capivari, 95% for Peixe, and 71% for Sorocaba. Conversely, Siluriformes were more representative in the Tietê River, representing 100% of abundance, except at Tie (II) due to the capture of only a single specimen of Characiformes. In the dry period, in addition to the previously mentioned orders, Cichliformes and Gymnotiformes were also sampled, and the predominance of Siluriformes occurred in all sampling sites, except in the Peixe River, where Characiformes represented 66% of the individuals. Cichliformes were found only in the tributaries.

Table 6. List of ichthyofauna species captured in the Tietê River and tributaries during the rainy (R) and dry (D) periods.

Specie	Tie (I)	Tie (II)	Tie (III)	Cap	Sor	Pei
CHARACIFORMES						
Acestrorhynchidae						
Acestrorhynchus lacustris (Lütken, 1875)						D
Anostomidae						
Leporinus friderici (Bloch, 1794)					D	R/D
Leporinus lacustris (Amaral Campos, 1945)						R/D
Megaleporinus obtusidens (Valenciennes, 1837)						R
Schizodon intermedius (Garavello and Britski, 1990)						R/D
Characidae						
Astyanax lacustris (Lütken, 1875)				R/D	R/D	R/D
Psalidodon schubarti (Britski 1964)				R		
Roeboides descalvadensis (Fowler, 1932)				R/D	R	R/D
Curimatidae						
Cyphocharax nagelii (Steindachner, 1881)					D	R
Steindachnerina insculpta (Fernández-Yépez, 1948)				D	R	
Erythrinidae						
Hoplias malabaricus (Bloch, 1794)					D	R/D
Parodontidae						
Apareiodon sp.				D		
Prochilodontidae						
Prochilodus lineatus (Valenciennes, 1837)	D	R/D	D	R/D	D	R/D
Serrasalmidae						
Serrasalmus maculatus (Kner, 1858)						D

Table 6. *Cont.*

Specie	Tie (I)	Tie (II)	Tie (III)	Cap	Sor	Pei
CICHLIFORMES						
Cichlidae						
Geophagus brasiliensis (Quoy and Gaimard, 1824)				D	D	
Oreochromis niloticus (Linnaeus, 1758)				D		
Saxatilia britskii (Kullander 1982)						D
GYMNOTIFORMES						
Gymnotidae						
Gymnotus cuia (Craig, Malabarba, Crampton and Albert, 2018)			D			D
SILURIFORMES						
Callichthyidae						
Hoplosternum littorale (Hancock, 1828)	R/D	D	R/D	D		R/D
Loricariidae						
Hypostomus ancistroides (Ihering, 1911)		D		D	D	
Hypostomus hermanni (Ihering, 1905)					R/D	
Hypostomus regani (Ihering, 1905)					D	
Hypostomus strigaticeps (Regan, 1908)				D		
Proloricaria prolixa (Isbrücker and Nijssen, 1978)					R/D	/
Pterygoplichthys ambrosettii (Holmberg, 1893)	R		R			
Total richness	3	3	4	11	12	14

Figure 4 shows the relative numerical abundance per order in the distinct sampling sites and periods.

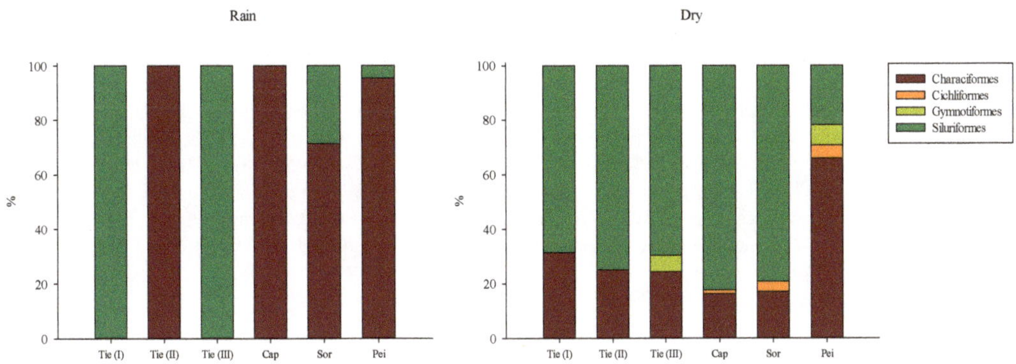

Figure 4. Relative abundance per order of the ichthyofauna in the Tietê River and tributaries, during the rainy and dry periods.

Considering the taxonomical families, we observed the predominance of Callichthyidae in the Tietê River, except at Tie (II) during the rainy period, explained by the capture of only a single representative of Prochilodontidae, *Prochilodus lineatus* (Valenciennes, 1837). In the rainy period, this family was also predominant in the tributary Capivari, corresponding to 61% of the fish abundance. The tributaries Sorocaba and Peixe were better represented by Characidae (57%) and Anostomidae (36%), respectively. In the dry period, Loricariidae represented 79% of the abundance in Sorocaba and 82% in Capivari. In Peixe River, Erythrinidae (29%) was the dominant family. Prochilodontidae was the only family found in all sampling sites, considering both sampling periods (Figure 5).

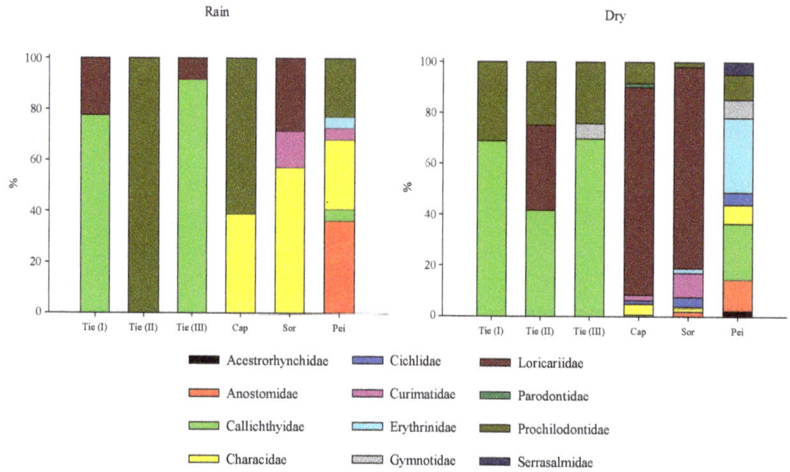

Figure 5. Relative abundance per family of the ichthyofauna in the Tietê River and tributaries during the rainy and dry periods.

The total species richness was higher in the tributary rivers, with 11 species in Capivari, 12 in Sorocaba, and 14 in Peixe, while in the Tietê River, only 3, Tie (I) and Tie (II), and 4 species were found, Tie (III). Three species were only captured in the rainy period, while ten were exclusively found in the dry period. *Prochilodus lineatus* was the only species common to all sampling sites and periods. Species richness was higher in the dry period for all sites, except for Tie (I), where the same richness was observed in both periods (Table 6).

In terms of CPUE (ind./1000 m^2/16 h), 947.3 individuals were captured, totalizing a biomass of 183.3 kg (Supplementary Material: Tables SA–SD). In the rainy period, the CPUEn mean values were higher at Tie (III), 2.6, and for CPUEb at Capivari, 805.7 g. The lowest mean value in the rainy period, for CPUEn, was obtained at Tie (II), 0.10, and for CPUEb at Sorocaba, 45.6 g. In the dry period, the Capivari River exhibited the highest mean values of CPUEn and CPUEb, with 14.3 individuals and 2928.1 g, respectively. In the same period, Tie (II) exhibited the lowest mean values for both, with 1.30 (CPUEn) and 119.69 g (CPUEb) (Figure 6).

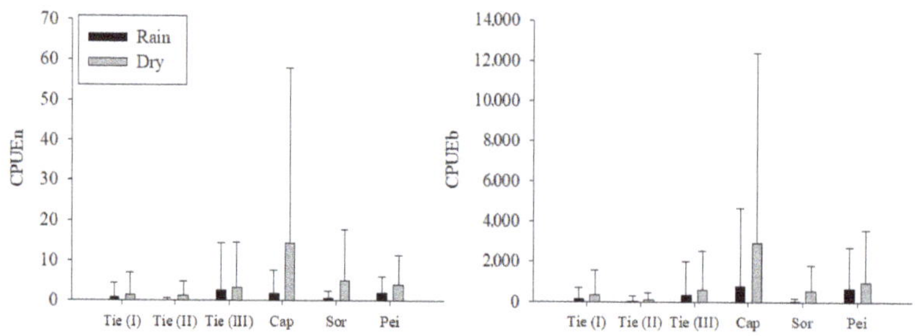

Figure 6. Values (means and standard deviation for the set of nets per site) of catch per unit effort in number of individuals (CPUEn) and in biomass (g) (CPUEb) for the sampled ichthyofauna in the Tietê River and tributaries during the rainy and dry period.

The Peixe River showed the highest diversity values with the Shannon-Wiener index (H′) (bits ind.$^{-1}$), for both sampling campaigns, with values of 2.9 and 3.0 in the rainy and dry periods, respectively. The lowest value for the rainy period was found in Tie (II), 0, and for the dry period in Tie (I), 0.90. For Pielou's equitability (J), the Sorocaba River had the highest value, 0.92, in the rainy period, and Tie (III) the lowest, 0.41. It was not possible to calculate the J value for Tie (II) in the rainy period due to the capture of only a single specimen of fish. In the dry period, Tie (II) had the highest J value, 0.98, and Capivari the lowest, 0.54 (Figure 7).

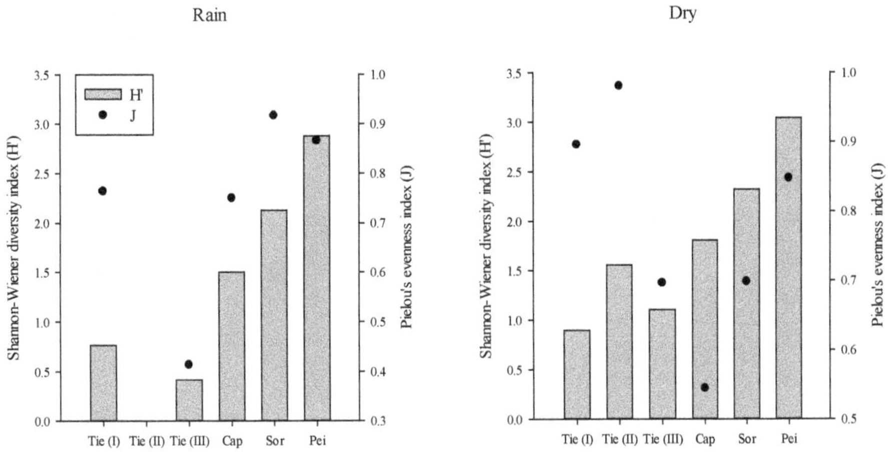

Figure 7. Pielou equability values (J′) and Shannon–Wiener index (H′) for the ichthyofauna of the Tietê River and tributaries during the rainy and dry periods.

The beta diversity values for the rivers are summarized in Table 7. The composition of species in the rainy period, comparing the sites, did not differ between Tie (I) and Tie (III), but there was a high difference between these environments and the other analyzed sites. Intermediate values of beta diversity were observed between the tributaries Sorocaba and Capivari, 0.56, and between the Peixe and Capivari, 0.57.

Table 7. Ichthyofauna beta diversity (Whittaker) values among the sampling sites of the Tietê River and tributaries during the rainy and dry periods.

		Dry					
	Site	**Tie (I)**	**Tie (II)**	**Tie (III)**	**Cap**	**Sor**	**Pei**
Rainy	Tie (I)	-	0.2	0.2	0.66667	0.83333	0.71429
	Tie (II)	1	-	0.3333	0.53846	0.69231	0.73333
	Tie (III)	0	1	-	0.69231	0.84615	0.6
	Cap	1	0.6	1	-	1	1
	Sor	1	1	1	0.55556	-	0.63636
	Pei	0.8333	0.81818	0.83333	0.57143	0.73333	-

In the dry period, the Tietê River sites showed low beta diversity, ranging from 0.20 to 0.33. The tributaries presented a maximum beta diversity, 1, Sorocaba/Capivari and Peixe/Capivari, or a relatively high beta diversity, 0.64, Sorocaba/Peixe.

Regarding functional diversity, fish were classified into eight different trophic guilds (i.e., piscivore, omnivore, herbivore, invertivore, detritivore, algivore, insectivore, and iliophagous) (Table 8). In the Tietê River, most species were detritivores (50–67%). In the tributaries, eating habits were more diversified, especially for the Peixe River, with a predominance of omnivorous (37%) and piscivorous (21%) species. In the Sorocaba River, there was a high percentage of detritivorous species (42%), but a lower percentage than that found in the main river, while in the Capivari River omnivorous species predominated (46%) (Figure 8).

Table 8. Trophic guilds of the fish species of the Tietê River and tributaries.

Specie	Trophic Guild	Reference(s)
CHARACIFORMES		
Acestrorhynchidae		
Acestrorhynchus lacustris (Lütken, 1875)	Piscivore	[33,34]
Anostomidae		
Leporinus friderici (Bloch, 1794)	Omnivore	[35,36]
Leporinus lacustris (Amaral Campos, 1945)	Omnivore	[36]
Megaleporinus obtusidens (Valenciennes, 1837)	Omnivore	[37]
Schizodon intermedius (Garavello and Britski, 1990)	Herbivore	[38]
Characidae		
Astyanax lacustris (Lütken, 1875)	Omnivore	[37]
Psalidodon schubarti (Britski 1964)	Omnivore	[30]
Roeboides descalvadensis (Fowler, 1932)	Invertivore	[33]
Curimatidae		
Cyphocharax nagelii (Steindachner, 1881)	Iliophagous	[39]
Steindachnerina insculpta (Fernández-Yépez, 1948)	Detritivore	[33]
Erythrinidae		
Hoplias malabaricus (Bloch, 1794)	Piscivore	[35–37,40,41]
Parodontidae		
Apareiodon sp. *	Algivore	[37]
Prochilodontidae		
Prochilodus lineatus (Valenciennes, 1837)	Detritivore	[33,36]
Serrasalmidae		
Serrasalmus maculatus (Kner, 1858)	Piscivore	[33,36,40]
CICHLIFORMES		
Cichlidae		
Geophagus brasiliensis (Quoy and Gaimard, 1824)	Omnivore	[34,41]
Oreochromis niloticus (Linnaeus, 1758)	Omnivore	[37,42]
Saxatilia britskii (Kullander 1982)	Insectivore	[43]
GYMNOTIFORMES		
Gymnotidae		
Gymnotus cuia (Craig, Malabarba, Crampton and Albert, 2018) *	Invertivore	[43]
SILURIFORMES		
Callichthyidae		
Hoplosternum littorale (Hancock, 1828)	Omnivore	[42]
Loricariidae		
Hypostomus ancistroides (Ihering, 1911)	Detritivore	[43,44]
Hypostomus hermanni (Ihering, 1905)	Detritivore	[43]
Hypostomus regani (Ihering, 1905)	Detritivore	[43]
Hypostomus strigaticeps (Regan, 1908)	Detritivore	[43,44]
Proloricaria prolixa (Isbrücker and Nijssen, 1978)	Herbivore	[43]
Pterygoplichthys ambrosettii (Holmberg, 1893)	Detritivore	[33]

Note: * Trophic guild for the genus.

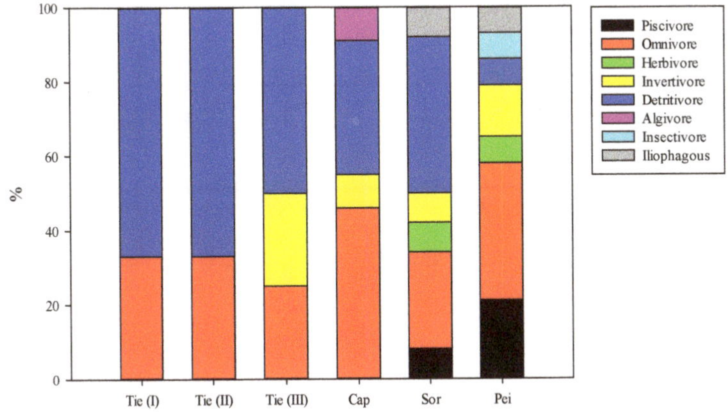

Figure 8. Relative abundance of the trophic guilds of the fish species of Tietê River and tributaries.

The plotted results of the nMDS analysis show a clear difference between the ichthyofauna of the main river compared to the tributaries (Figure 9), with very few species directly associated with the Tietê. Differences between the sampling periods were not evident.

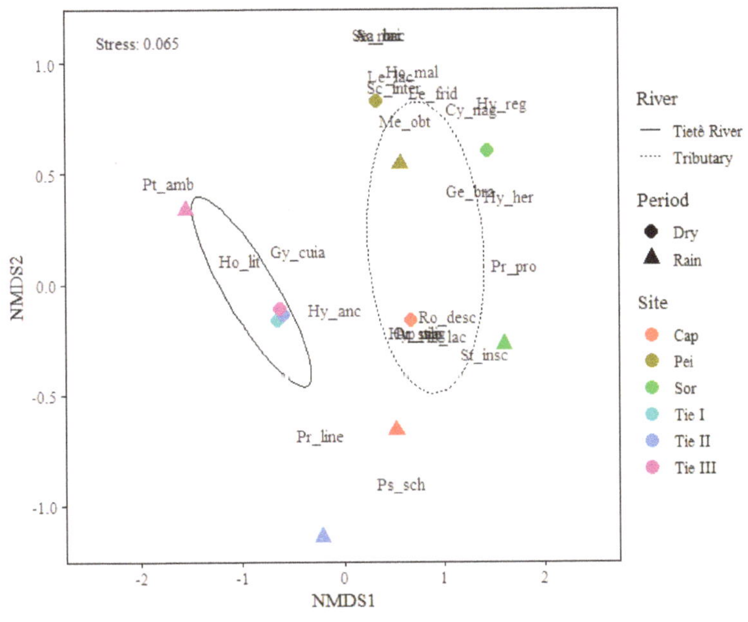

Figure 9. Non-metric multidimensional scaling (nMDS) for the ichthyofauna of the Tietê River and tributaries during the rainy and dry periods. Ac_lac: *Acestrorhynchus lacustris*; Le_frid: *Leporinus friderici*; Le_lac: *Leporinus lacustris* Amaral; Me_obt: *Megaleporinus obtusidens*; Sc_inter: *Schizodon intermedius*; As_lac: *Astyanax lacustris*; Ps_sch: *Psalidodon schubarti*; Ro_desc: *Roeboides descalvadensis*; Cy_nag: *Cyphocharax nagelii*; St_insc: *Steindachnerina insculpta*; Ho_mal: *Hoplias malabaricus*; Ap_sp: *Apareiodon* sp.; Pr_line: *Prochilodus lineatus*; Se_mac: *Serrasalmus maculatus*; Ge_bra: *Geophagus brasiliensis*; Or_nilo: *Oreochromis niloticus*; Sa_bri: *Saxatilia britskii*; Gy_cuia: *Gymnotus cuia*; Ho_lit: *Hoplosternum littorale*; Hy_anc: *Hypostomus ancistroides*; Hy_her: *Hypostomus hermanni*; Hy_reg: *Hypostomus regani*; Hy_strig: *Hypostomus strigaticeps*; Pr_pro: *Proloricaria prolixa*; and Pt_amb: *Pterygoplichthys ambrosettii*.

3.3. *Influence of Environmental Variables on Fish Assemblage Structure*

The DistLM analysis indicated significant relationships ($p < 0.05$) between the structure of the fish assemblages and the environmental variables, but only for the rainy period. Dissolved oxygen (DO) contributed with the highest percentage of explained variance (48.1%), followed by TDS (41.0%), conductivity (40.1%), total phosphorus (TP) (40.0%) chlorophyll a (36.3%), and temperature (33.9%) (Table 9).

Table 9. Values of Pseudo-F, coefficient of significance (p), and proportion of explanation (Prop. %) obtained by DistLM analysis between environmental variables and fish assemblage structure (abundance per species per site) for the Tietê River and tributaries, during rainy and dry periods. In bold, the significant variables ($p < 0.05$).

Variable	Rainy			Dry		
	Pseudo-F	p	Prop. %	Pseudo-F	p	Prop. %
Temperature	2.0507	**0.032**	33.89	1.1573	0.391	22.44
pH	0.30005	0.937	6.98	1.3611	0.271	25.39
ORP	1.5062	0.124	27.35	0.90929	0.573	18.52
Condutivity	2.6735	**0.039**	40.06	1.7201	0.16	30.07
Turbidity	1.0828	0.401	21.30	1.7254	0.142	30.14
DO	3.7083	**0.006**	48.11	1.5284	0.23	27.65
TDS	2.7779	**0.017**	40.99	2.0837	0.118	34.25
Chlorophyll *a*	2.2769	**0.017**	36.28	1.5465	0.176	27.88
TP	2.6657	**0.046**	39.99	1.9401	0.096	32.66
TN	2.4214	0.071	37.71	1.5972	0.201	28.54

4. Discussion

4.1. *Environmental Conditions*

The Tietê River receives, daily, a significant amount of domestic and industrial effluents from the largest Brazilian megalopolis, the city of São Paulo, with an enormous and well-known negative impact on the water quality [15,45]. Our results demonstrated that even 300 km downstream from the major sources of contamination, a severe eutrophic condition still persists.

The principal component analysis (PCA) explained satisfactorily the environmental data variance (>84% for both seasons) and evidenced, graphically, the distinctiveness of the Tietê River in relation to the tributaries. The Tietê sites' positioning was determined by the higher values of temperature, conductivity, total dissolved solids, chlorophyll, and nutrients (phosphorus and nitrogen) found in the rainy season. In the dry season, these sites correlated with conductivity, total dissolved solids, and nutrients.

The analyzed tributaries are less impacted by urban centers, and their courses flow predominantly along typical countryside landscapes. Particularly, the Peixe River has no city located in its watershed.

An eventual longitudinal dilution process could be expected, due to the entrance of the tributaries. Despite their better water quality conditions, the volumes (tributary water flow) are probably insufficient to cause a significant positive effect on the Tietê River. However, this is not a reason to neglect the potential contribution of the small and middle-size tributaries widely distributed along the Tietê River basin [46,47]. This process is especially important for observation in the near future, due to the present trend of sanitation measures implementation (sewage treatments plants) all over São Paulo State, after decades without effective actions.

Concentrations of dissolved oxygen, total phosphorus, and total nitrogen in the Tietê River, during both rainy and dry periods, were not in conformity with the reference values of the Brazilian legislation. Turbidity, in the rainy season, and chlorophyll, in the dry season, also exceed the established limits. For the tributaries, turbidity was very high (above the limit) in the rainy season, indicating intensive soil erosion and a reduced riparian forest in the three sub-basins. In the dry season, oxygen was below the limit in

Sorocaba and Peixe and chlorophyll and nutrients (TP, TN) above the limits in Capivari and Sorocaba, respectively.

From a temporal series of ten years (2007–2017), as part of the environmental monitoring program carried out in the Barra Bonita hydropower reservoir, a high level of non-conformities for total phosphorus, dissolved oxygen, thermotolerant coliforms, and biochemical demand of oxygen was found [18]. This reservoir is located in the Tietê River, immediately downstream of our study area. Another study along the middle and lower stretches of the Tietê River also observed high concentrations of nutrients, as well as of heavy metals (cadmium, chromium, copper, and zinc) [17]. These authors performed tests (bioassays) with water and sediment samples that revealed acute toxicity for zooplankton and chronic toxicity for zooplankton and fish species.

Nutrient loads favor the primary productivity in rivers and this explains the high levels of eutrophication indicated by the trophic state index [19]. Using data of total phosphorus and chlorophyll, the Tietê River sampling sites varied from supereutrophic to hypereutrophic. The trophic conditions increased in the dry period, indicating that rainfall may act as a dilution factor for nutrients and phytoplankton biomass. A seasonal trend of trophy was unclear for the tributaries: all classified as mesotrophic in the rainy period and from ultraoligotrophic to hypertrophic in the dry period. It is worth mentioning that the downstream site of the Tietê River (Tie III) was classified as hypereutrophic in both seasons, demonstrating that an expected longitudinal improvement does not occur, which was corroborated by the DO values. In this case, the temporal variation was different, getting worse in the rainy season, when concentrations were even lower in the three Tietê sites (water column means from 0.18 to 1.50 mg L^{-1}). A considerable DO decrease in the rainy season may be associated with previous (a few days earlier) storm events on the São Paulo metropolis, resulting in the downstream displacement of a highly polluted water mass. In another emblematic case of contamination in South America, Riachuelo River in the Buenos Aires city area, a study using a statistical approach analysis on temporal variation (eight consecutive years) of six water quality variables observed that increasing storm water runoff may have determined the observed trends (e.g., non-biodegradable organic matter) [48].

4.2. *Ichthyofauna*

The Tietê River fish fauna has been deeply affected (e.g., disappearance of migrant carnivorous species) by human actions since the nineteen century, with the incidental and deliberated introduction of non-native species, damming for hydropower generation, fisheries, and eutrophication [11].

Our study, focused on the effects of eutrophication, corroborated the initial hypothesis, showing considerable differences in the composition of the ichthyofauna when comparing the main river (Tietê) with the tributaries (Sorocaba, Capivari, and Peixe). This was clearly evidenced through analyses such as nMDS and Whittaker's beta diversity.

The number of captured species in the Tietê was very low. Considering the two seasonal periods (rainy and dry), the species richness was much higher in the tributaries, 11 to 14, compared to the Tietê River, only 3 to 4. A good correspondence occurred between the number of species and the values of the Shannon–Wiener index.

In a study carried out in the late 1990s [20], with sampling performed throughout the Tietê basin, the dominance of only one or two species was seen in the most polluted stretches; in fact, in stretches coincident with our study area. Probably, at that time the river condition was even worse in terms of water quality [49], as many sewage treatment plants have started to operate over the last three decades, or some other species have become further adapted to the ecosystem degradation.

An explanation for the very low number of species captured in the Tietê River main channel is that during episodes of poor water quality conditions, such as anoxia in the whole water column, fish may search for refuge in the marginal lagoons, which we did not sample. Local anglers (personal communication) frequently report this behavior. An investigation in the Sorocaba River, next to the urban area of Sorocaba city during the early 1990s (at that time with ~500,000 inhabitants), when most industrial and domestic sewage was discharged into the river almost with no treatment, has already reported that the lagoons were an escape for fish from the river pollution [49]. The lagoons exhibited more stable conditions, including higher oxygen concentrations.

The capture method is also an important factor that influences the richness results. In our study, we employed gillnets, which are effective for sampling the fish fauna that explores the main river channels, but have limited efficacy for small species living in marginal habitats.

In terms of feeding habits, we observed that most species in the Tietê River were detritivores, while in the tributaries the fish have a higher variety of food resources. Evaluation though isotopic signatures demonstrated that different fish trophic guilds change their diets in degraded environments and the detritivores, particularly, exhibit even higher trophic plasticity [50].

It is worth mentioning that *P. lineatus*, a detrivorous fish, was the only species found in all studied sites. Its distribution evidences a wide resilience to the observed contrasting water quality conditions, including the ingestion of considerable amounts of microplastics [51]. *Prochilodus lineatus* is an abundant species in the middle Tietê River basin, representing the main fishery resource of the region, with a socio-economics relevance. As a detritivorous/iliophagous (bottom-dwelling) fish, it feeds on detritus and other organisms that live in the mud deposited on the riverbed [20]. Certainly, it benefits from the large amount of organic matter (including huge populations of tolerant macroinvertebrates) present in the substrate, originating from untreated or partially treated organic sewage. A comparative analysis of the condition factor based on the weight/length relationship showed a better performance (feeding efficiency) of *P. lineatus* in the highly eutrophic Tietê River than in the Peixe River [52]—as already mentioned, a tributary not affected by human settlements. It is also important to consider that as a long-distance migrator, i.e., an efficient swimmer, it can escape from sudden decreases in dissolved oxygen, moving into to marginal lagoons and tributaries.

Other important detritivorous fish in the Tietê River were the Siluriformes *Hypostomus* spp., *H. littorale*, and *P. ambrosetti*. These fish are also known for their high tolerance to poor environmental conditions, such as anoxia or hypoxia, as they have accessory breathing mechanisms [49,53].

In terms of fish abundance, values in the tributary Capivari during the dry period, CPUEn and CPUEb, were remarkably high. This was due to the capture of more than 100 specimens of *Hypostomus strigaticeps* and *Hypostomus ancistroides*. On that occasion, when these two Loricariidae corresponded to approximately 80% of the fish dominance, this river was classified as hypereutrophic due to a very high value of chlorophyll concentration (100 $\mu g\ L^{-1}$).

Correlation between the structure of fish communities and environmental variables was significant for the rainy season. As previously discussed, during this period, the concentration of dissolved oxygen reached zero or close to zero in the Tietê River, while in the tributaries values were higher than 5 mg L^{-1}. In recurrent situations like that, only hypoxic-tolerant species survive (or remain) in the main river channel. Oxygen is a key variable, which contributed to the highest percentage of variance shown by the DistLM analysis (48.11%). In a long-term analysis carried out in a river with a long history of environmental degradation, followed by effective mitigation measures, the Seine (France), the level of tolerance to low oxygen values was considered a determinant factor for the recolonization of some fish species [54].

4.3. Final Considerations

The excessive eutrophication of the Tietê River acts as a chemical barrier to the dispersion of sensitive, or less tolerant, fish species. The concept of a chemical barrier applied to fish distribution has been previously considered, based on either laboratory studies [55] or literature reviews [54,56]. Our study addresses this problem directly in the field.

Our data evidence that the considered 40 km stretch of the Tietê River is highly degraded in terms of water quality, without any signals of resilience to the destructive impact caused by the huge loads of pollution it receives continuously. The official Class 2 classification is far from the river's present condition, with recurrent non-conformities in relation to the standard limits, and should be revised by the authorities.

Detrivorous feeding habits, especially when associated with accessory breathing or efficient swimming capability, is an essential fish characteristic that assures resilience and adaptability for living in degraded environments, such as the Tietê River.

Conservation of the tributaries is fundamental, given their higher diversity. That is the best guarantee of future recolonization processes in the basin, as long as restoration (urgently required) becomes effective in the main river.

Future studies are necessary for a better comprehension of the impact on the local and regional distribution of other biological communities. Temporal variability should also be addressed, considering both seasonality, alternation between rainy and dry periods, and the almost-instantaneous effects of random displacements of heavily polluted water masses (upstream storm events in the São Paulo metropolis).

Supplementary Materials: The following supporting information can be downloaded at: https://www.mdpi.com/article/10.3390/w16101383/s1.

Author Contributions: Conceptualization, B.U. and M.N.; Methodology, B.U.; Formal analysis, B.U. and M.N.; Investigation, B.U. and M.N.; Resources, M.N.; Data curation, B.U. and M.N.; Writing—original draft, B.U. and M.N.; Writing—review & editing, B.U. and M.N.; Visualization, B.U. and M.N.; Funding acquisition, M.N. All authors have read and agreed to the published version of the manuscript.

Funding: This research was partially funded by FAPESP (Fundação de Amparo à Ciência do Estado de São Paulo) (Process 2019/00177-4) and Coordenação de Aperfeiçoamento de Pessoal de Nível Superior (CAPES) (Processes 88887.663615/2022-00 and 88887.826892/2023-00).

Data Availability Statement: Data are contained within the article.

Acknowledgments: The authors would like to thank Limnética Consultant in Hydric Resources Ltd., for support in the fieldwork activities, and Welber Senteio Smith for valuable suggestions.

Conflicts of Interest: The authors declare no conflicts of interest.

References

1. Ronce, O. How Does It Feel to Be Like a Rolling Stone? Ten Questions About Dispersal Evolution. *Annu. Rev. Ecol. Evol. Syst.* **2007**, *38*, 231–253. [CrossRef]
2. Gibbs, M.; Saastamoinen, M.; Coulon, A.; Stevens, V.M. Organisms on the move: Ecology and evolution of dispersal. *Biol. Lett.* **2010**, *6*, 146–148. [CrossRef] [PubMed]
3. Radinger, J.; Wolter, C. Patterns and predictors of fish dispersal in rivers. *Fish Fish.* **2014**, *15*, 456–473. [CrossRef]
4. Lucas, M.C.; Baras, E. The Stimulus and Capacity for Migration. In *Migration of Freshwater Fishes*, 1st ed.; Lucas, C.M., Baras, E., Thom, T.J., Duncan, A., Slavík, O., Eds.; Blackwell Science Ltd.: London, UK, 2001; pp. 14–65.
5. Bowler, D.E.; Benton, T.G. Causes and consequences of animal dispersal strategies: Relating individual behaviour to spatial dynamics. *Biol. Rev.* **2005**, *80*, 205–225. [CrossRef] [PubMed]
6. Vianna, N.C.; Nogueira, M.G. Ichthyoplankton and limnological factors in the Cinzas River—An alternative spawning site for fishes in the middle Paranapanema River basin, Brazil. *Acta Limnol. Bras.* **2008**, *20*, 139–151.
7. Agostinho, A.A.; Pelicice, F.M.; Gomes, L.C. Dams and the fish fauna of the Neotropical region: Impacts and management related to diversity and fisheries. *Braz. J. Biol.* **2008**, *68*, 1119–1132. [CrossRef] [PubMed]
8. Britto, S.G.C.; Carvalho, E.D. Reproductive migration of fish and movement in a series of reservoirs in the Upper Paraná River basin, Brazil. *Fish. Manag. Ecol.* **2013**, *20*, 426–433. [CrossRef]
9. Pelicice, F.M.; Pompeu, P.S.; Agostinho, A.A. Large reservoirs as ecological barriers to downstream movements of Neotropical migratory fish. *Fish Fish.* **2015**, *16*, 697–715. [CrossRef]
10. Ferrareze, M.; Casatti, L.; Nogueira, M.G. Spatial heterogeneity affecting fish fauna in cascade reservoirs of the Upper Paraná Basin, Brazil. *Hydrobiologia* **2014**, *738*, 97–109. [CrossRef]
11. Smith, W.S.; Stefani, M.S.; Espíndola, E.L.G.; Rocha, O. Changes in fish species composition in the middle and lower Tietê River (São Paulo, Brazil) throughout the centuries, emphasizing rheophilic and introduced species. *Acta Limnol. Bras.* **2018**, *30*, 310. [CrossRef]

12. Jacquin, L.; Petitjean, Q.; Côte, J.; Laffaille, P.; Jean, S. Effects of Pollution on Fish Behavior, Personality, and Cognition: Some Research Perspectives. *Front. Ecol. Evol.* **2020**, *8*, 86. [CrossRef]
13. Instituto Brasileiro de Geografia e Estatística—IBGE (Brazilian Institute of Geography and Statistics). Censo Demográfico. 2022. Available online: https://censo2022.ibge.gov.br (accessed on 20 March 2024).
14. Tundisi, J.G. A crise hídrica e a qualidade da água na Região Metropolitana de São Paulo. In *Livro Branco da Água. A Crise Hídrica na Região Metropolitana de São Paulo em 2013-2015: Origens, Impactos e Soluções*, 1st ed.; Buckeridge, M., Ribeiro, W.C., Eds.; Instituto de Estudos Avançados: São Paulo, Brazil, 2018; pp. 39–45.
15. Buckeridge, M.; Ribeiro, W.C. *A Crise Hídrica na Região Metropolitana de São Paulo em 2013–2015: Origens, Impactos e Soluções*; Instituto de Estudos Avançados: São Paulo, Brazil, 2018; 175p.
16. Tundisi, J.G.; Matsumura-Tundisi, T.; Abe, D.S. The ecological dynamics of Barra Bonita (Tietê River, SP, Brazil) reservoir: Implications for its biodiversity. *Braz. J. Biol.* **2008**, *68*, 1079–1098. [CrossRef] [PubMed]
17. Rodgher, S.; Espíndola, E.L.G.; Rocha, O.; Fracácio, R.; Pereira, R.H.G.; Rodrigues, M.H.S. Limnological and ecotoxicological studies in the cascade of reservoirs in the Tietê River (São Paulo, Brazil). *Braz. J. Biol.* **2005**, *65*, 697–710. [CrossRef] [PubMed]
18. Silva, F.L.; Fushita, A.T.; Cunha-Santino, M.B.; Bianchini, I., Jr. Adopting basic quality tools and landscape analysis for applied limnology: An approach for freshwater reservoir management. *Sustain. Water Resour. Manag.* **2022**, *8*, 65. [CrossRef]
19. Dorgham, M.M. Effects of Eutrophication. In *Eutrophication: Causes, Consequences and Control*, 1st ed.; Ansari, A.A., Gill, S.S., Eds.; Springer: Dordrecht, The Netherlands, 2014; pp. 29–44.
20. Barrella, W.; Petrere, M., Jr. Fish community alterations due to pollution and damming in Tietê and Paranapanema rivers (Brazil). *River Res. Appl.* **2003**, *19*, 59–76. [CrossRef]
21. American Public Health Association—APHA. *Standard Methods for the Examination of Water and Wastewater*, 3rd ed.; Baird, R.B., Eaton, A.D., Rice, E.W., Eds.; American Public Health Association: Washington, DC, USA, 2017; 1546p.
22. Lamparelli, M.C. Grau de Trofia em Corpos D'água do Estado de São Paulo: Avaliação dos Métodos de Monitoramento. Ph.D. Thesis, University of São Paulo, São Paulo, Brazil, 2004.
23. Conselho Nacional do Meio Ambiente—CONAMA. (National Environment Council). Resolução nº 357, de 17 de Março de 2005. Dispõe Sobre a Classificação dos Corpos de Água e Diretrizes Ambientais para o seu Enquadramento, bem como Estabelece as Condições e Padrões de Lançamento de Efluentes. Available online: http://www.mma.gov.br/port/conama/res/res05/res35705.pdf (accessed on 15 March 2024).
24. Britski, H.A. *Peixes de Água Doce do Estado de São Paulo*; Comissão Interestadual da Bacia do Paraná—Uruguai. Poluição e Piscicultura: São Paulo, Brasil, 1972; pp. 79–108.
25. Graça, W.J.; Pavaneli, C.S. *Peixes da Planície de Inundação do Alto Rio Paraná e Áreas Adjacentes*, 1st ed.; EDUEM: Maringá, Brazil, 2007; 241p.
26. Langeani, F.; Rego, A.C.L. *Guia Ilustrado dos Peixes da Bacia do Rio Araguari*; Grupo de Mídia Brasil Central: Uberlândia, Brazil, 2014; 141p.
27. Ota, R.R.; Deprá, G.C.; Graça, W.J.; Pavaneli, C.S. Peixes da planície de inundação do alto rio Paraná e áreas adjacentes: Revised, annotated and updated. *Neotrop. Ichthyol.* **2018**, *16*, 170094. [CrossRef]
28. Clarke, K.R.; Warwick, R.M. *Change in Marine Communities: An Approach to Statistical Analysis and Interpretation*, 2nd ed.; Primer-E Ltd.: Plymouth, UK, 2001; 172p.
29. Hammer, O.; Harper, D.A.; Ryan, P.D. PAST—Paleontological Statistics Version 1.28. 2001. Available online: https://palaeo-electronica.org/2001_1/past/past.pdf (accessed on 11 February 2024).
30. Froese, R.; Pauly, D. FishBase. World Wide Web Electronic Publication. Version (10/2023). 2023. Available online: http://www.fishbase.org (accessed on 10 December 2023).
31. Anderson, M.; Gorley, R.N.; Clarke, K. *PERMANOVA+ for primer: Guide to Software and Statistical Methods*; Primer-E: Plymouth, UK, 2008.
32. R Core Team. R: A Language and Environment for Statistical Computing. R Foundation for Statistical Computing. 2023. Available online: https://www.R-project.org (accessed on 24 January 2024).
33. Lopes, T.M.; Muniz, C.M.; Schmitz, M.H.; Dias, R.M.; Rodrigues, A.C.; Buzo, M.G.; do Couto, E.V.; Agostinho, A.A. Drivers of fish trophic guild composition in lakes of the Upper Paraná River floodplain. *Aquat. Sci.* **2022**, *84*, 27. [CrossRef]
34. Carvalho, D.R.; Alves, C.B.M.; Moreira, M.Z.; Pompeu, P.S. Trophic diversity and carbon sources supporting fish communities along a pollution gradient in a tropical river. *Sci. Total Environ.* **2020**, *738*, 139878. [CrossRef]
35. Brazil-Sousa, C.; Soares, B.E.; Svanbäck, R.; Albrecht, M.P. Individual specialization is the highest in generalist populations from intermediary to high trophic positions in tropical freshwater fishes. *Aust. Ecol.* **2024**, *49*, 13368. [CrossRef]
36. Barbosa, A.P.D.; Oliveira-Junior, E.S.; Muniz, C.C.; Lázaro, W.L.; Silva, D.J.; Santos-Filho, M.d. Dynamic of fish trophic guilds in the plateau-plain gradient in the Paraguay River, Northern Pantanal. *Pap. Avulsos Zool.* **2022**, *62*, 202262041. [CrossRef]
37. Carvalho, D.R.; Alves, C.B.M.; Pompeu, P.S. Uncertainty in estimating fish trophic positions and food web structure in highly polluted river basins. *Environ. Biol. Fishes* **2022**, *105*, 119–137. [CrossRef]
38. Lima, F.P.; Nobile, A.B.; Freitas-Souza, D.; Carvalho, E.D.; Vidotto-Magnoni, A.P. Can dams affect the trophic structure of ichthyofauna? A long-term effect in the Neotropical region. *Iheringia Ser. Zool.* **2018**, *108*, e2018030. [CrossRef]
39. Gandini, C.V.; Boratto, I.A.; Fagundes, D.C.; Pompeu, P.S. Estudo da alimentação dos peixes no rio Grande à jusante da usina hidrelétrica de Itutinga, Minas Gerais, Brasil. *Iheringia Ser. Zool.* **2012**, *102*, 56–61. [CrossRef]

40. Ramos, J.K.K.; Silva, N.L.; Bonfim, V.C.; Fornari, B.Y.; Kliemann, B.C.K.; Pagliarini, C.D.; Brandão, H.; Ramos, I.P. Characterization of wild fish diet and trophic guild in a protected area. *Acta Limnol. Brasil* **2022**, *34*, 15. [CrossRef]
41. Souza, F.B.; Santos, A.C.A.; Silva, A.T. Trophic structure of ichthyofauna in streams of the Contas River basin, Brazil. *Stud. Neotrop. Fauna Environ.* **2022**, *57*, 29–42. [CrossRef]
42. Azevedo-Santos, V.M.; Gonçalves, G.R.L.; Manoel, P.S.; Andrade, M.C.; Lima, F.P.; Pelicice, F.M. Plastic ingestion by fish: A global assessment. *Environ. Pollut.* **2019**, *255*, 112994. [CrossRef] [PubMed]
43. Ganassin, M.J.M. Spatial Distribution Patterns of Fish Assemblages in Neotropical Reservoirs. Ph.D. Thesis, State University of Maringá, Maringá, Brazil, 2021.
44. Bonato, K.O.; Delariva, R.L.; da Silva, J.C. Diet and trophic guilds of fish assemblages in two streams with different anthropic impacts in the northwest of Paraná, Brazil. *Zoologia* **2012**, *29*, 27–38. [CrossRef]
45. Abraham, W.R.; Macedo, A.J.; Gomes, L.H.; Tavares, F.C.A. Occurrence and resistance of pathogenic bacteria along the Tietê River downstream of São Paulo in Brazil. *CLEAN—Soil Air Water* **2007**, *35*, 339–347. [CrossRef]
46. Moretto, E.M.; Nogueira, M.G. Physical and chemical characteristics of Lavapés and Capivara rivers, tributaries of Barra Bonita Reservoir (São Paulo, Brazil). *Acta Limnol. Bras.* **2003**, *15*, 27–39.
47. Oliveira, P.C.R.; Nogueira, M.G.; Sartori, L.P. Differential environmental impacts on small and medium size rivers from center of São Paulo State, Brazil, and regional management perspectives. *Acta Limnol. Bras.* **2014**, *26*, 404–419. [CrossRef]
48. Casares, M.V.; De Cabo, L.I. Análisis de tendencias de variables indicadoras de calidad de agua para el Riachuelo (cuenca Matanza-Riachuelo, Argentina). *Rev. Int. Contam. Ambie.* **2018**, *34*, 651–665. [CrossRef]
49. Smith, W.S.; Barrella, W. The ichthyofauna of the marginal lagoons of the Sorocaba River, SP, Brazil: Composition, abundance and effect of the anthropogenic actions. *Braz. J. Biol.* **2000**, *60*, 627–632. [CrossRef] [PubMed]
50. Prado, M.R.; de Carvalho, D.R.; Alves, C.B.M.; Moreira, M.Z.; Pompeu, P.S. Convergent responses of fish belonging to different feeding guilds to sewage pollution. *Neotrop. Ichthyol.* **2020**, *18*, 190045. [CrossRef]
51. Urbanski, B.Q.; Denadai, A.C.; Azevedo-Santos, V.M.; Nogueira, M.G. First record of plastic ingestion by an important commercial native fish (*Prochilodus lineatus*) in the middle Tietê River basin, Southeast Brazil. *Biota Neotrop.* **2020**, *20*, 20201005. [CrossRef]
52. Urbanski, B.Q.; Brambilla, E.M.; Nogueira, M.G. Length-weight relationship and condition factor for *Prochilodus lineatus*, an important commercial fish, in contrasting water-quality environments of the middle Tietê River basin, Southeast Brazil. *Biota Neotrop.* **2023**, *23*, 20231467. [CrossRef]
53. Hochachka, P.W.; Lutz, P.L. Mechanism, origin, and evolution of anoxia tolerance in animals. *Comp. Biochem. Physiol. B Biochem. Mol. Biol.* **2001**, *130*, 435–459. [CrossRef]
54. Le Pichon, C.; Lestel, L.; Courson, E.; Merg, M.-L.; Tales, E.; Belliard, J. Historical Changes in the Ecological Connectivity of the Seine River for Fish: A Focus on Physical and Chemical Barriers Since the Mid-19th Century. *Water* **2020**, *12*, 1352. [CrossRef]
55. Araújo, C.V.M.; Silva, D.C.V.R.; Gomes, L.E.T.; Acayaba, R.D.; Montagner, C.C.; Moreira-Santos, M.; Ribeiro, R.; Pompêo, M.L.M. Habitat fragmentation caused by contaminants: Atrazine as a chemical barrier isolating fish populations. *Chemosphere* **2018**, *193*, 24–31. [CrossRef]
56. Noatch, M.R.; Suski, C.D. Non-physical barriers to deter fish movements. *Environ. Rev.* **2012**, *20*, 1–12. [CrossRef]

water

MDPI

Article

Macrobenthic Community Structure and Water Quality Evaluation in Ulungu River Basin (Northwest China)

Yong Song [1,†], Qiang Huo [1,†], Fangze Zi [1], Jianmin Ge [2], Xuelian Qiu [1], Long Yun [1], Gulden Serekbol [2], Liting Yang [1], Baoqiang Wang [3] and Sheng'ao Chen [1,*]

1 College of Life Sciences and Technology, Tarim Research Center of Rare Fishes, Tarim University, Alar 843300, China; songyongdky@126.com (Y.S.); 10757223082@stumail.taru.edu.cn (Q.H.); 10757213076@stumail.taru.edu.cn (F.Z.); 10757231124@stumail.taru.edu.cn (X.Q.); 10757232147@stumail.taru.edu.cn (L.Y.); 10757231127@stumail.taru.edu.cn (L.Y.)
2 College of Animal Sciences and Technology, Tarim University, Alar 843300, China; 10757212140@stumail.taru.edu.cn (J.G.); 10757231098@stumail.taru.edu.cn (G.S.)
3 Institute of Hydrobiology, Chinese Academy of Sciences, Wuhan 430072, China; wangbq@ihb.ac.cn
* Correspondence: shengao@taru.edu.cn; Tel.: +86-136-5757-6029
† These authors contributed equally to this work.

Abstract: In this study, the composition and structure of the macrobenthic community in the Ulungu River Basin was investigated and the water quality status of the basin was evaluated. In May and August 2022, and October 2023, the macrobenthic and water environmental factors at 11 sample sites in the basin were investigated, and the water quality of the basin was evaluated by the Shannon–Wiener index, FBI, and BMWP index. A total of 6101 macrobenthic organisms were collected; these organisms belonged to 3 phyla, 7 classes, 14 orders, 57 families, and 117 genera. Arthropod species accounted for the largest number of species (87.9%). A total of nine dominant species were found: *Micronecta* sp., *Eukiefferiella* sp., *Baetis* sp., *Polypedilum* sp., *Saetheria* sp., *Ephemerella* sp., *Limnodrilus* sp., *Ephemera* sp., and *Hydropsyche* sp. At the temporal level, the average density and biomass of macrobenthos were in the order of August > October > May; at the spatial level, the average density was greater in the tributaries than in the main stream, and the average biomass was greater in the main stream than in the tributaries. The mean values of the Shannon—Wiener index and Margalef richness index were as follows: August > October > May at the temporal level; the mean values of the Pielou evenness index were as follows: May > October > August at the temporal level. At the spatial level, the overall mean value of each diversity index indicated that the tributaries had larger diversity indices than the main stream. The water quality evaluation results revealed that the overall water quality level of the Ulungu River Basin ranges from light pollution to poor quality. Human interference activities greatly impact the water quality of the basin. To restore the ecology of the basin, it is necessary to strengthen the management and control of pollution sources.

Keywords: Ulungu River; macrobenthos; community structure; evaluation of water quality

1. Introduction

Freshwater ecosystems are fragile [1], and water quality is an important factor related to freshwater ecological conditions and biodiversity [2]. At present, watershed development and anthropogenic pressure are increasingly causing the degradation of aquatic ecosystems [3], and changes in freshwater ecosystems will cause changes in the communities of fish, algae, and invertebrates in the water. Therefore, it is necessary to use reliable methods and indicators to regularly monitor the water quality status of freshwater ecosystems. In fact, the freshwater resources available to humans are very limited, and precious freshwater resources should be protected.

The Ulungu River is an important inland river in the southern part of the Altai Mountains in the Junggar Basin in Northwest China. It originates in Qinghe County, flows from east to west through Fuyun County and Fuhai County, and finally flows into Ulungu Lake, the second largest freshwater lake in Xinjiang. The Ulungu River Basin has a total length of 821 km and a drainage area of 61,400 km^2 [4]. It is in a cold temperate zone with an arid continental climate. It relies mostly on mountain rainfall and ice and snow meltwater to replenish water [5]. It is the main water source for Fuhai and Qinghe Counties. The Ulungu River and the Irtysh River were originally two independent water systems. However, due to the new construction of the "Diversion of the Irtysh River" project in 1986~1987, a 3-km-long canal was dug, causing the Ulungu River to become a tributary of the Irtysh River. Ulungu Lake also became an affiliated water body of the Irtysh River. The Ulungu River Basin is rich in fish resources, especially the Ulungu Lake as one of the important fishing grounds in inland China. Commercial fish such as Pond Smelt (*Hypomesus olidus*), Common Bream (*Abramis brama*), and Northern Pike (*Esox lucius*) live in the basin and are being captured [6]. In recent years, due to the continuous growth of social and economic development, the increase in population, the continuous expansion of urbanization and agriculture and animal husbandry, and the poor stability of surface runoff in the Ulungu River Basin, residents near the basin have experienced water shortages, agricultural irrigation, and the interruption of the Ulungu River [7]. These conditions seriously threaten the survival and reproduction of aquatic organisms in the basin.

Macrobenthos refers to aquatic animals that live at the bottom of water bodies for all or most of their life history. They are individual benthic animals that cannot pass through a 0.5 mm aperture mesh [8]. Macrobenthos are an important part of freshwater bodies. They are in the middle link in food webs of the ecosystem and are crucial in material exchange and energy flow in the ecosystem. Changes in their community structure will affect fish and other important aquatic organisms through the food chain [9–11]. In aquatic environments, macrobenthos have limited mobility and may also respond to changes in water quality through changes in species composition or population dynamics [12]. Macrobenthos are among the most important groups of organisms used to determine river water quality and ecological conditions [13]. Compared with zooplankton and fish, because they live on the bottom of the water for a long time, they are key links in the ecosystem and have strong regional characteristics and weak migratory ability [14]; therefore, they are sensitive to changes in the ecological environment. The quality of the environment directly affects their growth, reproduction, species distribution, and community composition [15]; most of these species are large and easy to identify and collect, so they can be used as indicator species for evaluating water quality. Macrobenthos have been widely used as biological indicators of river water quality in temperate regions [16]. Using macrobenthos to assess river ecological health is the mainstream method for the biological assessment of water quality [17], and it is also an indispensable prerequisite for river ecosystem restoration and management [18].

At present, there are few studies on macrobenthos in the Ulungu River. Existing studies have reported only macrobenthos in affiliated lakes and reservoirs in the Ulungu River Basin [19]. In this study, the composition and structure of the macrobenthic community in the Ulungu River Basin were investigated to determine the basin's community changes. The Shannon–Wiener index, FBI, and BMWP were used to conduct a biological evaluation of the water quality of the Ulungu River Basin to determine its water pollution status. The impacts of various environmental factors on macrobenthos were also analyzed. This study is expected to provide an important reference for the protection of the Ulungu River Basin ecosystem and provide a reference and scientific basis for the protection of important commercial fish and other aquatic biological resources in the basin.

2. Materials and Methods

2.1. Study Area and Sample Collection

According to the geographical and water system characteristics of the Ulungu River Basin, 11 sampling points were set up in the basin: tributaries—the Daqing River Bridge (S1), Qinggri River Twin Bridges (S2), Agashi Aobao Township (S3), Tucker Shiken Town

(S4), and Ihigen Village (S5); and main streams—the Ertai Bridge (S6), Qiakultu Town (S7), Yerk Kala (S8), Dure Town (S9), Kalamagai Town (S10), and Fuhai Bridge (S11). The specific distribution of sampling points is shown in Figure 1. The sampling times were May, August 2022, and October 2023. A D-type net with a 60-mesh aperture and a net port width of 25 cm was used to collect macrobenthos. The collection was repeated three times, and 0.25 m^2 was quantitatively collected at each sampling point. The collected samples were placed on a white disk, clean water was added, and various types of macrobenthos were selected on site. The selected macrobenthos were placed into 50 mL specimen bottles filled with a fixative (7% formalin solution). Most of the macrobenthos were identified to the genus level, and a few species were identified to the family level or higher. The identification methods used were *Economic Animals of China* [20] and *Chironomid Larvae of Northern China* [21]. The identified samples were counted and weighed to calculate the density and biomass per square meter.

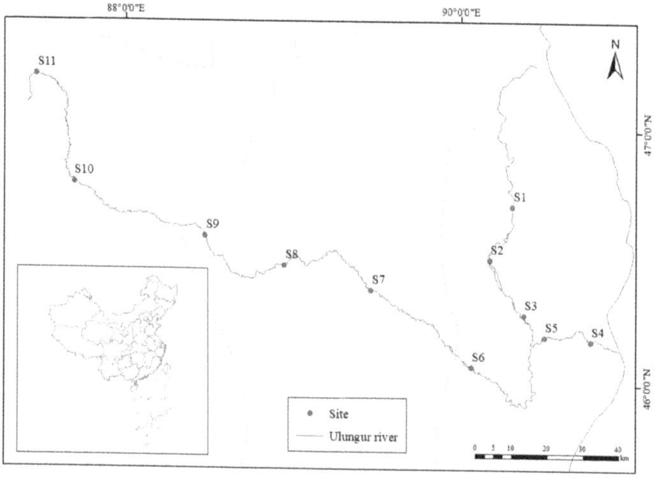

Figure 1. Distribution of sampling points in the Ulungu River.

2.2. Determination of Environmental Factors

The American Jinquan Multi-Parameter Water Quality Sonde (YSI 556MPS) (YSI Inc., Yellow Springs, OH, USA) was used to measure water temperature (WT), dissolved oxygen (DO), conductivity (Cond), salinity (SAL), and pH.

2.3. Data Analysis

1. Dominant species

It was determined according to the dominance value (Y). $Y > 0.02$ was considered the dominant species. The calculation formula [22] is as follows:

$$Y = (N_i/N) \times F_i \quad (1)$$

In the formula, N_i is the number of individuals of the i-th species, N is the total number of individuals of all species, and F_i is the frequency of occurrence of the i-th species.

2. Calculation of the biological indices

For the Shannon—Wiener diversity index, the calculation formula [23] is as follows:

$$H' = -\sum_{i=1}^{S} (N_i/N) \times \log_2(N_i/N) \quad (2)$$

For the Margalef richness index, the calculation formula [23] is as follows:

$$d = (S - 1) / \log_2 N \tag{3}$$

For the Pielou uniformity index, the calculation formula [23] is as follows:

$$J = H' / \log_2 S \tag{4}$$

In the formula, N_i is the number of individuals of the i-th species, N is the total number of individuals of all species, and S is the number of species.

For the family-level biotic index (FBI), the calculation formula [24] is as follows:

$$\text{FBI} = \sum_{i=1}^{F} n_i t_i / N \tag{5}$$

In the formula, n_i is the number of individuals in the i-th family, t_i is the pollution resistance value of the i-th family, and N is the total number of individuals in each family.

With respect to the biological monitoring working party (BMWP) score system, the calculation formula [25] is as follows:

$$\text{BMWP} = \sum_{i=1}^{n} t_i \tag{6}$$

In the formula, t_i is the sensitivity value of the i-th family.

The water quality evaluation standards for the various biological indices are shown in Table 1 [9,25,26].

Table 1. Water quality evaluation standards for various indices.

H′		FBI		BMWP	
Value	**Level**	**Value**	**Level**	**Value**	**Level**
>3	Clean	0.00~3.75	Healthy	≥81	Healthy
2~3	Light pollution	3.51~5.00	Good	51~80	Good
1~2	Moderate pollution	5.01~5.75	General	25~50	General
		5.76~7.25	Poor	10~24	Poor
0~1	Heavy pollution	7.26~10.00	Very poor	0~9	Very poor

2.4. Data Processing

The statistics and calculations of the species composition, abundance, biomass, dominance value, and various biological indices of macrobenthos were completed through Microsoft Excel 2010. A visual analysis of density, biomass, and various indices was completed using Origin 2018. A one-way analysis of variance and nonparametric tests were completed using IBM SPSS Statistics 25.

3. Results

3.1. Environmental Factors

In May, Kalamagai Town (S10) had the highest water temperature and the lowest dissolved oxygen; the water temperature of the Qinggri River Twin Bridges (S2) was the lowest, but the dissolved oxygen was the highest. The conductivity of Tucker Shiken Town (S4) was the highest, and that of the Qinggri River Twin Bridges (S2) was the lowest. The salinity of Tucker Shiken Town (S4) was the highest, and the Daqing River Bridge (S1) and the Qinggri River Twin Bridges (S2) were the lowest. The pH of the Ertai Bridge (S6) was the highest, and the Qinggri River Twin Bridges (S2) were the lowest.

In August, the water temperature of Fuhai Bridge (S11) was the highest, and that of Ihigen Village (S5) was the lowest. Dissolved oxygen was the highest in Ihigen Village (S5) and the lowest at the Ertai Bridge (S6). The conductivity and salinity were the highest in Kalamagai Town (S10) and the lowest in Tucker Shiken Town (S4). The pH was the highest in Kalamagai Town (S10) and the lowest at the Qinggri River Twin Bridges (S2).

In October, the water temperature in Kalamagai Town (S10) was the highest and the dissolved oxygen was the lowest. The Daqing River Bridge (S1) had the lowest water temperature, but the highest dissolved oxygen. The conductivity was the highest in Agashi Aobao Township (S3) and the lowest at the Daqing River Bridge (S1). Salinity was the highest in Kalamagai Town (S10) and the lowest at the Daqing River Bridge (S1). The pH was the highest in Dure Town (S9) and the lowest at the Daqing River Bridge (S1).

According to the average value of environmental factors in different months, the water temperature was higher in August than in May than in October. Dissolved oxygen was higher in October than in May than in August, which was contrary to the trend of water temperature. The conductivity and salinity were higher in August than in October than in May, and the pH value was higher in August than in May than in October. The specific changes in environmental factors are shown in Table 2.

Table 2. Environmental factors of the Ulungu River Basin.

Site/Parameters		Water Temperature/°C	Dissolved Oxygen/mg/L	Conductivity	Salinity/ppt	pH
May	S1	10.4	13.63	64.6	0.04	7.38
	S2	7.1	14.21	58.6	0.04	7.15
	S3	12.8	12.27	227.8	0.14	7.45
	S4	15.6	9.84	326.0	0.19	7.57
	S5	15.2	10.47	306.5	0.18	7.79
	S6	13.7	11.26	146.1	0.09	7.90
	S7	13.8	10.85	163.3	0.10	7.65
	S8	14.0	8.79	192.4	0.12	7.89
	S9	15.2	8.43	294.2	0.17	7.74
	S10	20.8	5.37	279.6	0.14	7.62
	S11	20.6	6.70	303.5	0.16	7.83
	M ± SD	14.5 ± 3.75	10.2 ± 2.60	214.8 ± 92.51	0.1 ± 0.05	7.6 ± 0.22
August	S1	21.4	5.09	224.5	0.11	8.42
	S2	19.4	5.11	1426.0	0.81	7.37
	S3	22.1	5.09	327.8	0.17	8.29
	S4	22.6	7.01	93.6	0.05	8.17
	S5	18.0	7.31	293.0	0.16	7.98
	S6	23.7	5.08	369.9	0.18	8.46
	S7	24.3	5.94	394.6	0.19	8.29
	S8	24.3	6.91	465.0	0.23	8.32
	S9	23.7	6.92	920.0	0.46	8.30
	S10	29.4	6.47	2105.0	0.98	8.61
	S11	37.2	6.21	764.0	0.32	8.48
	M ± SD	24.2 ± 4.98	6.1 ± 0.85	671.2 ± 579.96	0.3 ± 0.29	8.2 ± 0.32
October	S1	7.9	21.62	103.0	0.07	6.15
	S2	9.4	12.75	329.2	0.23	7.17
	S3	11.4	20.97	1167.0	0.80	7.19
	S4	9.8	11.54	174.2	0.12	6.30
	S5	11.3	10.49	223.8	0.15	7.13
	S6	13.7	12.55	362.3	0.22	6.78
	S7	14.7	10.73	435.4	0.26	6.92
	S8	12.6	16.00	444.4	0.28	7.07
	S9	14.4	15.15	469.7	0.29	8.08
	S10	15.8	7.04	561.0	0.33	7.00
	S11	14.1	14.95	229.0	0.14	6.66
	M ± SD	12.3 ± 2.40	14.0 ± 4.20	409.0 ± 274.25	0.3 ± 0.19	7.0 ± 0.49

3.2. Macrobenthos

A total of 6101 macrobenthos were collected in the three surveys; these organisms belonged to 3 phyla, 7 classes, 14 orders, 57 families, and 117 genera. The proportion of the number of individuals in different taxa was as follows: arthropods had the largest number, accounting for 89.7%, followed by annelids, accounting for 8.4%, and mollusks had the smallest number, accounting for 1.9%. Insecta accounted for the largest proportion of arthropods, among which Diptera (51.4%) accounted for the highest proportion, followed by Hemiptera (14.5%) and Ephemeroptera (12.5%) (Figure 2). The larvae of Chironomidae in Diptera were common in almost all stations in different months. The EPT species were mainly distributed in the tributaries in May, and the density tended to increase in August, and its distribution spread to the middle of the basin. Hemiptera was mainly composed of Corixidae and Aphelocheiridae. In May, it is mainly distributed in the main stream, and the density increased as a whole in August and October, and it was evenly distributed in the whole basin (Figure 3). Ephemeroptera was mainly distributed in tributaries in May, and its density increased in August. It was distributed at all sites except Fuhai Bridge (S11), but still accounted for a relatively high proportion in tributaries. In October, its density decreased to lower than that in May, and it was mainly distributed in the main stream. Decapoda was only discovered on the Fuhai Bridge in August and October. Amphipods were mainly Gammaridae, found in Tucker Shiken Town (S4) and Dure Town (S9) in May, only in Agashi Aobao Township (S3) in August, and distributed at the Ertai Bridge (S6) and Dure Town (S9) in October. Odonata were mainly distributed in the main stream in May and October, and were distributed at the Daqing River Bridge (S1), Tucker Shiken Town (S4), and Kalamagai Town (S10) in August, but in small numbers. A total of 9 dominant species ($Y > 0.02$) were found in the survey, and the dominant species in May were *Micronecta* sp., *Baetis* sp., and *Ephemerella* sp. The dominant species in August were *Eukiefferiella* sp., *Polypedilum* sp., *Micronecta* sp., *Saetheria* sp., and *Baetis* sp. The dominant species in October were *Limnodrilus* sp., *Micronecta* sp., *Ephemera* sp., and *Hyclropsyche* sp. The common dominant species in the three-season survey was *Micronecta* sp.

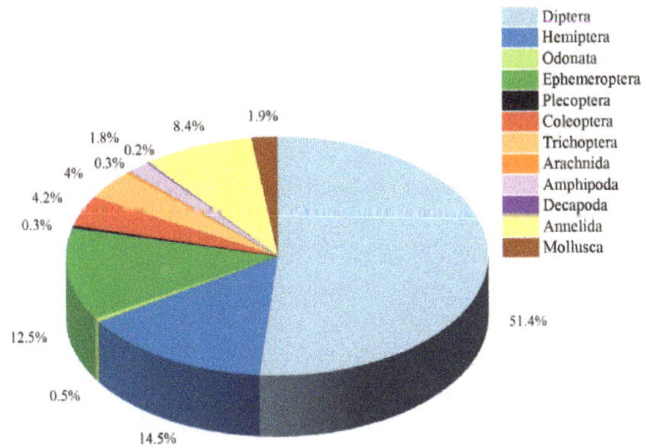

Figure 2. Proportions of individual numbers of different taxa of macrobenthos.

In May, the density of macrobenthos ranged from 52 to 1120 ind/m^2, with the lowest point occurring in Qiakultu Town (S7) and the highest point occurring in Agashi Aobao Township (S3). The average value was 283 ind/m^2. The biomass range was 0.056~5.897 g/m^2, with the lowest point occurring at Fuhai Bridge (S11) and the highest point occurring at Ertai Bridge (S6). The average value was 2.17 g/m^2 (Figure 3).

In August, the density of macrobenthos ranged from 36 to 8448 ind/m^2, with the lowest point occurring at Fuhai Bridge (S11) and the highest point occurring at Tucker Shiken Town (S4). The average value was 1426 ind/m^2. The biomass range was 0.6404~14.5832 g/m^2, with the lowest point occurring at Ertai Bridge (S6) and the highest point occurring at Fuhai Bridge (S11). The average value was 4.39 g/m^2.

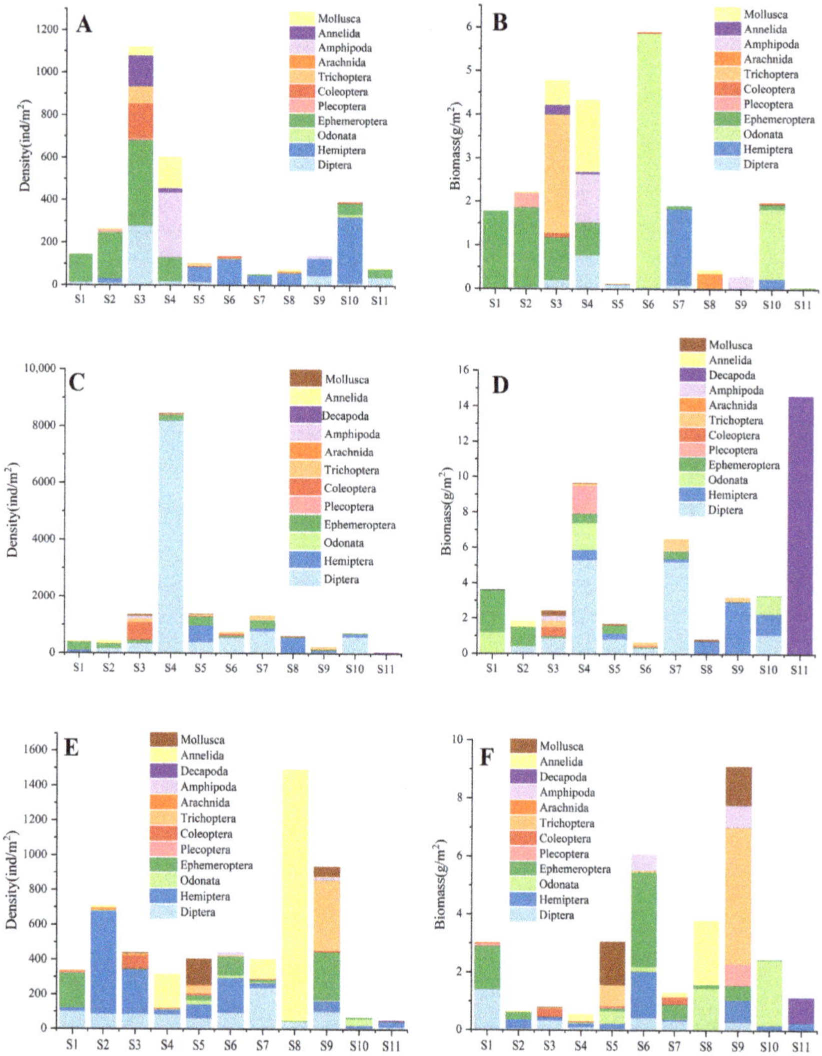

Figure 3. Density and biomass of macrobenthos. (**A**) Macrobenthic density in May, (**B**) macrobenthic biomass in May, (**C**) macrobenthic density in August, (**D**) macrobenthic biomass in August, (**E**) macrobenthic density in October, and (**F**) macrobenthic biomass in October.

In October, the density of macrobenthos ranged from 52 to 1492 ind/m^2, with the lowest point occurring at Fuhai Bridge (S11) and the highest point occurring at Yerk Kala (S8). The average value was 509 ind/m^2. The biomass ranged from 0.5504 to 9.1056 g/m^2, with the lowest point occurring in Tucker Shiken Town (S4) and the highest point occurring in Dure Town (S9). The average value was 2.90 g/m^2.

The average density and biomass of macrobenthos in the Ulungu River were in the order of August > October > May at the temporal level; at the spatial level, the average density was greater in the tributaries than in the main stream, and the average biomass was greater in the main stream than in the tributaries (Table 3). After nonparametric testing, there was a significant difference in macrobenthic density between seasons ($p = 0.038$), but there was no significant difference in biomass between seasons ($p = 0.234$). The changes in macrobenthic density and biomass at various points in the Ulungu River Basin are shown in Figure 3.

Table 3. The average value of spatial horizontal density and biomass of macrobenthos.

Site		Density (ind/m²)			Biomass (g/m²)		
		May	August	October	May	August	October
tributaries	S1	144	416	336	1.77	3.60	3.02
	S2	264	440	708	2.21	1.84	0.64
	S3	1120	1360	440	4.78	2.42	0.79
	S4	604	8448	316	4.33	9.64	0.55
	S5	104	1380	404	0.13	1.69	3.06
	Mean		1099			2.70	
main stream	S6	136	748	444	5.90	0.64	6.08
	S7	52	1324	404	1.90	6.52	1.32
	S8	76	600	1492	0.45	0.82	3.79
	S9	140	228	936	0.30	3.22	9.11
	S10	392	708	68	1.98	3.30	2.45
	S11	84	36	52	0.06	14.58	1.14
	Mean		440			3.53	

The Shannon—Wiener index (H') of the Ulungu River in May ranged from 0.57 to 3.25, with an average of 1.91; the Margalef richness index (d) ranged from 0.47 to 2.46, with an average of 1.35; and the Pielou evenness index (J) ranged between 0.29 and 0.83, with an average value of 0.63. The maximum values of H' and d both occurred in Agashi Aobao Township (S3), and the maximum value of J occurred at the Fuhai Bridge (S11). The minimum values of H' and J both occurred at Ertai Bridge (S6), and the minimum value of d occurred at Yerk Kala (S8) (Figure 4).

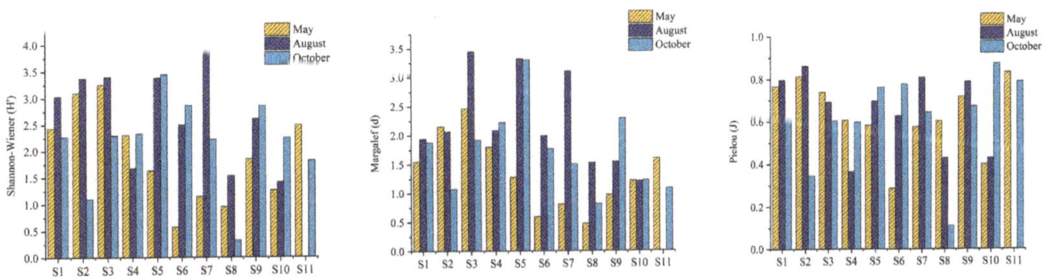

Figure 4. Spatial changes in the macrobenthic diversity index.

In August, the Shannon—Wiener index ranged from 0 to 3.83, with an average value of 2.43; the Margalef richness index ranged from 0 to 3.45, with an average value of 2.02; and the Pielou evenness index ranged from 0 to 0.86, with an average value of 0.59. The maximum values of H', d, and J occurred in Qiakultu Town (S7), Agashi Aobao Township (S3), and Qinggri River Twin Bridge (S2), respectively. During the investigation of Fuhai Bridge (S11), only species of Palaemonidae were found at this site, and all diversity indices were zero.

In October, the Shannon—Wiener index ranged from 0.33 to 3.44, with an average value of 2.16; the Margalef richness index ranged from 0.82 to 3.30, with an average value of 1.73; and the Pielou evenness index ranged from 0.11 to 0.87, with an average value of 0.62. The maximum values of H' and d both occurred in Ihigen Village (S5), and the maximum value of J occurred in Kalamagai Town (S10). The minimum values of H′, d, and J were all in Yerk Kala (S8).

The mean values of the Shannon—Wiener index and Margalef richness index followed the order August > October > May at the temporal level, and the mean values of the Pielou evenness index followed the order May > October > August at the temporal level. At the spatial level, the overall mean value of each diversity index indicated that the tributaries had larger diversity indices than the main stream. The specific changes in the diversity indices are shown in Figure 4.

3.3. Water Quality Evaluation Results

The Shannon—Wiener index water quality evaluation results show (Table 4) that the Ulungu River Basin was experiencing light pollution overall, with the largest proportion of clean sites occurring in August. The water quality conditions were better in August than in October, followed by May. The water quality of the Qinggri River Twin Bridge (S2) and Agashi Aobao Township (S3) was clean in May and August, and the water quality of Ihigen Village (S5) was clean in August and October. The water quality of the Daqing River Bridge (S1) was poor in August. According to the average value of the Shannon–Wiener index at different sites, all the tributaries had light pollution, and the water quality was better than that of the main stream, and Yerk Kala (S8) had heavy pollution in the main stream.

Table 4. Results of the Shannon–Wiener index for evaluating water quality.

Site	May		August		October		Mean	Level
	H′	Level	H′	Level	H′	Level		
S1	2.43	Light pollution	3.03	Clean	2.28	Light pollution	2.58	Light pollution
S2	3.09	Clean	3.37	Clean	1.10	Moderate pollution	2.52	Light pollution
S3	3.25	Clean	3.40	Clean	2.29	Light pollution	2.98	Light pollution
S4	2.31	Light pollution	1.68	Moderate pollution	2.33	Light pollution	2.11	Light pollution
S5	1.63	Moderate pollution	3.38	Clean	3.44	Clean	2.82	Light pollution
S6	0.57	Heavy pollution	2.51	Light pollution	2.86	Light pollution	1.98	Moderate pollution
S7	1.15	Moderate pollution	3.83	Clean	2.22	Light pollution	2.40	Light pollution
S8	0.95	Heavy pollution	1.53	Moderate pollution	0.33	Heavy pollution	0.94	Heavy pollution
S9	1.85	Moderate pollution	2.61	Light pollution	2.85	Light pollution	2.44	Light pollution
S10	1.26	Moderate pollution	1.41	Moderate pollution	2.25	Light pollution	1.64	Moderate pollution
S11	2.49	Light pollution	0	Heavy pollution	1.82	Moderate pollution	1.44	Moderate pollution
Mean	1.91	Moderate pollution	2.43	Light pollution	2.16	Light pollution	2.17	Light pollution

The average FBI index shows (Table 5) that the water quality grade of the Ulungu River Basin was poor, with the water quality conditions in May being better than those in August, followed by those in October. In May, the water quality in Qiakultu Town (S7) was healthy, and the water quality at Daqing River Bridge (S1) and Qinggri River Twin Bridge (S2) was good; in August, the water quality in Dure Town (S9) was good; and in October, the water quality levels at the Daqing River Bridge (S1) and Dure Town (S9) were good. During the investigation period, the remaining sites were polluted to varying degrees, and the water quality level of Yerk Kala (S8) was very poor in May, August, and October. According to the average value of the FBI index at different sites, the water quality of Daqing River Bridge (S1) was good, and Yerk Kala (S8) was very poor. The water quality of the tributaries and the main stream was similar.

The average BMWP index shows (Table 6) that the water quality level of the Ulungu River Basin was moderate, with the water quality conditions in August being better than those in October, and May being the worst. In May, the water quality levels of the Qinggri River Twin Bridge (S2), Agashi Aobao Township (S3), and Tucker Shiken Town (S4) were good; in August, the water quality levels of Agashi Aobao Township (S3) and Ihigen Village (S5) were healthy, and the water quality levels of Tucker Shiken Town (S4), Ertai Bridge (S6), and Qiakultu Town (S7) were good; in October, the water quality level of the Daqing River Bridge (S1) was good. According to the average value of BMWP scores at different sites, the water quality of Agashi Aobao Township (S3) and Tucker Shiken Town (S4) and

Ihigen Village (S5) in the tributary was good, Fuhai Bridge (S11) was poor, and the other sites were general. The water quality of tributaries was better than that of the main stream.

Table 5. FBI results for water quality evaluation.

Site	May		August		October		Mean	Level
	FBI	Level	FBI	Level	FBI	Level		
S1	4.03	Good	5.24	General	4.29	Good	4.52	Good
S2	4.76	Good	6.83	Poor	7.62	Very poor	6.40	Poor
S3	6.80	Poor	6.53	Poor	6.95	Poor	6.76	Poor
S4	6.38	Poor	6.98	Poor	8.28	Very poor	7.21	Poor
S5	6.77	Poor	6.83	Poor	5.78	Poor	6.46	Poor
S6	7.31	Very poor	6.94	Poor	6.27	Poor	6.84	Poor
S7	2.01	Healthy	6.30	Poor	7.53	Very poor	5.28	General
S8	7.52	Very poor	7.47	Very poor	9.08	Very poor	8.02	Very poor
S9	7.57	Very poor	4.32	Good	4.83	Good	5.57	General
S10	7.19	Poor	6.26	Poor	5.29	General	6.25	Poor
S11	6.58	Poor	6.00	Poor	7.42	Very poor	6.67	Poor
Mean	6.08	Poor	6.34	Poor	6.67	Poor	6.36	Poor

Table 6. BMWP water quality evaluation results.

Site	May		August		October		Mean	Level
	BMWP	Level	BMWP	Level	BMWP	Level		
S1	25.1	General	35.9	General	57.0	Good	39.3	General
S2	54.7	Good	34.2	General	17.0	Poor	35.3	General
S3	71.4	Good	87.7	Healthy	33.0	General	64.0	Good
S4	55.2	Good	77.7	Good	32.5	General	55.1	Good
S5	34.9	General	95.1	Healthy	93.4	Healthy	74.5	Good
S6	22.6	Poor	54.8	Good	39.0	General	38.8	General
S7	22.9	Poor	77.1	Good	40.2	General	46.7	General
S8	6.1	Very poor	38.1	General	30.1	General	24.8	General
S9	7.5	Very poor	29.0	General	102.8	Healthy	46.4	General
S10	44.5	General	38.3	General	20.5	Poor	34.4	General
S11	18.0	Poor	4.0	Very poor	12.0	Poor	11.3	Poor
Mean	33.0	General	52.0	Good	43.4	General	42.8	General

Taken together, according to the respective overall average values of H', FBI, and BMWP indices, overall, the water quality of the Ulungu River Basin ranged from light pollution to poor. The water quality of the tributary was better than that of the main stream (Table 7).

Table 7. Water quality assessment results of the average values of Shannon–Wiener, FBI, and BMWP indexes in tributaries and main streams.

River Basin	H′		FBI		BMWP	
	Mean	Level	Mean	Level	Mean	Level
tributaries	2.6	Light pollution	6.27	Poor	53.65	Good
main stream	1.81	Moderate pollution	6.44	Poor	33.75	General

4. Discussion

4.1. Environmental Factors

Aquatic ecological factors are potential indicators for predicting changes in aquatic biological communities. The water temperature of natural water bodies will experience seasonal and diurnal changes [27]. The highest water temperature monitored during the survey was at Fuhai Bridge (S11) in August. This may be due to the basin's proximity to towns and human activity areas, coupled with the nearby terrain being flat, and the absence of tall trees blocking light. The average pH value of the three surveys (7.61 ± 0.65) indicated that the water body was weakly alkaline, which is consistent with the pH range of most natural water bodies in the range of 5–9 [28]. This weak alkalinity may be related to the alkaline wastewater discharged into the water and human activities occurring around the riverbanks. DO was at a low level in August, and DO could cause changes in community structure by affecting the respiration and diffusion of macrobenthos [29–31]. DO is often related to water temperature, with generally higher water temperatures causing DO levels to decrease. The results of this study found that the trend of water temperature and dissolved oxygen in the Ulungu River Basin was opposite at the temporal level. The low DO level in August was accompanied by an increase in the abundance of Chironomidae species, which is similar to previous findings [32].

4.2. Macrobenthos

The survey results of the macrobenthos in the Ulungu River Basin showed that aquatic insects accounted for the largest proportion of the population in the region. These findings are similar to the results of previous studies in the Irtysh River Basin, which has a similar geographical location. The number of dominant species is less than that in the Irtysh River Basin [33], though the species are similar. The density survey results are higher than those in the Irtysh River Basin, indicating that the average density of macrobenthos in the Ulungu River is greater than that in the Irtysh River Basin. A possible reason is that the water environmental factors in the Ulungu River Basin are more favorable than those in the Irtysh River [19]. The Ulungu River Basin accounts for the largest proportion of arthropod species, especially Diptera, and this is similar to the results of studies on macrobenthos in lakes in the southern Ob-Irtysh basin [34]. However, the Chinese urban rivers with similar latitudes also have the largest number of arthropods, but the proportion is lower than that of the Ulungu River Basin [35]. Compared with the African Ganges with a similar longitude, the Ganges has the largest proportion of mollusks, and there are only a few species of Ephemeroptera, which is significantly different from the macrobenthos community structure in the Ulungu River Basin [36]. The density and biomass of macrobenthos in the Ulungu River Basin are much greater than those in the Tarim River Basin, which is also an inland river, and the dominant species are quite different from those in the Tarim River Basin [37]. This may be because Ulungu River is close to the Altai Mountains, and Tarim River is close to the edge of the Taklimakan Desert. There are obvious differences in the ecological environment between the two regions. The habitats of the Ulungu River Basin and the Ili River Basin are complex. The dominant species all include Ephemeroptera and Chironomid species, and the average density and biomass are not much different [38].

At the spatial level, the overall density of tributaries is higher than that of the main stream, but the biomass is lower than that of the main stream. This is mainly because the main stream is mostly distributed with some large-bodied but small-number macrobenthic animals, such as Amphipoda, Decapoda, Odonata, etc. The average diversity index showed that the diversity in the tributaries was greater than that in the main stream, indicating that the overall macrobenthic diversity in the tributary reaches was greater than that in the main stream. This is similar to the results of the study on the Jarama River in Madrid, Europe, where tributaries close to mountains and at high altitudes will have higher diversity of macrobenthic animals [39]. A possible reason is that the main stream is close to the downstream reaches, and less runoff will reduce the benthos diversity [40]. Because the tributaries are close to the Altai Mountains and receive a large amount of rainfall, the downstream reaches of the main stream are close to the Junggar Basin and receive less rainfall. In addition, the overexploitation and utilization of land in tributary reaches and the increase in new industrial water use have resulted in the runoff of the downstream reaches of the main stream being much smaller than that of the tributaries [41]. In addition, during the investigation, it was found that the substrate in the tributary river sections consisted mostly of pebbles and gravel, which are more suitable for the survival of macrobenthos [42,43]. The diversity indices of the Fuhai Bridge (S11), located in the main stream of the Ulungu River in August, were all zero. The main reason may be the impact of the flow interruption in the Fuhai section of Ulungu River's lower reaches of the main stream during the August survey. Since the 1990s, due to the overexploitation of land in the tributaries of the Ulungu River and a surge in industrial water use, the number of days without flow in the Fuhai section of the downstream main stream has increased annually. As a result, the habitats and breeding grounds of aquatic life in this section have almost disappeared, and the aquatic ecological environment has been severely damaged.

4.3. Water Quality Evaluation

In the Ulungu River Basin, the biological evaluation of the water quality of each index had different results due to differences in evaluation standards. The Shannon–Wiener index and BMWP evaluation results were relatively similar, and the water quality of the tributaries was better than that of the main stream. The water quality evaluation results of the FBI of Yerk Kala (S8) in the three surveys were all very poor, which is consistent with the conditions observed during the survey. The water body at this site was turbid during the survey, the surrounding water body was mostly rocky, the vegetation was sparse, and the biodiversity was low. Most of the species were pollution-resistant species, such as oligochaetes, and the water quality was poor. Overall, the water quality of the Ulungu River Basin ranges from light pollution to poor quality. The possible reason is that the Ulungu River Basin is dominated by agriculture and animal husbandry, and rainfall and snowmelt will bring a large amount of harmful substances produced by agriculture and animal husbandry into the river. In addition, a large amount of human interference, such as an increase in organic matter and an increase in chironomid larvae and oligochaetes in the water body [44], has also impacted the water quality of the basin. Chironomidae taxa found in tropical rivers may be pollution-resistant taxa. Compared with tropical rivers, the proportion of Chironomidae taxa in the Ulungu River Basin is larger, which is consistent with the results of the presence of pollution in water bodies in this study. In areas where organic pollution is present, the proportion of chironomid taxa is associated with a higher abundance of chironomid larvae [45]. At the spatial level, there are many types of EPTs in tributaries, and the water quality is better than that in the main stream. The main reason may be that the tributaries are more rural and group farms and receive more rainfall. The vegetation coverage near the basin is high, the water flow rate is fast, and the self-purification ability is strong. The discharge of industrial and domestic wastewater and pollution caused by livestock farming flow into the main stream with the water flow. However, the downstream reaches of the main rivers are prone to flow interruptions, the discharged wastewater cannot be diluted, the erosion and siltation capacity of the river sections is reduced, the self-purification ability is lost, and the water quality is deteriorating annually.

As a tributary of the Irtysh River, the Ulungu River's polluted water may have a negative impact on the water quality of the Irtysh River and even the Ob River along with runoff. However, previous studies have found [46] that the pollution level of the upper reaches of the Irtysh River is low, and the cross-border runoff does not have the characteristics of serious pollution. This may be due to increased rainfall and reduced permafrost volume, coupled with the effective operation of large dams and reservoirs in the Irtysh River Basin in China and the implementation of water-saving technologies, which ensured the water level in the basin [47] and weakened the impact of polluted water in the Ulungu River, ensuring the health of the river. However, the water quality of the Irtysh River in East Kazakhstan is poor, which may be mainly due to heavy metal and nuclear pollution caused by previous industries in the area [48]. There is no obvious pollution in the water quality of the Irtysh River downstream after Pavlodar in Kazakhstan, which may be due to its high self-purification ability downstream. The Irtysh River eventually merges into the Ob River. Some studies have found that the Ob River leech diversity in Russia is relatively large [49], which may mean that the water quality of the river is good. This may be due to the flat nature of the branch water system, and many tributaries carry nutrients from the vast territory, combined with the Atlantic Multidecadal Variability (AMV), inducing an increase in normal rainfall in the Ob River [50], giving the river a high level of self-purification capacity. But, in fact, the Ob River is also suffering from some radionuclides and heavy metal pollution that may come from European nuclear power plants and some human factors [51,52]. In contrast, the impact of pollution from the Ulungu River may be diluted during the long runoff process, but after long-term accumulation, it is speculated that it may also have a potentially deeper impact on the Ob River Basin ecosystem. In addition, compared with the Shuya River in northern Russia, due to its lower degree of eutrophication, the water quality is better than that of the Ulungu River Basin. There are many residents and farms along the Ulungu River Basin, so it is speculated that the water body already has a certain degree of eutrophication [53].

5. Conclusions

As the second largest river in the Altay region of Xinjiang, China, the water quality of the Ulungu River is closely related to the ecological health of the basin and the lives of surrounding residents. The macrobenthos in the basin are mainly aquatic insect species with high density and biomass. According to the respective overall average values of H′, FBI, and BMWP indices, overall, the water quality level of the Ulungu River Basin ranges from light pollution to poor quality, and human interference activities such as basin development and livestock husbandry have a greater impact on the water quality of the basin. The water ecology of the Ulungu River Basin is relatively fragile and has little ability to withstand pollution. To protect and restore basin ecology, it is necessary to strengthen the management and governance of pollution sources, strengthen the management of the harm of livestock, irrigation, and other activities to basin ecology, and rationally plan the scale of basin development without harming the ecological environment.

Author Contributions: Conceptualization, Q.H., Y.S. and S.C.; Software, Q.H., X.Q. and L.Y. (Long Yun); Validation, Q.H. and F.Z.; Investigation, Q.H., F.Z. and J.G.; Data curation, Q.H. and F.Z.; Writing—original draft preparation, Q.H. and Y.S.; Writing—review and editing, S.C., Y.S., B.W., X.Q., L.Y. (Long Yun), G.S. and L.Y. (Liting Yang). All authors have read and agreed to the published version of the manuscript.

Funding: The authors declare that this study received funding from the Third Xinjiang Scientific Expedition Program (Grant No. 2021xjkk0604) and the Tianshan Talent Training Program of Xinjiang. The funder was not involved in the study design, collection, analysis, interpretation of data, writing of this article, or decision to submit it for publication.

Data Availability Statement: Because the project is not finalized, a link to the data has not been made public.

Acknowledgments: Thank you to our Fisheries Department colleagues and Yingcheng Studio.

Conflicts of Interest: All authors of this manuscript have no conflicts of interest to declare.

References

1. Medeiros, A.S.; Milošević, D. Progress in understanding the vulnerability of freshwater ecosystems. *Sci. Prog.* **2023**, *106*, 1–11. [CrossRef]
2. Tongnunui, S.; Beamish, F.W.H.; Sooksawat, T.; Wattanakornsiri, A.; Chotwiwatthanakun, C.; Supiwong, W.; Intacharoen, P.; Sudtongkong, C. Temporal Changes in Water Quality with Increasing Ambient Temperatures Affect the Distribution and Relative Abundance of 10 Species of Balitorid Fishes in Small Streams of Eastern Thailand. *Water* **2023**, *15*, 2791. [CrossRef]
3. Premke, K.; Wurzbacher, C.; Felsmann, K.; Fabian, J.; Taube, R.; Bodmer, P.; Attermeyer, K.; Nitzsche, K.N.; Schroer, S.; Koschorreck, M.; et al. Large-scale sampling of the freshwater microbiome suggests pollution-driven ecosystem changes. *Environ. Pollut.* **2022**, *308*, 119627. [CrossRef] [PubMed]
4. Li, X.; Jia, P.; Niu, C.; Xing, Y.; Liu, H.; Li, H.; Tang, W.; Zhao, Y. Species diversity of freshwater fish and assessment on watershed health in the Irtysh River and Ulungur River basins in Xinjiang, China. *Biodivers. Sci.* **2020**, *28*, 422–434. [CrossRef]

5. Gao, F.; He, B.; Xue, S.; Li, Y. Impact of landscape pattern change on runoff processes in catchment area of the Ulungur River Basin. *Water Supply* **2020**, *20*, 1046–1058. [CrossRef]
6. Ji, F.; Ma, X.; Qiu, L.; Kang, Z.; Shen, J. Quantifying the effects of introduced Bighead Carp (Cyprinidae; Aristichthys nobilis) stocking on dominant fish species in the Ulungur Lake, China. *Biol. Invasions* **2022**, *24*, 1253–1265. [CrossRef]
7. Xin, T. Study on Hydrochemical Field and Renewal of Different Water Bodies in Ulungur River Watershed. Master's Thesis, Xi'an University of Science and Technology, Xi'an, China, 2021. [CrossRef]
8. Yuan, Y. Research progress on the biological evaluation of freshwater water quality by Macrobenthic fauna. *Bull. Biol.* **2014**, *49*, 1–4.
9. Duan, X.; Wang, Z.; Xu, M. *Benthic Macroinvertebrate and Application in the Assessment of Stream Ecology*; Tsinghua University Press: Beijing, China, 2010; ISBN 978-7-302-21860-9.
10. Shen, H.; Shi, P.; Wu, Y.; Li, Y.; Huang, X.; Li, H.; Qiu, W.; Xie, P. Macrozoobenthos community structure and water quality bioassessment of the Liangtang River. *Acta Hydrobiol. Sin.* **2016**, *40*, 203–210.
11. Zhang, Y.; Peng, W.; Peng, S.; Zhang, M.; Zhang, H.; Xie, Y.; Ge, J.; Yu, Y.; Qu, X. Temporal-spatial distribution and ecological evaluation of macroinvertebrate functional feeding groups in Yongding River Basin. *Chin. J. Appl. Ecol.* **2022**, *33*, 3433–3440. [CrossRef]
12. Kalyoncu, H.; Yorulmaz, B.; Barlas, M.; Yildirim, M.Z.; Zeybek, M. Water Quality of Aksu Stream and Effect of Physicochemical Parameters on the Macroinvertebrate Diversity. *Sci. Eng. J. Firat Univ.* **2008**, *20*, 23–33.
13. Kazanci, N.; Türkmen, G.; Basoren, O.; Gültutan, Y.; Ekingen, P.; Öz, B. A research on water quality of Kelkit Stream using benthic macroinvertebrates and physicochemical variables. *Rev. Hydrobiol.* **2008**, *1*, 145–160.
14. Li, F.; Ma, Y.; Song, X.; Li, S.; Zhang, X.; Wang, X.; Wang, T.; Sun, Z. Community structure and ecological quality assessment of macrobenthos in the coastal sea areas of Northern Yantai, China. *Front. Mar. Sci.* **2022**, *9*, 989034. [CrossRef]
15. Xiong, J.; Mei, X.; Hu, C. Comparative Study on the Community Structure and Biodiversity of Zoobenthos in Lakes of Different Pollution States. *J. Lake Sci.* **2003**, *15*, 160–168.
16. Muluye, T.; Mengistou, S.; Fetahi, T. Assessing the ecological health of the upper and middle Awash River, Ethiopia, using benthic macroinvertebrates community structure and selected environmental variables. *Environ. Monit. Assess.* **2023**, *196*, 45. [CrossRef]
17. Wang, B. Water Quality Bioassessment Using Benthic Macroinvertebrates. Ph.D. Thesis, Nanjing Agricultural University, Nanjing, China, 2003.
18. Gezie, A.; Goshu, G.; Tiku, S. Ecological effect of a small dam on the macroinvertebrate assemblage and water quality of Koga River, Northwest Ethiopia. *Heliyon* **2023**, *9*, e17285. [CrossRef]
19. Tan, Y. Study on Benthic Ecology in the E-Er-Qi-Si River Basin and the Adjacent Inland River Basin. Master's Thesis, Shanghai Ocean University, Shanghai, China, 2010.
20. Liu, Y. *Economic Zoology of China: Freshwater Molluscs*; Science Press: Beijing, China, 1979; ISBN 9787100000000.
21. Wang, J.; Wang, X. *Chironomid Larvae of Northern China*; China Yanshi Press: Beijing, China, 2011; ISBN 7-80250-200-0.
22. Zhang, C.; Zhang, T.; Zhu, T.; Li, W.; Xie, Z. Community Structure of Macrozoobenthos and Its Relationship with Environmental Factors in Lake Hongze. *J. Hydrol.* **2012**, *33*, 27–33. [CrossRef]
23. Zhao, Y.; Shen, J.; Wang, T.; Li, H.; Wang, J.; Zhang, Y.; Ma, X. Biological Assessment of Water Quality of Ulungur Lake Using Benthic Macroinvertebrates. *J. Hydrol.* **2010**, *31*, 7–11. [CrossRef]
24. Hilsenhoff, W.L. Rapid Field Assessment of Organic Pollution with a Family-Level Biotic Index. *J. N. Am. Benthol. Soc.* **1988**, *7*, 65–68. [CrossRef]
25. Elliott, J.M.; Hellawell, J.M. Biological Surveillance of Rivers: A Biological Monitoring Handbook. *J. Anim. Ecol.* **1978**, *47*, 1030–1031. [CrossRef]
26. Gen, S.; Qu, X.; Zhang, Y.; Lin, K. Comparison and Application of Biological Indices of Macroinvertebrates in River Health Assessment. *Environ. Sci.* **2012**, *33*, 2281–2287. [CrossRef]
27. Liu, F.; Zi, F.; Wang, X.; Zeng, H.; Huo, B.; Wang, C.; Ge, J.; Chen, S.; Wang, B. Assessing the Impact of Anthropic Pressures on Aquatic Macroinvertebrates: A Functional Trait Approach in the Irtysh River Watershed. *Biology* **2023**, *12*, 1315. [CrossRef] [PubMed]
28. Kumar, H.; Naim, N.; Ali, D. Geochemical Survey and Estimation of Baseline Concentrations of Major and Trace Elements in Stream Water, Uttar Pradesh, India: Implication for Environmental Studies. *J. Geol. Soc. India* **2023**, *99*, 723–732. [CrossRef]
29. Calapez, A.R.; Serra, S.R.Q.; Rivaes, R.; Aguiar, F.C.; Feio, M.J. Influence of river regulation and instream habitat on invertebrate assemblage' structure and function. *Sci. Total Environ.* **2021**, *794*, 148696. [CrossRef]
30. Calapez, A.R.; Serra, S.R.Q.; Santos, J.M.; Branco, P.; Ferreira, T.; Hein, T.; Brito, A.G.; Feio, M.J. The effect of hypoxia and flow decrease in macroinvertebrate functional responses: A trait-based approach to multiple-stressors in mesocosms. *Sci. Total Environ.* **2018**, *637–638*, 647–656. [CrossRef]
31. Calapez, A.R.; Branco, P.; Santos, J.M.; Ferreira, T.; Hein, T.; Brito, A.G.; Feio, M.J. Macroinvertebrate short-term responses to flow variation and oxygen depletion: A mesocosm approach. *Sci. Total Environ.* **2017**, *599–600*, 1202–1212. [CrossRef]
32. Jolejole, M.E.; Cayetano, M.G.; Magbanua, F.S. Responses of benthic macroinvertebrate communities in tropical Asian streams passing through an industrial zone. *Chem. Ecol.* **2021**, *37*, 399–418. [CrossRef]
33. Wang, J.; Zhou, Q.; Xie, C.; Li, J.; Wei, L. The community structure of Macrozoobenthos and biological assessment of water quality in the Irtysh River of Xinjiang. *Chin. J. Ecol.* **2014**, *33*, 2420–2428. [CrossRef]

34. Vdovina, N.O.; Bezmaternykh, M.D. Peculiarities of Macrozoobenthos in Lakes of Different Mineralization of the Southern Section of the Ob-Irtysh Interfluve. *Hydrobiol. J.* **2016**, *52*, 65–73. [CrossRef]
35. Zhang, Y.; Yu, H.; Liu, M.; Liu, J.; Dong, W.; Xu, T.; Wang, Y.; Guo, Y. Analysis of Water Quality and Habitat Suitability for Benthic Macro-Invertebrates in the Majiagou Urban River, China. *Water* **2023**, *15*, 2269. [CrossRef]
36. Goel, A.; Sharma, J.; Durgapal, N.C. Assessment of biological water quality of river ganga in Patna (India) using benthic macro-invertebrates. *Appl. Biol. Res.* **2021**, *23*, 136–146. [CrossRef]
37. Zheng, Z.; Zhang, K.; Huo, B.; Shen, J.; Zhou, Q.; Wang, J. Community structure of macroinvertebrates after ecological water transportation in lower reaches of Tarim River Basin. *J. Huazhong Agric. Univ.* **2023**, *42*, 206–214. [CrossRef]
38. Yu, S.; Ahan, J.; Zhang, Z.; Li, K.; Sun, D.; Yang, H. Macrobenthos communities in different habitats of the Ili River in Xinjiang and their influencing factors. *Acta Hydrobiol. Sin.* **2017**, *41*, 1062–1070.
39. Elsa, G.; Vanesa, M.; Diego, J.D.G. Long-term Macrobentos Responses to Environmental Flows and Water Quality Improvement along River Jarama. *Ecohydrol. Hydrobiol.* **2023**, *23*, 304–315. [CrossRef]
40. Asmamaw, M.; Mereta, S.T.; Ambelu, A. Response of macroinvertebrates to changes in stream flow and habitat conditions in Dinki watershed, central highlands of Ethiopia. *Ecol. Indic.* **2021**, *133*, 108448. [CrossRef]
41. Nurlan, H. Hydrological Characteristics in the Ulungur River Basin. *Arid Zone Res.* **2014**, *31*, 798–802. [CrossRef]
42. Belmar, O.; Bruno, D.; Guareschi, S.; Mellado-Díaz, A.; Millán, A.; Velasco, J. Functional responses of aquatic macroinvertebrates to flow regulation are shaped by natural flow intermittence in Mediterranean streams. *Freshw. Biol.* **2019**, *64*, 1064–1077. [CrossRef]
43. Theodoropoulos, C.; Vourka, A.; Stamou, A.; Rutschmann, P.; Skoulikidis, N. Response of freshwater macroinvertebrates to rainfall-induced high flows: A hydroecological approach. *Ecol. Indic.* **2017**, *73*, 432–442. [CrossRef]
44. Zemo, M.A.T.; Menbohan, S.F.; Atchrimi, B.T.; Betsi, W.C.N.; Nwaha, M.; Dzavi, J.; Mavunda, C.A.; Lactio, N. Effect of Anthropogenic Pressure on the Biodiversity of Benthic Macroinvertebrates in Some Urban Rivers (Yaoundé). *Water* **2023**, *15*, 2383. [CrossRef]
45. Federico, S.; Manuel, C.; Lenin, V.; Edison, S.; Pablo, C. Determination of the ecological water quality in the Orienco stream using benthic macroinvertebrates in the Northern Ecuadorian Amazon. *Integr. Environ. Assess. Manag.* **2022**, *19*, 615–625. [CrossRef]
46. Burlibayeva, M.D.; Burlibayev, Z.M.; Opp, C.; Bao, A. Regime dynamics of hydrochemical and toxicological parameters of the Irtysh River in Kazakhstan. *J. Arid Land* **2016**, *8*, 521–532. [CrossRef]
47. Wang, D.; Zhang, S.; Wang, G.; Liu, Y.; Wang, H.; Gu, J. Reservoir Regulation for Ecological Protection and Remediation: A Case Study of the Irtysh River Basin, China. *Int. J. Environ. Res. Public Health* **2022**, *19*, 11582. [CrossRef]
48. Ivan, R.; Linus, Z.; Daulet, A.; Gulira, M.; Kamshat, T. A state-of-the-art and future perspectives of transboundary rivers in the cold climate—A systematic review of Irtysh River. *J. Hydrol. Reg. Stud.* **2022**, *42*, 101173. [CrossRef]
49. Fedorova, L.I.; Kaygorodova, I.A. First data on the Hirudinea fauna of lotic ecosystems of the Khanty-Mansi Autonomous Area (Russia). *ZooKeys* **2022**, *1082*, 73–85. [CrossRef]
50. Nicolì, D.; Bellucci, A.; Iovino, D.; Ruggieri, P.; Gualdi, S. The impact of the AMV on Eurasian summer hydrological cycle. *Sci. Rep.* **2020**, *10*, 14444. [CrossRef] [PubMed]
51. Standring, W.J.; Oughton, D.H.; Salbu, B. Potential remobilization of 137Cs, 60Co, 90Tc, and 90Sr from contaminated Mayak sediments in river and estuary environments. *Environ. Sci. Technol.* **2002**, *36*, 2330–2337. [CrossRef] [PubMed]
52. Guay, C.K.H.; Zhulidov, A.V.; Robarts, R.D.; Zhulidov, D.A.; Gurtovaya, T.Y.; Holmes, R.M.; Headley, J.V. Measurements of Cd, Cu, Pb and Zn in the lower reaches of major Eurasian arctic rivers using trace metal clean techniques. *Environ. Pollut.* **2010**, *158*, 624–630. [CrossRef] [PubMed]
53. Baryshev, I.A. Evaluating Water Quality in a Large River System in the Northern European Russia by Macrozoobenthos. *Water Resour.* **2021**, *48*, 774–781. [CrossRef]

water

MDPI

Article

Analyses of Morphological Differences between Geographically Distinct Populations of *Gymnodiptychus dybowskii*

Linghui Hu [1,†], Na Yao [2,†], Chengxin Wang [2], Liting Yang [2], Gulden Serekbol [1], Bin Huo [3], Xuelian Qiu [2], Fangze Zi [2], Yong Song [2] and Shengao Chen [2,*]

1 College of Animal Science and Technology, Tarim University, Alar 843300, China; 10757222068@stumail.taru.edu.cn (L.H.); 10757231098@stumail.taru.edu.cn (G.S.)

2 College of Life Science and Technology, Tarim Research Center of Rare Fishes, Tarim University, Alar 843300, China; yaona0212@hotmail.com (N.Y.); 10757203067@stumail.taru.edu.cn (C.W.); 10757231127@stumail.taru.edu.cn (L.Y.); 10757231124@stumail.taru.edu.cn (X.Q.); 10757213076@stumail.taru.edu.cn (F.Z.); songyongdky@126.com (Y.S.)

3 College of Fisheries, Huazhong Agricultural University, Wuhan 430070, China; huobin@mail.hzau.edu.cn

* Correspondence: shengao@taru.edu.cn; Tel.: +86-136-5757-6029

† These authors contributed equally to this work.

Abstract: To study the morphological differences between and the evolutionary mechanisms driving the differentiation of geographically distinct populations of *Gymnodiptychus dybowskii*, 158 fish were collected from the Turks River and the Manas River in Xinjiang from 2020 to 2021 with the approval of the Academic Ethics Committee. The morphological characteristics of the fish were assessed using classical fish ecology methods such as traditional morphometric measurements and the framework approach. The results showed that the morphological characteristics of the populations in the Turks River and Manas River were significantly different; a one-way ANOVA revealed 22 highly significant differences ($p < 0.01$) and 1 significant difference ($p < 0.05$) among the 33 morphological traits of the observed populations, and a principal component analysis revealed that there was no overlap between the two populations of *G. dybowskii*. The main characteristics associated with principle component 1 were the terminus of the dorsal fin to the ventral origin of the caudal fin (D—F), the dorsal origin of the caudal fin to the origin of the anal fin (E—H), and the insertion of the pectoral fin to the terminus of the pectoral fin (J—K); the main factors associated with principal component 2 were the body height (BD), the terminus of the dorsal fin to the insertion of the pelvic fin (D—I), the caudal peduncle height (CPH), and the tip of the snout to the last end of the frontal maxilla (A—B); and the main traits associated with principle component 3 were the terminus of the anal fin to the origin of the anal fin (G—H), the body width (BW), the insertion of the pelvic fin to the terminus of the pelvic (I—L), the insertion of the pectoral fin to the terminus of the pectoral fin (J—K), and the insertion of the pelvic fin to the insertion of the pectoral fin (I—J). An OPLS-DA revealed that the two populations could be wholly separated and that the intergroup growth traits of the Manas River population were different and significantly greater than those of the Turks River population. The discriminant functions of the Turks River and Manas River populations of *G. dybowskii* were as follows: $Y_T = -432.033 + 1787.748X_1 + 826.517X_2 + 249.002X_3 + 1183.050X_4 + 554.934X_5 + 999.296X_6 + 627.428X_7$; $Y_M = -569.819 + 2041.044X_1 + 344.942X_2 + 333.737X_3 + 940.512X_4 + 348.222X_5 + 1167.770X_6 + 1015.904X_7$. According to a coefficient of variation analysis, a total of nine traits, namely, EI/BL, C-D/BL, E-F/BL, F-H/BL, H-I/BL, C-J/BL, D-I/BL, D-H/BL, and D-F/BL, had a CD > 1.28, indicating that the differences in these nine traits had reached the subspecies level. The results showed that *G. dybowskii* significantly differed between the two geographically distinct populations in the Turks River and the Manas River and have differentiated to the subspecies level. This study provides a basis for a better investigation of the population structure of highland endemic fishes and the mechanisms by which they diverged and lays a foundation for developing and utilizing germplasm resources from endemic fishes in Xinjiang.

Keywords: *Gymnodiptychus dybowskii*; biological characteristics; morphological differences

1. Introduction

Gymnodiptychus dybowskii (Kessler, 1874) is a carp species belonging to Cypriniformes, Cyprinidae, Schizothoracinae, *Gymnodiptychus* (Herzenstein, 1892) [1]. Schizothoracine fishes originated from the primitive Barbinae subfamily of fishes on the Ti-betan Plateau [2]. After continuous differentiation, they evolved into a specialized group [3,4]; they have unique adaptations to special environments, such as high altitude, hypoxia, and low water temperature [5], and have important economic and scientific value [6]. In recent years, its ecological niche has been disturbed due to human activities [7–9] and the introduction of exotic species [1,10], which has triggered food competition and reproductive competition, leading to a very significant decline in its population. It has been listed as Class I key protected aquatic wildlife in the Xinjiang Uygur Autonomous Region since 2004 (Xinjiang Uygur Autonomous Region Key protected Wildlife List (Revised), Xinzhengfa (2022) No. 75). Now that *G. dybowskii* has become a rare and endangered indigenous fish in Xinjiang, the conservation of *G. dybowskii* is urgently needed, and studies related to its different geographic populations will provide crucial assistance for long-term biodiversity conservation.

G. dybowskii is mainly distributed in China in the waters of the Ili River basin and the Junggar Basin on the northern slopes of the Tianshan Mountains [1]. The Turks River is one of the largest tributaries of the Ili River. It originates from the northern slope of the central peak of the Tianshan Mountain, i.e., Khan Tengri Peak, and eventually flows into the Ili River; it has a length of approximately 408 km, an elevation of 800~4600 m, and an average annual runoff of 79.33×10^8 m^3, accounting for more than 60% of the annual runoff of the Ili River [10]. The Manas River originates from the northern Tianshan Mountains. The Manas River, which has a length of approximately 450 km, originates from the northern foothills of the Tianshan Mountains and eventually flows into Manas Lake in the north. It is an inland river with the most extensive glacier extent in the Junggar inland area, with an elevation of 256 m to 5242 m in the watershed and an average annual runoff of 11.90×10^8 m^3 [11]. The geographic differences between the Turks and Manas River basins are significant, and relevant studies have found that populations with significant geographic differences produce greater genetic variation [12]. Fish from different geographic populations can trigger changes in their morphology due to differences in factors such as water flow, predators, habitat modification, and food availability [13–15]. Studies to date have focused on comparing differences in fish morphology between reservoirs or lakes and rivers [16–18]. Fewer studies have analyzed morphological differences in fish populations in rivers from different catchments at high altitudes and *G. dybowskii* has been found to have a high degree of variability in its morphology across geographic populations in natural surveys. Therefore, it would be important to carry out studies related to morphological differences between the populations of the Turks River and the Manas River.

In this study, we collected samples of *G. dybowskii* from the Turks and Manas Rivers. We analyzed the morphological characteristics and quantitative traits of these fishes to identify the characteristics and adaptive evolutionary mechanisms of specialized *G. dybowskii* fishes in high-elevation areas as well as the effects of geographic isolation on the formation of these species. The aim is to provide a scientific basis for the evolution of highland fish fauna and the conservation of the germplasm.

2. Materials and Methods

2.1. Materials

From 2020 to 2021, 123 *G. dybowskii* (body length: 50.08–121.99 mm, 82.29 ± 13.76 mm; weight: 1.86–24.29 g, 8.32 ± 4.16.04 g) were collected from the Turks River (T, E80°55′ N42°81′) using gillnets and ground cages (2a = 2.0 cm), and 35 *G. dybowskii* (body length: 52.36–273.60 mm, 195.36 ± 11.36 mm; body weight: 2.20–340.00 g, 125.82 ± 15.13 g) were collected from the Manas River (M, E86°13′ N44°30′), Sampling locations are shown in Figure 1. After biological determination on site, the sampled fish were fixed in formalin and returned to the laboratory for subsequent processing. Permission for sampling and collection was provided by an aquatic wildlife license of the People's Republic of China (approval number: (new) Mizuno Catch Word [2019] No. 4). All experimental protocols were approved by the Ethics Committees of the Tarim University of Technology (approval code: TDDKYXF 20200426, approval date: 26 April 2020) and adhered to animal welfare laws, guidelines, and policies.

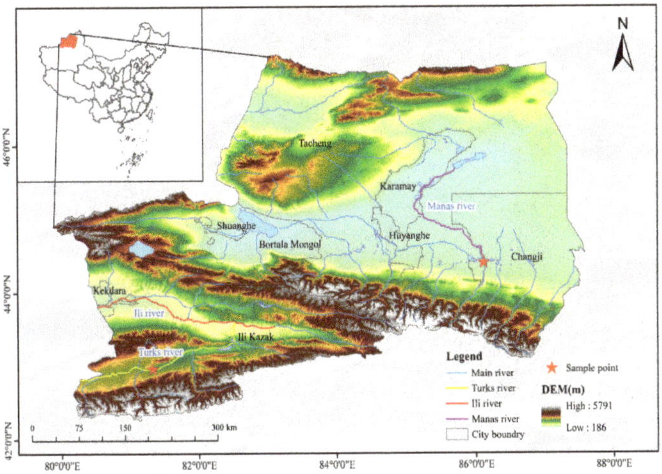

Figure 1. Sampling sites for *G. dybowskii*.

2.2. Methodology

2.2.1. Morphological Indicators

Traditional morphological measurements and the frame method were used to measure the total length (TL), body length (BL), body height (BD), body width (BW), head length (HL), snout length (SL), and eye diameter (ED). Eye interval (EI), caudal peduncle length (CPL), and caudal peduncle height (CPH) were also measured [19,20] (Figure 2).

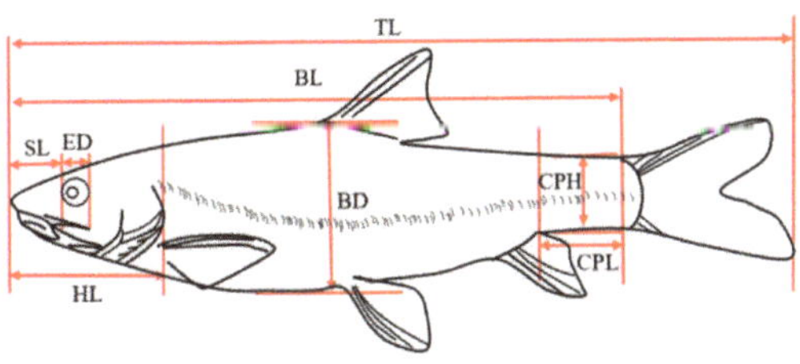

Figure 2. Traditional morphological measurements of *G. dybowskii*. TL: total length; BL: body length; BD: body depth; HL: head length; SL: snout length; ED: eye diameter; CPL: caudal peduncle length; CPH: caudal peduncle height.

The frame measurements were determined by 12 anatomical coordinate points [21] (Figure 3), and the 24 frame distances were as follows: tip of snout to the last end of the frontal maxilla (A—B), tip of snout to origin of dorsal fin (A—C), tip of snout to insertion of pelvic fin (A—I), tip of snout to insertion of pectoral fin (A—J), the last end of the frontal maxilla to origin of dorsal fin (B—C), the last end of the frontal maxilla to insertion of pelvic fin (B—I), the last end of the frontal maxilla to insertion of pectoral fin (B—J), origin of dorsal fin to terminus of dorsal fin (C—D), origin of dorsal fin to ventral origin of caudal fin (C—F), origin of dorsal fin to origin of anal fin (C—H), origin of dorsal fin to insertion of pelvic fin (C—I), origin of dorsal fin to Insertion of pectoral fin (C—J), terminus of dorsal fin to dorsal origin of caudal fin (D—E), terminus of dorsal fin to ventral origin of caudal fin (D—F), terminus of dorsal fin to origin of anal fin (D—H), terminus of dorsal fin to insertion of pelvic fin (D—I), dorsal origin of caudal fin to ventral origin of caudal fin (E—F), dorsal origin of caudal fin to origin of anal fin (E—H), ventral origin of caudal fin to origin of anal fin (F—H), terminus of anal fin to origin of anal fin (G—H), origin of anal fin to insertion of pelvic fin(H—I), insertion of pelvic fin to insertion of pectoral fin (I—J), insertion of pelvic fin to terminus of pelvic (I—L) and insertion of pectoral fin to terminus of pectoral fin (J—K).

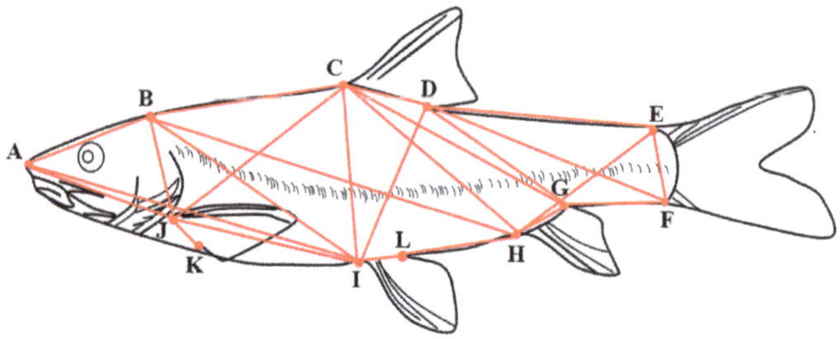

Figure 3. Frame structure of *G. dybowskii*. A. Tip of the snout; B. end of the frontal maxilla; C. origin of the dorsal fin; D. terminus of the dorsal fin; E. dorsal origin of the caudal fin; F. ventral origin of the caudal fin; G. terminus of the anal fin; H. origin of the anal fin; I. insertion of the pelvic fin; J. insertion of the pectoral fin; K. terminus of the pectoral fin; L. terminus of the pelvic fin; L. terminus of the pelvic fin; L. terminus of the pectoral fin; L. terminus of the pelvic fin.

A digital Vernier caliper (CD67-S15PS) was used to measure the length (accuracy: 0.01 mm), and an electronic balance (LE403E) was used to measure the weight (accuracy: 0.01 g).

2.2.2. Morphological Analysis

Based on the experience of Xie Congxin [22] and Guo Yan [1] et al., the color of the fish body was compared and evaluated by visual observation and a color contrast card test, and the error of color observation was reduced.

One-way ANOVA (at the 0.05 significance level) was used to identify the significance of differences between geographically distinct populations. The least significant difference (LSD) method was used for variables that were significant according to the ANOVA after multiple comparisons, and Tamhane's T2 method was used for variables that were not significant according to the ANOVA [23].

Principal component analysis (PCA) was used to extract the principle components from the covariance matrix. Contribution rates were calculated, and principal component scatter plots were constructed [24]. Orthogonal partial least squares discriminant analysis (OPLS-DA) was used to further distinguish differences among the data, thus improving the confidence level [25].

Discriminate analysis was used to assess all the proportional data [19]. The discrimination formula was established by identifying the parameters that played a significant role in determining the morphological differences among populations, and the discrimination accuracy and comprehensive discrimination rate were calculated as follows:

$$\text{Discrimination accuracy (\%)} = (\text{Ai}/\text{Bi}) \times 100\%$$

$$\text{Combined discrimination rate (\%)} = \sum_{i=1}^{k} \text{Ai} / \sum_{i=1}^{k} \text{Bi}$$

where k is the number of populations, Ai is the number of correctly discriminated individuals in the ith population, and Bi is the number of measured individuals in the ith population.

The extent of the morphological variation among the geographically isolated populations was analyzed by the coefficient of difference (CD) [26], which determines whether the differences among populations are due to geographic isolation within a species and is given by the following formula:

$$CD = (M1 - M2)/(SD1 + SD2)$$

where M1 and M2 are the standardized means of a parameter for populations 1 and 2, respectively, and SD1 and SD2 are the standard deviations of a parameter for populations 1 and 2, respectively.

2.3. Data Processing

The 34 sets of morphological data were standardized by dividing by body length to eliminate the effect of individual size differences on the morphological parameters, resulting in 33 sets of measured ratio data.

The data were statistically analyzed using SPSS 17.0 and ORIGIN 9.0 and are expressed as the mean ± standard deviation (M ± SD).

3. Results

3.1. Characteristics of *G. dybowskii*

G. dybowskii fish body: prolonged, slightly laterally compressed. The posterior surface of the head is slightly elevated, the head is conical, the muzzle is obtuse and protruding, the mouth is inferior and horseshoe-shaped, the upper and lower lips are very well developed. The postlabial groove is interrupted, 1 pair of jaw whiskers, gill rakers are short and sparse, and the gill covert membrane is connected to the isthmus. Most of the body is lacking scales, appearing only on the shoulders, buttock, and laterally; the upper part of the ventral fins have axillary scales.

Turks River population (Figure 4 (T)): a grayish-white abdomen from the waist up to the lateral line, and from the lateral line to the dorsum, the color fades from light brown to dark brown; the dorsal spine is partly dark brown to partially black, with black spots varying from small to large between the lateral line to the dorsum.

Figure 4. Morphological characteristics of *G. dybowskii* (T: Turks River population; M: Manas River population).

Manas River population (Figure 4 (M)): a creamy white abdomen from the waist upward to the back, and the color gradually deepens to grayish white; the grayish white of the back fades darker; the lateral line is light brown, and the dorsal spine is dark gray, accompanied by fine black spots.

3.2. One-Way ANOVA

There were highly significant differences ($p < 0.01$) in 22 proportional traits, such as BW/BL, HL/BL, SL/BL, ED/BL, and EI/BL, and significant differences ($p < 0.05$) in E—F/BL between the two populations, with differences mainly occurring in the posterior part of the carapace (Table 1). The proportional traits that are not significantly different are shown in Appendix A Table A1.

Table 1. One-way ANOVA results for measured traits of *G. dybowskii* from distinct geographical populations (fraction with significant differences).

Traits	T Mean ± SD	M Mean ± SD	p
BW/BL	0.139 ± 0.012	0.135 ± 0.019	<0.001 **
HL/BL	0.253 ± 0.011	0.228 ± 0.022	0.007 **
SL/BL	0.075 ± 0.007	0.080 ± 0.012	<0.001 **
ED/BL	0.055 ± 0.006	0.036 ± 0.010	<0.001 **
EI/BL	0.074 ± 0.006	0.081 ± 0.007	<0.001 **
CPL/BL	0.127 ± 0.016	0.173 ± 0.046	<0.001 **
A—B/BL	0.201 ± 0.010	0.187 ± 0.017	<0.001 **
B—C/BL	0.267 ± 0.014	0.274 ± 0.046	<0.001 **
C—D/BL	0.127 ± 0.010	0.110 ± 0.016	<0.001 **
D—E/BL	0.331 ± 0.029	0.403 ± 0.085	<0.001 **
E—F/BL	0.071 ± 0.006	0.114 ± 0.060	0.03 *
F—H/BL	0.205 ± 0.012	0.252 ± 0.037	<0.001 **
H—I/BL	0.195 ± 0.022	0.205 ± 0.029	<0.001 **
A—C/BL	0.455 ± 0.025	0.457 ± 0.030	<0.001 **
C—I/BL	0.180 ± 0.014	0.194 ± 0.022	<0.001 **
C—H/BL	0.319 ± 0.026	0.319 ± 0.045	<0.001 **
C—F/BL	0.482 ± 0.036	0.551 ± 0.031	<0.001 **
D—I/BL	0.144 ± 0.013	0.156 ± 0.026	<0.001 **
D—H/BL	0.202 ± 0.013	0.225 ± 0.025	<0.001 **
D—F/BL	0.358 ± 0.017	0.440 ± 0.026	<0.001 **
E—H/BL	0.227 ± 0.013	0.290 ± 0.016	<0.001 **
J—K/BL	0.037 ± 0.005	0.093 ± 0.062	<0.001 **
I—L/BL	0.043 ± 0.005	0.077 ± 0.050	0.004 **

Note: Superscript "*" indicates a significant difference ($p < 0.05$), and superscript "**" indicates a highly significant difference ($p < 0.01$).

3.3. Principal Component Analysis

The KMO test (KMO = 0.767) and Bartlett's test of sphericity ($p < 0.01$) were performed. The KMO test value was 0.767 (>0.600), and the correlation between the variables was strong, which indicated that the morphological differences between the geographically distinct groups of *G. dybowskii* were suitable for the principal component analysis. As shown in Figure 5, after the third principal component, the eigenvalues of the principle components tended to stabilize. Therefore, the first three principal components were selected for analysis (PC1~PC3).

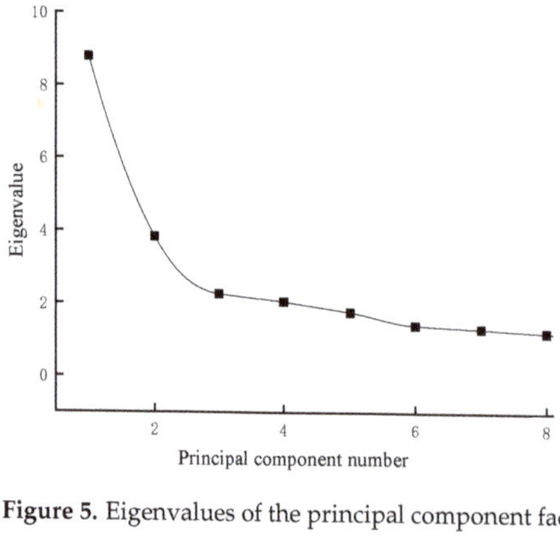

Figure 5. Eigenvalues of the principal component factors.

The contribution rates of these components were 25.9%, 11.50% and 6.70%, respectively, with a cumulative contribution rate of 44.10%. The main traits associated with principal component 1 were D—F, E—H, and J—K; those associated with principle component 2 were BD, D—I, CPH, and A—B; and those associated with principle component 3 were H—G, BW, I—L, J—K, and I—J. The first three principal components were associated with all of the measured traits of *G. dybowskii* (Table 2).

Table 2. Principal component analysis of *G. dybowskii* from geographically distinct populations.

Traits	Principal Components		
	PC1	PC2	PC3
TL/BL	0.0269	0.1811	0.2311
BD/BL	−0.0677	0.3448	−0.2049
BW/BL	0.0066	0.2441	−0.3475 [a]
HL/BL	0.2301	0.2124	0.0575
SL/BL	−0.0403	0.1062	−0.1167
ED/BL	0.2400	0.1326	0.1405
EI/BL	−0.1731	0.1897	−0.0684
CPL/BL	−0.2145	−0.0778	0.1633
CPH/BL	0.0750	0.3008 [a]	−0.0511
A—B/BL	0.1543	0.2680 [a]	0.2055
B—C/BL	−0.0835	0.0090	0.1412
C—D/BL	0.1552	0.1579	−0.1804
D—E/BL	−0.1876	−0.0088	−0.0190
E—F/BL	−0.1557	−0.0788	−0.1160
F—H/BL	−0.2374	−0.0484	−0.0071
H—I/BL	−0.0899	−0.0619	−0.0867
I—J/BL	−0.1884	0.1049	−0.2560 [a]
A—J/BL	0.2092	0.2177	0.2051
A—I/BL	−0.0276	0.2177	0.0015
B—J/BL	−0.0251	0.1918	0.1029
B—I/BL	−0.1060	0.2287	0.0059
A—C/BL	−0.0422	0.1484	0.0698
C—J/BL	−0.0631	−0.0088	−0.0920
C—I/BL	−0.1599	0.2499	−0.0863
C—H/BL	0.0029	0.1119	−0.1902
C—F/BL	−0.2216 [a]	−0.0659	−0.1287
D—I/BL	−0.1699	0.3295 [a]	−0.0890
D—H/BL	−0.2305	0.1111	−0.0782
D—F/BL	−0.2977	−0.0611	−0.0914
E—H/BL	−0.2821 [a]	−0.0496	−0.0523
J—K/BL	−0.2679 [a]	0.0541	0.3061 [a]
I—L/BL	−0.2425	0.0797	0.3426 [a]
H—G/BL	−0.2288	0.0614	0.3476 [a]
Eigenvalue	8.80	3.92	2.27
Percentage of Variance (%)	25.9	11.5	6.7
Cumulative (%)	25.9	37.4	44.1

Note: Superscript "a" indicates a load that exceeds 0.2500.

A three-dimensional scatter plot based on principal components 1, 2, and 3 (Figure 6) revealed no overlap between the geographically distinct populations. Each trait in the Turks River population was more compact and less discrete. In contrast, all the traits in the Manas River population were more dispersed, discrete, and significant than those in the Turks River population. Therefore, it was hypothesized that the Manas River population would exhibit more significant interspecific variation than the Turks River population, as assessed by an OPLS-DA.

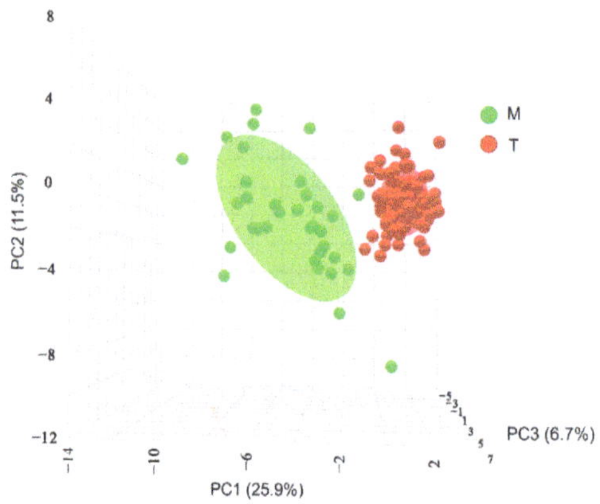

Figure 6. Three-dimensional spatial distribution of *G. dybowskii* from geographically distinct populations; PC1, PC2, and PC3 are principal components 1, 2, and 3, respectively; T is the population in the Turks River, and M is the population in the Manas River.

3.4. Orthogonal Partial Least Squares Discriminant Analysis (OPLS-DA)

Further assessment by an OPLS-DA (Figure 7) revealed that the Turks River population exhibited less intraspecific variation and more overlap than the Manas River population. In contrast, the Manas River population had more distinct intraspecific variation and almost no overlap between traits, and the Turks River population could be separated entirely from the Manas River population, which was consistent with the results of the 3D scatterplot of the principal components.

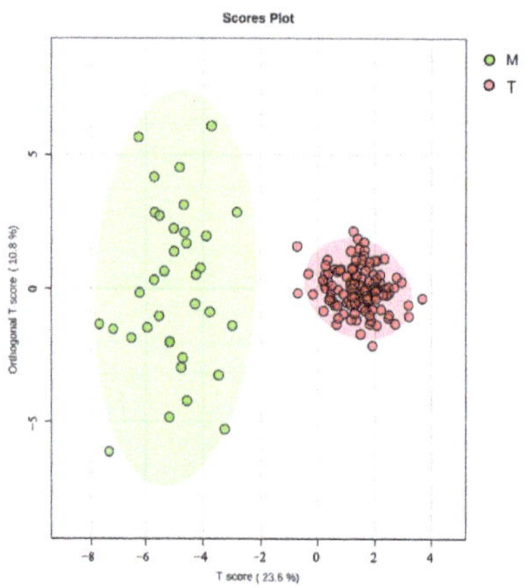

Figure 7. OPLS-DA of *G. dybowskii* from geographically distinct populations; T is the population from the Turks River, and M is the population from the Manas River; the vertical coordinate "Orthogonal T Score" indicates the orthogonal principal components test score, which responds to within-group differences. The horizontal coordinate "T Score" indicates the predictive principal component test score, which reflects the difference between groups.

3.5. Discriminant Analysis

The stepwise discriminant method was used to discriminate 33 morphometric proportional traits between the two geographically distinct populations, and seven eigenvalues with significant discriminant contributions were selected: SL/BL, ED/BL, CPL/BL, CPH/BL, CD/BL, DF/BL, and EH/BL. These identifiers were replaced by *X*1–*X*7, respectively, and discriminant formulas were established for the Turks River population (Y_T) and the Manas River population (Y_M):

$$Y_T = -432.033 + 1787.748X_1 + 826.517X_2 + 249.002X_3 + 1183.050X_4 + 554.934X_5 + 999.296X_6 + 627.428X_7;$$

$$Y_M = -569.819 + 2041.044X_1 + 344.942X_2 + 333.737X_3 + 940.512X_4 + 348.222X_5 + 1167.770X_6 + 1015.904X_7.$$

The combined discriminant value for both the Turks River population and the Manas River population was 100% (Table 3), indicating that the two populations were separated entirely and highly differentiated. These results are consistent with those of the OPLS-DA.

Table 3. Discriminant analysis of *G. dybowskii* from geographically distinct populations.

Method	Population	Predicted Group Membership		Identification Accuracy/Percent	Discriminant Accuracy/Percent
		T	M		
Stepwise discrimination	T	123	0	100	100
	M	0	35	100	

Note: T is the population from the Turks River, and M is the population from the Manas River.

3.6. Analysis of the Coefficient of Difference

The coefficients of difference (CD) of nine traits, including EI/BL, C—D/BL, and E—F/BL, exceeded 1.28, with D—F/BL having the highest CD value of 22.7899 and D—H/BL having the next highest value of 8.7849 (Table 4).

Table 4. Analysis of the coefficients of difference of *G. dybowskii* from geographically distinct populations.

Traits	T		M		CD
	Min~Max	Mean ± SD	Min~Max	Mean ± SD	
TL/BL	0.1106–2.2928	1.2354 ± 0.3158	0.9567–1.2606	1.1633 ± 0.0494	0.1490
BD/BL	0.1600–0.2300	0.1772 ± 0.0124	0.1200–0.2500	0.1797 ± 0.0292	−0.4488
BW/BL	0.1100–0.1700	0.1386 ± 0.0116	0.0900–0.1700	0.1354 ± 0.0187	−2.2707
HL/BL	0.2100–0.2700	0.2530 ± 0.0109	0.1400–0.2600	0.2280 ± 0.0219	0.8481
SL/BL	0.0600–0.1000	0.0754 ± 0.0069	0.0500–0.1000	0.0797 ± 0.0120	−4.7870
ED/BL	0.0400–0.0700	0.0554 ± 0.0060	0.0200–0.0600	0.0363 ± 0.0100	12.6230
EI/BL	0.0600–0.0900	0.0737 ± 0.0063	0.0700–0.0900	0.0814 ± 0.0069	1.5063 *
CPL/BL	0.1100–0.1900	0.1269 ± 0.0160	0.0700–0.3700	0.1726 ± 0.0463	−1.3499
CPH/BL	0.0600–0.0800	0.0675 ± 0.0049	0.0400–0.0800	0.0626 ± 0.0085	−1.0644
A—B/BL	0.1800–0.2200	0.2006 ± 0.0098	0.1500–0.2300	0.1874 ± 0.0169	0.2266
B—C/BL	0.2200–0.3000	0.2670 ± 0.0140	0.1800–0.4900	0.2743 ± 0.0462	−2.5602
C—D/BL	0.1000–0.1600	0.1267 ± 0.0097	0.0500–0.1400	0.1097 ± 0.0164	1.3048 *
D—E/BL	0.1000–0.3700	0.3305 ± 0.0289	0.0500–0.4600	0.4034 ± 0.0848	0.7900
E—F/BL	0.0600–0.0900	0.0707 ± 0.0055	0.0600–0.4500	0.1140 ± 0.0604	1.9013 *
F—H/BL	0.1800–0.2400	0.2051 ± 0.0123	0.0900–0.3000	0.2517 ± 0.0368	1.4903 *
H—I/BL	0.0200–0.2300	0.1949 ± 0.0222	0.0900–0.2600	0.2049 ± 0.0289	2.5655 *
I—J/BL	0.2500–0.3200	0.2842 ± 0.0151	0.2400–0.3600	0.3077 ± 0.0243	−8.3643
A—J/BL	0.2200–0.2900	0.2596 ± 0.0126	0.2000–0.2600	0.2371 ± 0.0153	−0.0874
A—I/BL	0.5000–0.5700	0.5367 ± 0.0117	0.3000–0.6000	0.5334 ± 0.0495	−0.3953
B—J/BL	0.1200–0.1600	0.1410 ± 0.0080	0.1200–0.1700	0.1400 ± 0.0106	0.2191
B—I/BL	0.3500–0.4100	0.3742 ± 0.0126	0.1200–0.4500	0.3831 ± 0.0532	0.4773
A—C/BL	0.2300–0.5000	0.4550 ± 0.0255	0.3800–0.5300	0.4571 ± 0.0299	0.2541
C—J/BL	0.1700–0.2800	0.2437 ± 0.0146	0.0100–0.3100	0.2520 ± 0.0473	1.8431 *
C—I/BL	0.1500–0.2800	0.1799 ± 0.0143	0.1500–0.2500	0.1940 ± 0.0220	0.0105
C—H/BL	0.2600–0.5100	0.3187 ± 0.0265	0.0800–0.3600	0.3189 ± 0.0454	−15.8851

Table 4. Cont.

Traits	T		M		CD
	Min~Max	Mean ± SD	Min~Max	Mean ± SD	
C—F/BL	0.1400–0.5500	0.4815 ± 0.0356	0.4400–0.6000	0.5506 ± 0.0312	0.9605
D—I/BL	0.1100–0.1800	0.1439 ± 0.0132	0.1200–0.6000	0.1563 ± 0.0261	1.8586 *
D—H/BL	0.1700–0.2400	0.2020 ± 0.0129	0.1600–0.2700	0.2246 ± 0.0250	8.7849 *
D—F/BL	0.3100–0.4100	0.3580 ± 0.0165	0.3600–0.5100	0.4397 ± 0.0258	22.7899 *
E—H/BL	0.2000–0.2700	0.2274 ± 0.0132	0.2400–0.3200	0.2903 ± 0.0160	0.9998
J—K/BL	0.0300–0.0500	0.0370 ± 0.0054	0.0300–0.1700	0.0934 ± 0.0618	0.7654
I—L/BL	0.0300–0.0500	0.0427 ± 0.0051	0.0200–0.1400	0.0769 ± 0.0498	0.7736
H—G/BL	0.0500–0.1100	0.0880 ± 0.0088	0.0600–0.2900	0.1300 ± 0.0631	0.2706

Note: T is the population from the Turks River, M is the population from the Manas River. SD represents the coefficient of difference, with a superscript "*" indicating that the data suggest the presence of two subspecies (CD > 1.28).

4. Discussion

In the present study, the carapace of *G. dybowskii* from the Turks River gradually deepened to a brownish gray color from the abdomen to the dorsum. The black spots between the lateral line and the dorsum became lighter and darker. In contrast, the carapace of fish from the Manas River population was grayish white, with tiny black dots only on the dorsal ridge, and these fish differed in body color. The morphological characteristics were the same as those reported in the study conducted by Zhang et al. [27]. A total of 22 of the 33 measured morphological traits were highly significantly different according to the one-way ANOVA ($p < 0.01$), and one was significantly different ($p < 0.05$). This finding is similar to the results of a previous study by Pashkov et al. [28], who reported significant differences ($p < 0.05$) in the morphometrics and body colors of geographically distinct populations. Compared with those of other vertebrates, the morphological characteristics of fish are more susceptible to the influence of the environment [29,30]. Morphological differences among the same fish species are closely related to habitat type and quality [31]. Dunn et al. [32] investigated intraspecific variation in the morphological traits of *Galaxias brevipinnis*, *G. gollumoides*, and *G. vulgaris* in different hydrological environments in rapid-flowing and slow-flowing habitats, and Langerhans et al. [33]. compared intraspecies morphological differences between *Bryconops caudomaculatus* and *Biotodoma wavrini* from two habitats in the Neotropical region: a river and a lagoon ($p < 0.05$). The morphological differences in *G. dybowskii* observed in this study were related to the different habitats in which the sampled fish were found.

A principal component analysis (PCA) provides adequate data degradation and feature extraction. Moreover, this approach can reveal the correlation, direction of change, and importance of the data [34,35]. In this study, the difference in the principal components between the populations of *G. dybowskii* from the Turks River and Manas River was noticeable. The samples of the two populations formed distinct groups, suggesting that the two populations were differentiated. The morphological traits associated with the first three principal components were mainly related to the body, tail, height, and head. Dunn et al. reported that fish in fast-flowing habitats tend to have shorter body lengths, slender caudal peduncles, small pointed craniums, and smaller muzzles, while fish in slow-flowing habitats tend to have longer bodies, shorter caudal peduncles, and larger craniums and muzzles [32]. These results are consistent with the results of the present study: the population in the Turks River was characterized by a short body, a slender caudal peduncle, and a small and pointed head, and the water volume in this river is sufficient and the water flow is rapid [10,36]. The population in the Manas River was characterized by having a large and long body, a thick caudal peduncle, and a large and rounded head, and the water flow in this river is gentle compared with that in the Turks River [37].

The PCA combined with the OPLS-DA further differentiated between the groups [38]. This study revealed significant differences between the Turks and Manas River populations of *G. dybowskii*. In this study, we found substantial interspecific differences between the Turks and Manas River populations of *G. dybowskii* by an OPLS-DA, and the results of the OPLS-DA were consistent with those of other studies using discriminant statistical approaches. The morphological differences between the Turks River and Manas River populations of *G. dybowskii* were significant ($p < 0.05$), and the degree of discrimination may suggest the presence of inter-subspecies or interspecies discrimination [39]. The accuracy of the discrimination between the two populations was 100%, indicating that this approach was suitable for discriminating individuals belonging to *G. dybowskii*.

According to the 75% subspecies identification and characterization criteria proposed by Mayr et al., subspecies are differentiated when the coefficient of difference (CD) is greater than 1.28, although the use of the coefficient of variation as a metric for distinguishing between subspecies needs to be considered in conjunction with other biological and biogeographical factors [26]. In this study, CD analyses between the Turks River population and the Manas River population showed that the CD values for all nine morphological characteristics of *G. dybowskii* were greater than 1.28. In combination with the geographic isolation and differences in the outward appearance between populations in the Turks River basin and the Manas River basin, as well as the results of the one-way ANOVA, PCA, OPLS-DA, and discriminant analyses, these findings demonstrated that the two populations had reached at least the subspecific level of differentiation. Meng Yanxiao et al. reported that only one trait of species the *Brachymystax lenok* and *Brachymystax lenok tsinlingensis* Li differed at the subspecies level (CD > 1.28); combined with the fact that *B. lenok* and *B. lenok tsinlingensis* Li are geographically isolated and because of differences in morphological indicators, these authors inferred that the degree of morphological differentiation between *B. lenok* and *B. lenok tsinlingensis* Li had reached the subspecies level [38], and the results of the present study are similar to theirs. Therefore, the results of this study support the conclusion that the degree of differentiation between the Turks River population and the Manas River population of *G. dybowskii* reaches at least at the subspecies level, although further research is needed to determine whether these two populations should be considered separate species.

5. Conclusions

This study shows that *G. dybowskii* is a typical cold-water plateau fish, exhibiting apparent differences in body color and morphology between the geographically distinct populations. A comprehensive multivariate analysis and one-way analysis of variance (ANOVA) revealed that 23 traits were significantly different between populations. The PCA, OPLS-DA, and discriminant analysis revealed that the two populations could be wholly distinguished from each other and that the differences in the morphological traits were significant ($p < 0.05$). The coefficient of difference analyses revealed a total of nine traits, with differences rising to the subspecies level (CD > 1.28). We hypothesize that the differentiation of *G. dybowskii* populations in the Turks and Manas Rivers has reached the subspecific level. This is due to the differences in the ecosystems and the complexity of the water system structure of the two rivers, leading to the deepening of the degree of differentiation. This study provides a scientific basis for the evolution of highland fish fauna and the conservation of the germplasm.

Author Contributions: Conceptualization, L.H. and N.Y.; software, L.H. and C.W.; validation, L.Y. and G.S.; investigation, B.H. and X.Q.; data curation, F.Z. and Y.S.; writing—original draft preparation, L.H. and N.Y.; writing—review and editing, S.C. All authors have read and agreed to the published version of the manuscript.

Funding: The Special Financial Project of the Ministry of Agriculture and Rural Affairs (Fishery Resources and Environment Survey in Key Waters of Northwest China); the Corps Science and Technology Bureau Key Areas of Science and Technology Public Relations Plan (2022DB019).

Institutional Review Board Statement: All experimental protocols were approved by the Ethics Committees of the Tarim University of Technology (approval code: TDDKYXF 20200426, approval date: 26 April 2020) and adhered to animal welfare laws, guidelines, and policies.

Data Availability Statement: Because the project is not finalized, a link to the data has not been made public.

Acknowledgments: Thank you to our fishery department colleagues and Yingcheng Studio.

Conflicts of Interest: The authors declare no conflicts of interest.

Appendix A

Table A1. One-way ANOVA results for measured traits of *G. dybowskii* from distinct geographical populations (fraction not significantly different).

Traits	T Mean ± SD	M Mean ± SD	*p*
TL/BL	1.235 ± 0.316	1.163 ± 0.049	0.46
BD/BL	0.177 ± 0.012	0.180 ± 0.029	0.217
CPH/BL	0.068 ± 0.005	0.063 ± 0.009	0.128
I—J/BL	0.284 ± 0.015	0.308 ± 0.024	0.495
A—J/BL	0.260 ± 0.013	0.237 ± 0.015	0.615
AI/BL	0.537 ± 0.012	0.533 ± 0.049	0.09
B—J/BL	0.141 ± 0.008	0.140 ± 0.011	0.695
B—I/BL	0.374 ± 0.013	0.383 ± 0.053	0.091
C—J/BL	0.244 ± 0.015	0.252 ± 0.047	0.984
H—G/BL	0.088 ± 0.009	0.130 ± 0.063	0.181

References

1. Guo, Y.; Zhang, R.M.; Cai, L.G. *Xinjiang of Fishery*; Xinjiang Science and Technology Press: Urumqi, China, 2012; ISBN 9787546615295.
2. Cao, W.X.; Chen, Y.Y.; Wu, Y.F. Origin and evolution of *schizothoracine* fishes in relation to the upheaval of the Xizang Plateau. In *Tibetan Expedition Team of the Chinese Academy of Science ed. Studies on the Period, Amplitude and Type of the Uplift of the Qinghai-Xizang Plateau*; Science Press: Beijing, China, 1981; pp. 118–130.
3. He, D.K.; Chen, Y.F.; Chen, Y.Y.; Chen, Z.M. Molecular phylogeny of the specialized *Schizothoracine* fishes (*Teleostei: Cyprinidae*), with their mplications for the uplift of the Qinghai-Tibetan Plateau. *J. Chin. Sci. Bull.* **2004**, *49*, 39–48. [CrossRef]
4. Qi, D.; Chao, Y.; Guo, S.; Zhao, L.; Li, T.; Wei, F.; Zhao, X. Convergent, parallel and correlated evolution of trophic morphologies in the subfamily *Schizothoracinae* from the Qinghai-Tibetan Plateau. *PLoS ONE* **2012**, *7*, e34070. [CrossRef]
5. Guan, L.; Chi, W.; Xiao, W.; Chen, L.; He, S. Analysis of ypoxia-inducible factor alpha polyploidization reveals adaptation to Tibetan plateau in the evolution of *Schizothoracine* fish. *BMC Evol. Biol.* **2014**, *14*, 192. [CrossRef]
6. Zhu, X.F.; Chen, Y.F. Preliminary study on the age and growth characteristics of *Schizothorax macropogon*. *Chin. J. Zool.* **2009**, *44*, 76–82.
7. Wang, Z.; Yan, Y.C.; Jiang, P.G.; Yan, A. Effects of different factors on soil moisture in farmland in Manas River Basin. *Xinjiang Agric. Sci.* **2013**, *50*, 1879–1886.
8. Li, Y.Y.; Zhang, H.L.; Zhang, F.H.; Chen, F.; Lai, X.Q. Analysis on the Potential of the Available Agricultur e Water Resour ces in Manas River Valley, Xinjiang. *J. Nat. Resour.* **2007**, *22*, 44–50. [CrossRef]
9. Sui, F.G. Tectonic evolution and its relationship with hydrocarbon accumulation in the northwest margin of Junggar Basin. *J. Geol.* **2015**, *89*, 779–793. [CrossRef]
10. Ren, M.L.; Guo, Y.; Zhang, Q.L.; Zhang, R.M.; Li, H.; A, D.K.; Cai, L.G.; Yong, W.D.; Ren, B.; Gao, H.; et al. *Fish Resources and Fishery of Ili River*; Heilongjiang Science and Technology Press: Harbin, China, 1998; pp. 13–310. ISBN 7-5388-3369-2.
11. Chen, L.P. Study on the Ecological Environment of Manas River Basin. Master's Thesis, Shihezi University, Xinjiang, China, 2020. [CrossRef]
12. Robinson, W.B.; Parsons, J.K. Changing times, spaces, and faces: Tests and implications of adaptive morphological plasticity in the fishes of northern postglacial lakes. *Can. J. Fish. Aquat. Sci.* **2002**, *59*, 1819–1833. [CrossRef]
13. Langerhans, R.B.; Layman, C.A.; Shokrollahi, A.M.; DeWitt, T. Predator-driven phenotypic diversification in Gambusia affinis. *Evolution* **2004**, *58*, 2305–2318.
14. Lostrom, S.; Evans, J.P.; Grierson, P.F.; Collin, S.P.; Davies, P.M.; Kelley, J.L. Linking stream ecology with morphlogical variabilityin a native freshwater fish from semi-arid Australia. *Ecol. Evol.* **2015**, *5*, 3272–3287. [CrossRef]
15. Scharnweber, K. Morphological and trophic divergence of lake and stream minnows (*Phoxinus phoxinus*). *Ecol. Evol.* **2020**, *10*, 8358–8367. [CrossRef]

16. Franssen, N.R. Anthropogenic habitat alteration induces rapid morphological divergence in a native stream fish. *Evol. Appl.* **2011**, *4*, 791–804. [CrossRef]
17. Lang, J.J.V.; Snyder, R.J.; Clapsadl, M.D.; Michalak, P.; Kang, L.; Pérez-Fuentetaja, A. Morphometric differentiation and gene flow in emerald shiner (*Notropis atherinoides*) from the lower Great Lakes and the Niagara River. *J. Great Lakes Res.* **2018**, *45*, 324–332. [CrossRef]
18. Ramler, D.; Palandacic, A.; Delmastro, G.B.; Wanzenbock, J.; Ahnelt, H. Morphological divergence of lake and stream Phoxinus of northern Italy and the Denube basin based on geometric morphometric analysis. *Ecol. Evol.* **2016**, *7*, 572–584. [CrossRef]
19. Li, S.F.; Li, C.H.; Li, J.L. Analysis of morphological differences among strains of Nile tilapia. *J. Zool.* **1998**, *44*, 75–82.
20. Yin, M.C. *Fish Ecology*; China Agricultural Press: Beijing, China, 1995; ISBN 9787109031432.
21. Wang, W.; Chen, L.Q.; Gu, Z.M.; Peng, S.M.; Li, Y.K. Analysis of Morphological Variations among Seven Populations of *Erythroculter ilishaeformis*. *Freshw. Fish.* **2007**, *37*, 40–44.
22. Xie, C.X.; Zhang, J.B.; Diao, X.M.; Kong, X.Y.; Li, S.A.; Wu, H.H.; Zhang, J.P.; Wu, Y.F.; Qin, W.; Huang, Q.; et al. *Ichthyology*; China Agricultural Press: Beijing, China, 2010; ISBN 978-7-109-13579-6.
23. Wang, J.L.; Chen, Q.H.; Lu, W.Z.; Yuan, S.Q.; Hu, Q.X.; Cen, S.S.; Zhou, C.J.; Meng, X.L.; Nie, G.X.; Gu, Q.H. Morphological differences among five species in rhodeinae in Huaihe River basin in Henan Province. *J. Aquat. Biol.* **2019**, *43*, 123–132.
24. Dray, S.; Dufour, A. The ade4 Package: Implementing the Duality Diagram for Ecologists. *J. Stat. Softw.* **2007**, *22*, 1–20. [CrossRef]
25. Trygg, J.; Wold, S. Orthogonal projections to latent structures (O-PLS). *J. Chemom.* **2002**, *16*, 119–128. [CrossRef]
26. Mayr, E.; Gorton, L.E.; Usinger, R.L. *Methods and Principles of Systematic Zoology*; Mc-Graw-Hill Book Company: New York, NY, USA, 1953; pp. 125–154.
27. Zhang, Y.J.; Chen, S.A.; Wang, C.X.; Wang, X.Y.; Li, J.L.; Xie, C.X.; Wei, Q. Morphological differences among different geographic populations of the naked heavy-lipped fishes in Xinjiang. *Xinjiang Agric. Sci.* **2023**, *60*, 511–520. [CrossRef]
28. Pashkov, N.A. Morphological characteristics of the populations of black-striped pipefish *Syngnathus abaster* Risso, 1827 (Pisces, Actinopterygii, Syngnathidae) in some North Caucasian water bodies. *Mar. Biol. J.* **2017**, *2*, 55–69.
29. Vincent, R.; Marc, P.; Pierre, M. Parallel evolution of morphological traits and body shape in littoral and pelagic brook charr, Salvelinus fontinalis, along a gradient of interspecific competition. *Oecologia* **2021**, *197*, 421–436.
30. Hetzel, C.; Forsythe, P. Phenotypic plasticity of a generalist fish species resident to lotic environments: Insights from the Great Lakes region. *Ecol. Evol.* **2023**, *13*, e10715. [CrossRef]
31. Caiger, P.E.; Croq, C.; Clements, K.D. Environmentally induced morphological variation in the temperate reef fish, *Forsterygion lapillum* (F. Tripterygiidae). *Mar. Biol.* **2021**, *168*, 131. [CrossRef]
32. Dunn, N.R.; O'Brien, L.K.; Burridge, C.P.; Closs, G.P. Morphological Convergence and Divergence in Galaxias Fishes in Lentic and Lotic Habitats. *Diversity* **2020**, *12*, 183. [CrossRef]
33. Langerhans, R.B.; Layman, C.A.; Langerhans, A.K.; Dewitt, T.J. Habitat-associated morphological divergence in two Neotropical fish species. *Biol. J. Linn. Soc.* **2003**, *80*, 689–698.
34. Ola-Oladimeji, F.A.; Oso, J.A.; Oladimeji, T.E.; Idowu, E.O.; Adeleke, K.; Urihe, F.O. Phenotypic Diversities of Four Populations of Clarias Gariepinus (Siluriformes, Clariidae) Obtained from Ogun and Ondo State Waterbodies in South-Western Nigeria. *Vestn. Zool.* **2017**, *51*, 285–294.
35. Iqbal, M.Z.; Chamily, F.A.; Rahman, M.M.; Tasnim, R.; Mohiuddin, M.; Sultana, F.; Rahman, S.M.; Ali, M.M. Habitat salinity and source-induced variation in body shape of euryhaline long whiskers catfish (*Mystus gulio*). *Reg. Stud. Mar. Sci.* **2024**, *69*, 103308.
36. Wang, Y.; Ding, J.; Li, X.; Zhang, J.; Ma, G. Impact of LUCC on ecosystem services values in the Yili River Basin based on an intensity analysis model. *J. Ecol.* **2022**, *42*, 3106–3118.
37. Jiang, L. Analysis on Groundwater Characteristics at Mosuowan Irrigation District of the ManasRiver Basin. *Groundwater* **2016**, *38*, 1–3.
38. Jiang, E.; Yang, Y.; Pan, M.; Tao, Y. Efficacy evaluation and metabolomics analysis of raw and salt-processed Achyranthes bidentata Radix in zebrafish larvae for osteoporosis treatment. *J. Pharm. Biomed. Anal.* **2024**, *237*, 115774. [CrossRef]
39. Mong, Y.X.; Wang, G.H.; Xiong, D.M.; Liu, H.X.; Zhang, J.L.; Wang, J.L.; Wang, L.X.; Liu, X.L. Exploring the validity of subspecies of Qinling Chinook salmon based on morphological differences. *J. Aquat. Biol.* **2018**, *42*, 550–560.

water

MDPI

Article

Use of Hydroacoustic Methods to Assess Ecological Status Based on Fish: A Case Study of Lake Dejguny (Poland)

Andrzej Hutorowicz

Hydroacoustic Laboratory, National Inland Fisheries Research Institute, Oczapowskiego 10, 10-719 Olsztyn, Poland; a.hutorowicz@infish.com.pl

Abstract: The idea of the article was to verify the thesis that acoustic studies of ichthyofauna in lakes provide sufficient information to determine the necessity (or lack thereof) of carrying out a full procedure for assessing the ecological status based on fish. The manuscript compares acoustic data collected in 2008 and 2021 in Lake Dejguny (Poland) as a result of acquisition carried out with the same EY-500 echosounder, on the same routes (transects), and the echogram analyses were carried out using software dedicated to this echosounder. Target strength (TS) distribution was determined in the lake and in individual water layers with a thickness of 2 m, as well as, using the mapping method, changes in the distribution of fish in selected water layers. Large changes in TS distribution in the vertical profile were demonstrated between both years of the study, as well as a clear reduction in the space available for fish in waters below 24 m in 2021, which was associated with a reduction in the concentration of dissolved oxygen below this depth to a value lower than 2.5 mg L^{-1}. It was suggested that it could be possible to develop a simple index enabling the assessment of changes in fish structure, based, similarly to the Large Fish Index, on the ratio of the number of large fish to small fish. The obtained results confirmed that the proposed methods of acoustic data analysis can be the basis for the decision on the necessity (or lack thereof) of conducting an assessment using the Polish national LFI-EN method based on the results of one-off fishing with Nordic multimesh gillnets.

Keywords: ecological status assessment; freshwater hydroacoustics; target strength TS

1. Introduction

In the second half of last year, Colby et al. [1] wrote that "oligotrophic lakes respond to progressive eutrophication by a sequence of predictable events." Higher nutrient loads result in increased plant production and, consequently, changes in the abiotic environment, including changes in water color and transparency, increased turbidity, oxygen depletion in the hypolimnion, and increased chemical stratification. This accelerates biotic changes in phytoplankton, zooplankton, benthos, and coastal algae (increase in food). In subsequent stages, the abundance of submerged macrophytes and benthos decreases and causes the main changes in the fish populations [1–5]. Salmonids dominating in very clean lakes disappear as their trophy grow (including due to inhibition of natural reproduction) and are replaced by others that can survive in the changed environment [1]. Smelt fish begin to dominate, and then perch and pike fish become more dominant, and finally cyprinid fish [2,3]. At the same time, biomass and fish production generally increase, but eventually they collapse [3]. Due to their relatively long lifespan, fish respond with changes in composition and structure to long-term changes in the lake environment [1,2].

It is widely recognized that fish enable detection and, above all, quantification of the impact of human activity on lakes and reservoirs, which has contributed to the creation of a number of indices, including the Index of Biotic Integrity (IBI) for streams in Illinois, USA [6], which was then extended to the water bodies of Central and North America [7]. They were a combination of taxonomic and/or functional indicators [7]. The IBI index was also applied for the assessment of water quality in the European part of Russia [8]. In Europe, as a result of the obligation arising from the provisions of the Water Framework Directive, a number of national indicators based on lake fish communities have been created since the beginning of the 21st century [9–13]. However, simpler indicators were also created, e.g., based on the number of a single species and the importance of fish diseases, enabling the assessment of the ecological quality of Catalan water reservoirs by Argillier et al. [7], or an index of the abundance of coastal fish species (perch was proposed as a key species) enabling the assessment of the coastal zone and transitional Baltic Sea, or using the share of large fish (larger than 30 cm) and the Large Fish Index (LAFI) [14]. This indicates that it is possible to attempt to develop a method for assessing changes in the ecological state that takes into account only some of the features describing ichthyofauna in lakes.

In Poland, two methods have been developed to assess the ecological condition of lakes based on ichthyofauna. The basis of the LFI+ method is the response of the structure of ichthyofauna (expressed in the weight shares of species and their assortments in the total fish catch) to positive and negative changes in the environment [3]. In this method, the ecological status is determined based on point assessments from ten years of commercial fishing catches. The basis for the assessment in the LFI-EN method is single catches with a set of Nordic multimesh gillnets in accordance with the EN 14757 standard (a fish sampling technique based on the European standard CEN using multi-mesh gill nets) [15]. Their number depends on the depth and surface of the lake and may vary from 8 in shallow lakes (<20 ha and depth up to 5.9 m) up to 68 wontons in large and deep lakes (over 1000 ha and depth over 75 m) [4]. The assessment is based on the weight shares (%) of individual fish species caught in the total fish caught. In stratified lakes, the weight shares (%) of bream (*Abramis Brama* L.), silver bream (*Abramis bjoerkna* L.), roach (*Rutilus rutilus* L.), bleak (*Alburnus alburnus* L.), and ruffe (*Gymnocephalus cernua* L.) show an increase with increasing environmental pressure indicators and the weight share of tench (*Tinca tinca* L.), rudd (*Scardinius erythrophthalmus* L.), and perch (*Perca fluviatilis* L.) show a decrease with increasing environmental pressure indicators [3].

Hydroacoustic makes it possible to collect a lot of spatial data with high resolution (in the order of cm) from a large area in a short time. It also enables separate analyses to be performed in any selected water layers (at any depth and in a layer of any thickness). The important advantages of this method are a coupling receiver with a GPS signal, which allows easy mapping, the lack of impact on the tested object and the environment (non-contact methods), as well as significant time savings and low costs when performing surveys. However, an undoubted disadvantage is the inability to determine the species composition of ichthyofauna, especially the number of individual species.

Hydroacoustic surveys allow continuous monitoring of both the size and the distribution of fish. In fact, the fish target strength (*TS*) is measured, which can be converted into weight or total length (*TL*). Over the hundred-year history of research in the field of fish detection using acoustic methods (they began to be carried out in the late 1920s), many relationships have been developed describing the relationship between the target strength of detected fish and the total length or weight of individual fish species or clusters formed by several species (e.g., [16–21]). One of the analyzed issues is the correspondence in fish length data obtained from vertical hydroacoustics and gillnetting. Research, including by Tušer et al. [22] and Emmrich et al. [23], showed a general good agreement between the distributions of quantities obtained by both methods. This confirmed the possibility of using vertical hydroacoustics to obtain the size structure of fish communities in lakes. Tušer et al., however, expressed the view that acoustic surveys cannot completely replace gillnets [22].

Therefore, the aim of the present study was to explore how hydroacoustic data can be used to conduct a preliminary assessment of changes in the ecological state of lakes that will confirm or exclude the need to carry out a full assessment procedure using the LFI-EN method. The search for a method of analyzing acoustic data has greater potential—it may enable systematic monitoring of changes in ichthyofauna in the reclamation of water reservoirs and, at the same time, omit the time-consuming and burdensome procedures of obtaining the necessary fishing permits.

2. Materials and Methods

The study was carried out the mesotrophic, dimictic Lake Dejguny (total area 765.3 ha, volume 92.6 × 10^6 m^3, mean depth 12.0 m, maximum depth 45.0 m, length max. 6.5 km, width max. 2.4 km) which is located in the Great Masurian Lake District (north-eastern Poland; latitude 54.0383°, longitude 21.6067°). The shape of the lake bowl allows us to distinguish three sections: the north-western part (maximum depth—29 m), which is separated by a not very narrow isthmus, the central part (maximum depth—45 m), and the south-eastern part of the lake (maximum depth > 30 m), which is separated by an island (Figure 1). The lake has been classified as the vendace type [24].

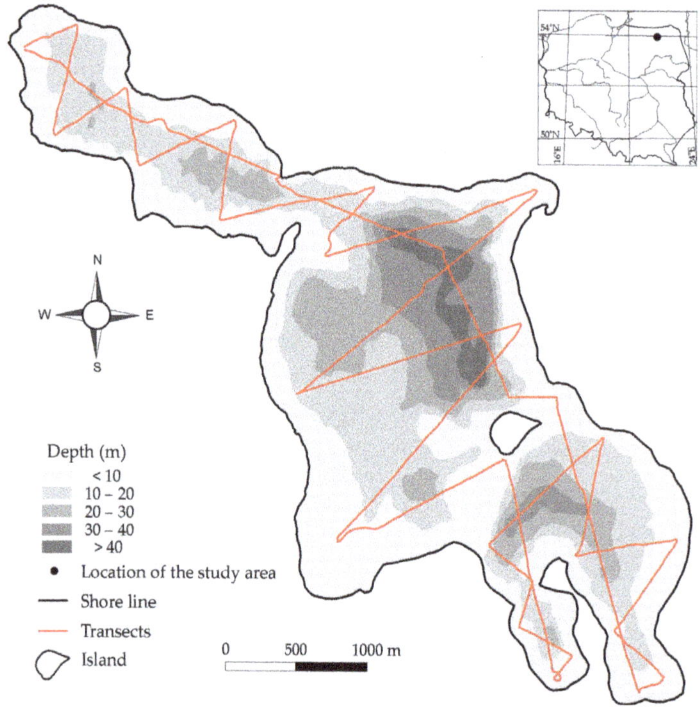

Figure 1. The contour of the Lake Dejguny with bathymetry (after bathymetric plan of the National Inland Fisheries Institute in Olsztyn) and survey transects in October 2008 and October 2021.

Lake Dejguny was not covered by systematic monitoring studies. In 1989, as part of the State Environmental Monitoring, the state of water purity was assessed on the basis of physico-chemical and several biological indicators, classifying the water as class II purity (on a three-level scale) [25]. According to the "Collective assessment of waters" for the years 1989–2013, available on the website of the Chief Inspectorate of Environmental Protection in Warsaw, Lake Dejguny was not examined during this period, and based on an extrapolated assessment on pressure, it was found that the lake had a very good ecological status [26]. Based on research conducted in 2021 as part of the State Environmental Monitoring, the value of the LFI-EN index was determined (0.53) and the ecological condition of the lake was determined to be moderate [27].

2.1. Hydroacoustic Surveys

The analysis was carried out on the basis of the results obtained on 13 October 2008 and 7 October 2021. Baseline data (from 2008) came from surveys conducted between 6:00 p.m. and 9:00 p.m. using Simrad EY500 120 kHz echosounder (Simrad A Konsberg Company, Bergen, Norway) with a 4 × 10° elliptical transducer oriented vertically downwards. Transducer was mounted onto a custom made frame to stabilize their position. Pulse duration was set to 0.3 ms and pulse interval was set to as fast as possible within the options of the system's controlling software. Hydroacoustic studies were carried out along closely separated zigzag transects covering the entire area of the lake (Figure 1), at a speed of 2.5 km h^{-1}. Data for comparison were collected on 7 October 2021 between 8:00 p.m. and 0:00 a.m. according to the same transects, pulse duration and constant speed of 2 km h^{-1}, using the same splitbeam echosounder.

The total length of the transects was approximately 8.4 km, and the degree of coverage, defined according to Aglen [28], as the ratio between the surveyed distance, that is, the cumulative length of the hydroacoustic transects (km) and the square root of the lake area (km^2) was 6.5. The degree of coverage should be at least 3.0 and preferably near to or above 6.0 [23]. This coverage was sufficiently high to expect the coefficient of variation for the estimated abundance to be well below 10% [29].

During the research, dissolved oxygen concentration in water and temperature were also measured in a vertical profile from surface to bottom using the OXI 196 (WTW) in 2008 and in 2021 the YSI 5740 Dissolved Oxygen and Temperature Probe (Yellow Springs Instruments Co, Inc., Ohio, USA).

2.2. Data Analysis

Hydroacoustic data were processed using the Simrad EP500 ver. 5.3 echo processing system dedicated to this echosounder. Signals were analyzed in the range from −50 dB to −17 dB with a *TS* class resolution of 3 dB. This range was adopted to take into account the low effectiveness of gillnets in catching small fish [30]. According to research by Prchalova et al. [30], gillnets are unable to catch roach—*Rutilus rutilus* (L.), perch—*Perca fluviatilis* L., and rudd—*Scardinius erythrophtalmus* (L.) with a length of less than about 40 mm. A similar total length of fish—less than 5 cm—was indicated by Tušer et al. [22].

The number of fish (single-fish echoes) was estimated in sections of 120 pings, which corresponded to a length of approximately 140 m, separately in 2 m thick layers from a depth of 2 m to the bottom. On the basis of these data, the target strength (*TS*) distribution in the depth gradient in Lake Dejguny at night on 13 October 2008 and 7 October 2021 were compared.

Based on the relationship between the target strength and the total length of the fish (*TL*), the *TL* limit values were indicated. According to previous research [31] for data from Lake Dejguny in 2021, the most appropriate relationship for estimating the total length of fish (*TL*) is the equation according to Frouzova et al. [20]:

$$TL_F = 10^{\left(\frac{TS+84.95}{21.15}\right)} \quad (1)$$

where *TS* is the maximum target strength in dB, and TL_F is the total length of the fish in mm for a given target strength.

In order to determine the changes caused by the disturbance, i.e., a decrease in the concentration of dissolved oxygen in the waters of the lake's hypolimnion, *XYZ* data visualization was performed by mapping (where *XY* determined the location in the rectangular coordinate system and *Z* the number of detected fish). The Surfer program was used. The kriging method with a linear variogram was used to create a regular grid of values. Then, maps were generated showing areas surrounded by isolines with a given number of detected single-fish echoes *n*, starting from *n* = 1, through *n* = 10, 20 to *n* = 40. The database used for these analyses also contained information about the lake contour and bathymetric lines (e.g., 20 m, 22 m, etc., respectively), defined by a series of points with appropriate x and y coordinates, in a rectangular coordinate system. Comparison of the areas where fish echoes could be encountered ($n \geq 1$) made it possible to indicate the degree of reduction in the space available for fish and, consequently, fragmentation or even disappearance of habitats. This type of analysis was carried out on layers selected on the basis of comparison of the distribution of *TS* in water layers from a depth of 2 to 34 m.

Normality of distributions was tested using the Shapiro–Wilk test. The null hypothesis of this test assumes that the population is normally distributed. Therefore, a data set with a *p*-value greater than an alpha value of 0.05 does not reject the null hypothesis that the data come from a normally distributed population. The skewness of the distributions was also assessed. However, to assess the consistency of the *TS* distribution in both years, Kendall's tau correlation coefficient (τ) was used, which assesses the similarity of the arrangement of a set of objects from the periods r and s. This coefficient expresses the correlation of features measured on an ordinal scale and allows us to determine the degree of change in the arrangement of objects over time [32]. Calculations were performed in R v 3.6.3, also using the "moments" package.

3. Results and Discussion

In October 2008, 20,100 single-fish echoes were detected, while in October 2021 there was an increase of 19%, i.e., 24,794 single-fish echoes. The size structure of the recorded fish echoes was also different on both dates. In 2008, the target strength (*TS*) distribution did not differ significantly from the normal distribution ($p = 0.08359$, Shapiro–Wilk normality test), and the maximum number of single-fish echoes was in the range $-44 < TS < -41$ dB (Figure 2). The share of fish with $TS > -38$ dB was approximately 25%. In 2021, the peak number of single-fish echoes was recorded in the range of $-50 < TS < -47$ dB. The *TS* of most fish echoes was lower than the average TS, and the target strength distribution was right-skewed (The skewness was 1.05, which indicated that the data were highly skewed). The share of fish with $TS > -38$ dB (which corresponds to $LT_F = 16.6$ cm according to Equation (1)), i.e., the largest fish that are the target of fishermen, decreased to 8%.

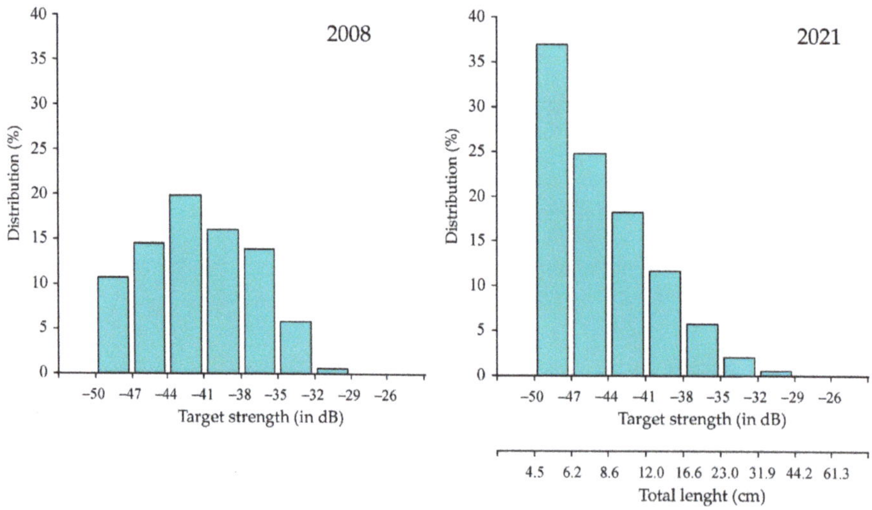

Figure 2. Target strength (*TS*) distribution in Lake Dejguny at night on 13 October 2008 and 7 October 2021. The predicted total length of fish (TL_F) in 2021 was calculated according to the formula of Frouzova et al. [20].

The *TS* distribution of single-fish echoes in water layers at depths from 2 to 34 m also showed very clear differences between both years of research (Figure 3). In 2008, a peak was observed in the water layer at a depth of 22 to 28 m, consisting of [illegible] (at a depth of 22–24 m) and 4047 (24–26 m) echoes of individual fish, i.e., 21% and 20% of all detected fish. At a depth of 20–24 m, larger fish with $TS \geq -41$ dB dominated (2179 and 2838 echoes of individual fish, which constituted 71% and 67% of fish detected in a given layer, respectively). The layer from 2 to 14 m was dominated by small fish with $TS < -41$ dB (142–777 echoes of individual fish), constituting 82–91% of fish detected in a given layer. Also, below 28 m (up to 30 m) depth, as few as 15 to 880 individual fish echoes accounted for 87% to 93% of the fish detected in a given layer.

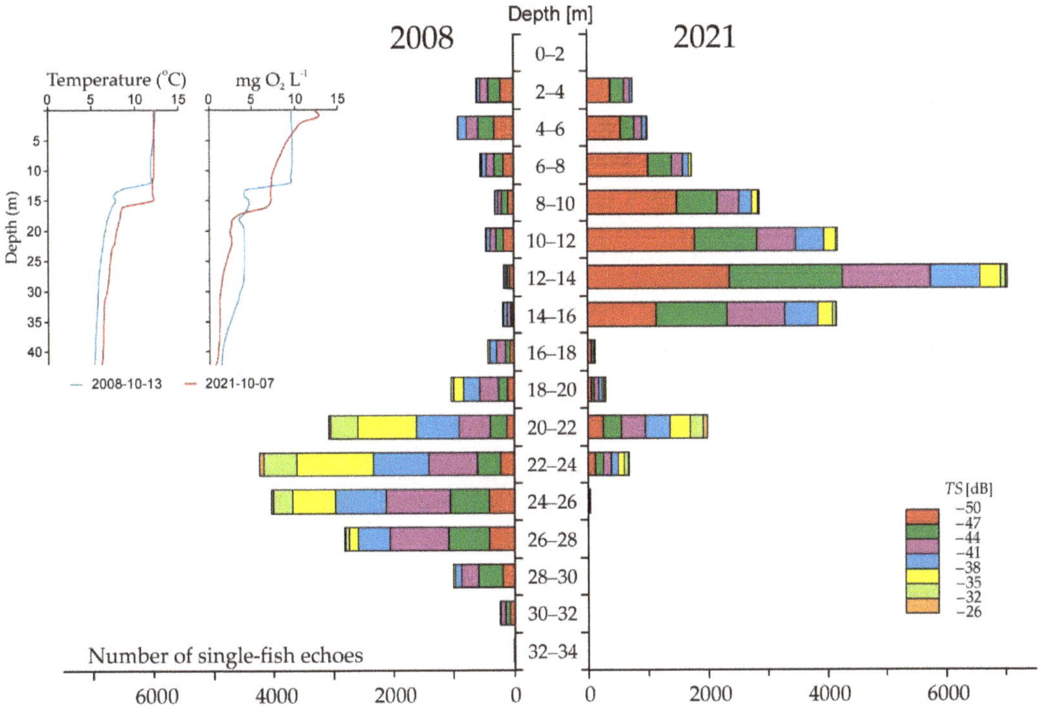

Figure 3. Temperature and oxygen profiles and distribution of target strength (*TS* in dB) at a depth of 2–34 m in Lake Dejguny on the night of 13 October 2008 (n = 20,100) and 7 October 2021 (n = 24,794).

In the fall of 2021, the vertical distribution of single-fish echoes and the *TS* distribution were different (Figure 3). The largest number of fish (i.e., 7035 single-fish echoes., approximately 28% of all detected) was recorded in the layer from 12 to 14 m. The second, more than three times smaller, peak occurred at a depth of 20–22 m. It consisted of 1993 single-fish echoes, which accounted for only 8% of all fish detected. From 2 m to 18 m, fish with $TS < -41$ dB dominated (which corresponds to $LT_F < 12.0$ cm according to Equation (1)). They accounted for 79–94% of the fish detected in these layers. On both compared dates, a significant decrease in the number of fish detected was observed near the thermocline. In 2008, a temperature drop of 3.1 °C m^{-1} was recorded between the 12 and 13 m depth and 1.1 °C m^{-1} in the 13–14 m depth layer. However, in the water layer of 12–16 m, only 174 to 188 single-fish echoes were found, respectively (i.e., <2% of all fish detected). In 2021, a sharp thermocline (3.3 °C m^{-1}) was recorded between 15 and 16 m deep. Just below this layer (16–18 m), only 119 single-fish echoes were found (i.e., approximately 0.5% of fish detected; Figure 3).

TS distributions in lake water layers between 2008 and 2021 differed significantly (Table 1). Kendall's tau coefficient (τ) for the single-fish echoes (n) series had the values

$\tau < 0.3$ with $p > 0.05$, except for the n series for $TS = -29$ dB. For these target strength values, the test showed quite high consistency of the n series ($\tau \approx 0.7$ with $p < 0.05$).

Table 1. The values of Kendall's τ with p values when testing correlations between series of single-fish echoes (n_x) in target strength (TS) intervals in 2008 and 2021 in Lake Dejguny.

	$\sum_{TS=-50}^{TS=-26} n$	n_{-50}	n_{-47}	n_{-44}	n_{-41}	n_{-38}	n_{-35}	n_{-32}	n_{-29}	n_{-26}	n_{-23}
τ	−0.177	−0.093	−0.220	−0.318	−0.042	0.112	0.133	0.106	**0.699**	−0.067	- *
p	0.355	0.624	0.254	0.108	0.823	0.554	0.484	0.577	**0.002**	0.724	- *

Note: *—in 2021 $n = 0$. Statistically important correlations with p values of less than 0.01 are shown in bold.

Based on the target strength distribution in the depth gradient (shown in Figure 3), water layers were selected in which mapping of the distribution of the total number of fish and the number of fish with $TS > -35$ dB, which was recorded on route sections with a length of 120 pings. According to Equation (1), the TL_F in 2021 is assumed to be >23.3 cm. Two layers of 20–22 m and 22–24 m were taken into account. In 2021, the presence of fish was still recorded in these layers; although, compared to 2008, a clear decrease in the number of fish larger with $TS > -35$ dB was observed.

Mapping showed that in 2008, the area in the central part of the lake where fish were recorded ($n \geq 1$ individuals/120 pings) covered approx. 122 ha (Figure 4a, Table 2). This area was almost equal to the area surrounded by the 20 m isobath in the central part of the lake (130 ha) and constituted 63% of the total bottom area surrounded by the 20 m isobath (182 ha). At this depth, in the north-western and south-eastern parts of the lake, fish appeared sporadically. However, no fish with $TS > -35$ dB were observed in these places (Figure 4c). In the central part, they were still recorded on 89% of the area available in this part of the lake and at a density of $10 \leq n < 20$ ind./120 pings in a similar area (72 ha) as the total number of fish (83 ha). In 2021, the area of fish occurrence ($n \geq 1$ ind./120 pings) at a depth of 20–22 m was much smaller (by 26%) than in 2008 (Figure 4c, Table 2). The area where a greater number of fish was recorded (n > 20) covered practically only the deepest parts of the water column—only 15% of the area with a depth of at least 20 m (Table 2). Fish with $TS > -35$ dB ($TL_F > 16.6$ cm) were occasionally found here. No fish were observed in the north-western and south-eastern parts of the lake (Figure 4d).

Table 2. An area (according to the maps in Figures 4 and 5) in the water layer 20–24 m deep in Lake Dejguny, where single-fish echoes with a TS value from −50 to −23 dB were identified in acoustic search sections of 120 pulses between the night of 13 October 2008 and 7 October 2021.

Layer	Number of Single-Fish Echoes (ind./120 Pings)	Area (ha)			
		$TS > -50$ dB		$TS > -35$ dB	
		2008	2021	2008	2021
20–22 m	1	121.89	89.85	115.29	53.79
	10	82.83	54.92	71.82	1.76
	20	59.64	27.43	38.52	0
	40	4.44	8.41	0.41	0
	60	0	0.75	0	0
22–24 m	1	108.63	68.98	104.61	30.58
	10	81.9	27.94	73.13	0.09
	20	68.3	11.61	53.87	0
	40	40.87	0.69	5.59	0
	60	0.13	0	0	0

In 2008, in a layer 22–24 m deep, fish were found ($n \geq 1$ ind./120 pings) only in the central and northern parts of the lake. They occurred in an area (108.6 ha) covering 75% of the bottom area surrounded by a 22 m isobath. In the central part, they were found in 97%

of the available area (Figure 5a). A higher density of fish ($10 < n \leq 20$) was recorded only in the central part—in 65% of the area available in this part of the lake. Densities greater than $n > 40$ ind./120 pings were observed in 37% of the available area. Fish with $TS > -35$ dB were found in an area of 104.6 ha, mainly in the central part of the lake (Figure 5b), and in slightly higher densities ($n > 20$ pieces/120 ping) in 48% of the area available in this part of the lake.

In 2021, the area of fish occurrence at a depth of 22–24 m ($n \geq 1$ ind./120 pings) was 37% smaller than in 2008 (Table 2). However, the area where fish occurred in larger numbers (n 10 $\geq$ ind./120 pulses) decreased by as much as 66%. Fish with $TS > -35$ dB ($TL_F > 16.6$ cm) were found in an area representing 27% of the available area in this part of the lake (Figure 5d). It was 71% lower than in 2008. Slightly higher densities of these fish ($n > 10$) were recorded in an area of only 0.1 ha (Figure 5d, Table 2).

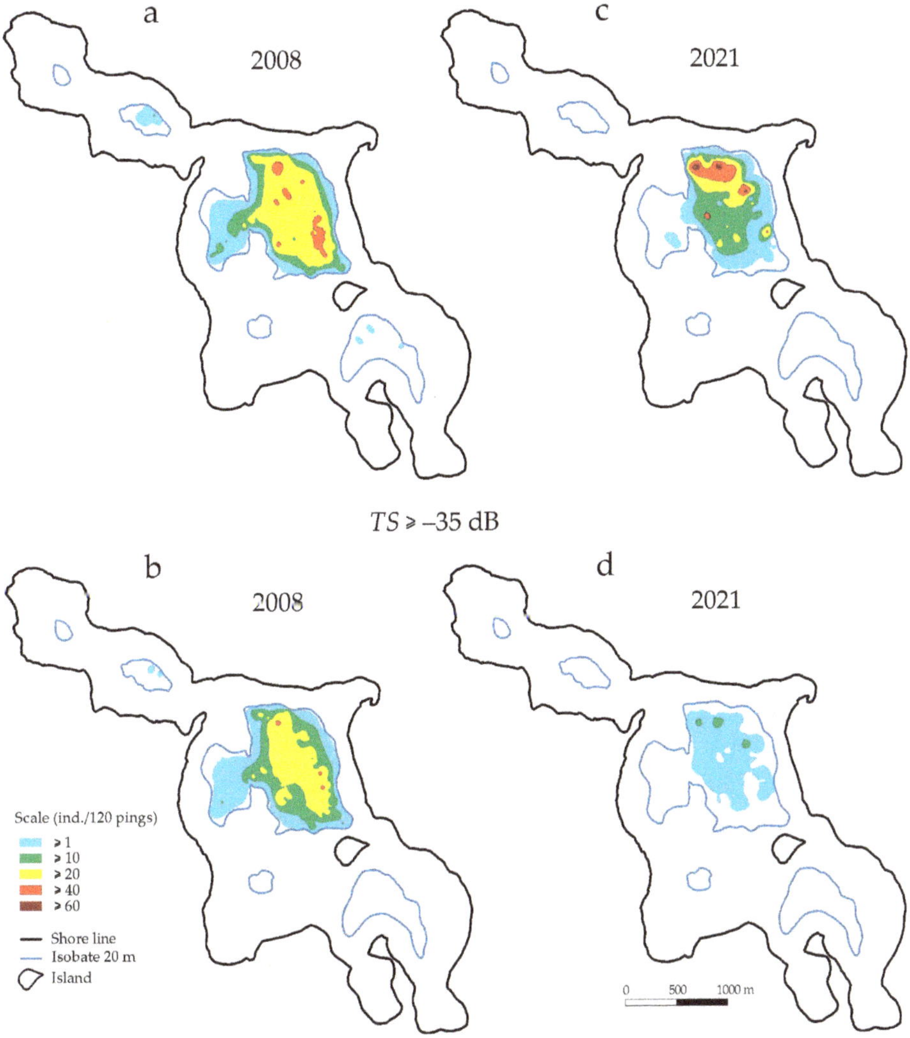

Figure 4. Mapping the distribution of the number of fish (**a**,**c**) and the number of fish with $TS > -35$ dB (**b**,**d**) in sections of the route with a length of 120 pings at a depth of 20–22 m in Lake Dejguny between the night of 13 October 2008 and 7 October 2021.

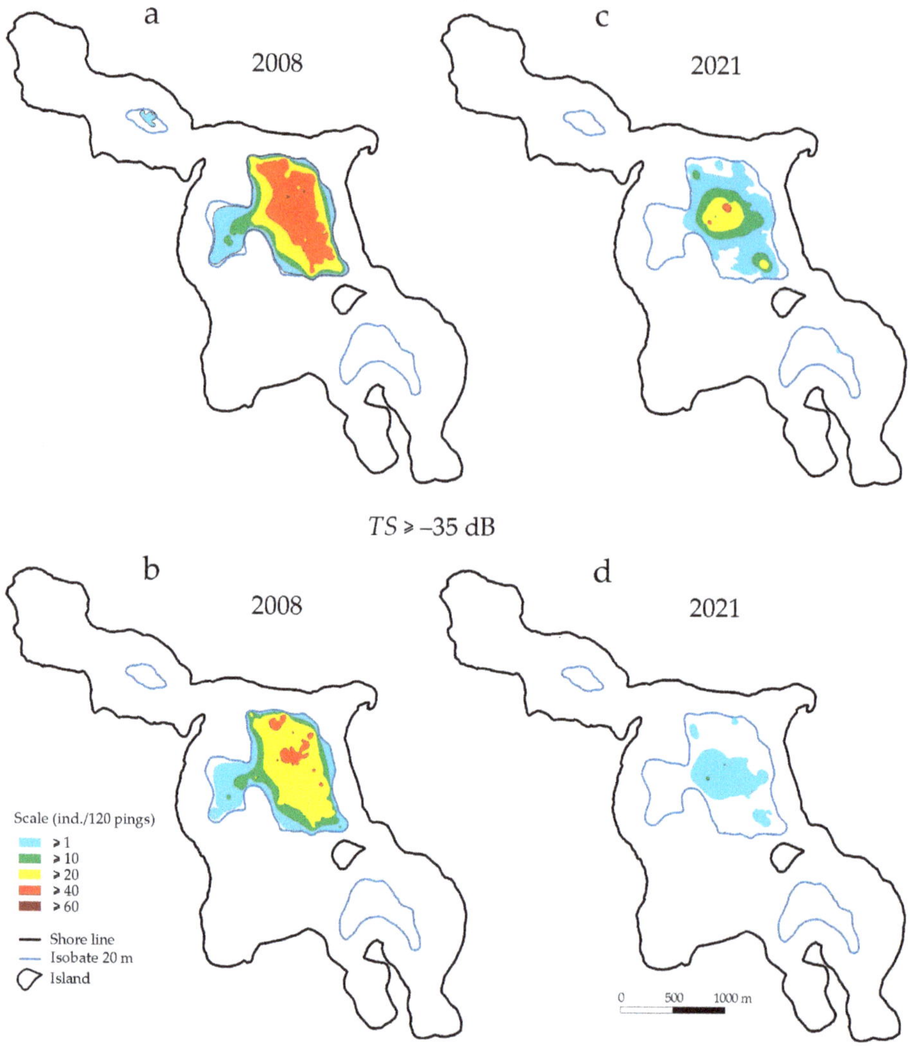

Figure 5. Mapping the distribution of the number of fish (**a**,**c**) and the number of fish with $TS > -35$ dB (**b**,**d**) in sections of the route with a length of 120 pings at a depth of 22–24 m in Lake Dejguny between the night of 13 October 2008 and 7 October 2021.

Acoustic studies of Lake Dejguny showed significant changes in the distribution of *TS* fish communities throughout the lake and in the vertical profile. However, mapping allowed us to demonstrate a reduction in the space available for fish in waters below 24 m deep (habitat loss).

Size is the main structural feature in fish assemblages. Size strongly influences organismal movement, predation, growth, mortality, and reproduction of fish [33]. It is well documented that most ecological and physiological processes are strongly dependent on body size [34]. Barange et al. [35] confirmed that *TS* information in studies of the structure and dynamics of fish populations can be effective in the case of low-density, multi-species assemblages. Emmerlich et al. [36] showed that non-taxonomic variables describing size-structured communities may be consistent with descriptors of lake morphometry, lake productivity (expressed by total phosphorus and chlorophyll a concentration), lake use intensity, and the taxonomic and functional composition of fish communities. Because the results presented in this study were obtained using the same method (same equipment, and same echogram analysis and mapping methods), the observed changes in the *TS* structure and the spatial distribution of fish in Lake Dejguny are the result of structural changes in the lake ecosystem.

This conclusion is confirmed by the observed changes in the distribution of oxygen in the lake. In October 2021, in the lower parts of the lake, below 23 m depth, the concentration of dissolved oxygen in the water dropped below 2.5 mg $O_2 \cdot L^{-1}$, which is over 10 m shallower than in October 2008 (Figure 3). This value limited the occurrence of fish in the lake.

Conducting a series of analogous studies on many lakes should enable the development of a simple indicator assessing changes in the ecological state. It seems that this future indicator could be based on the Large Fish Index (LAFI), which is used in marine ecology research. LAFI is based on the number of large fish as a percentage of all acoustically recorded fish echoes. LAFI is sensitive to the effects of fishing and relatively insensitive to environmental variability. However, it can be easily linked to a specific ecological quality status based on a reference period [29]. Therefore, until the possible calibration of this method of assessing changes in the ecological state over time, it is necessary to determine the value of the indicator on two dates (the initial one and the one being compared).

Therefore, it was assumed that:

- The target strength *TS* distribution sufficiently describes precisely the changes in the size spectrum of the fish community;
- An indicator based on the essence of the Large Fish Index can effectively assess changes in ecological status by comparing values at the beginning and end of a specific period;
- Percentage share (%), calculated as the number of large fish divided by the number of smaller fish, can be as effective as the percentage of biomass of fish species or fish assortment, used in LIF-EN and LAFI, to assess the structure of ichthyofauna.

The latter assumption was based on Murry & Farrell's statement that the distribution of the number *N* as a function of unit length *L* can determine the spectrum of community size [37].

In this study, these proportions (*IND*) of the number of fish with $TS > -38$ dB (i.e., when $TL_{F\ 2021} > 16.6$ cm) were determined in relation to the number of fish with -47 dB $< TS < -41$ dB, expressed as

$$IND = \frac{\sum_{TS=-38dB}^{TS=-32dB} n_{TS}}{\sum_{TS=-38dB}^{TS=-32dB} n_{TS}} \quad (2)$$

The IND_{2008} value was 0.404, IND_{2021}—0.154. IND_{2021} was therefore 2.6 times smaller than IND_{2008}. The index value for the 2008 data was also higher than the LAFI = 0.3 value, which corresponded to the range of values in the period 1925–1983 in the North Sea, when most of the exploited stocks were above the precautionary reference limit for spawning stock biomass [14]. Therefore, the conducted research does not exclude the possibility of conducting analyses to develop an indicator based on acoustic data.

To sum up, hydroacoustic studies carried out in Lake Dejguny (analysis of *TS* distribution and maps of fish distribution in water layers) indicate that they can be a complementary method to the intercalibrated national method for assessing the ecological state of lakes based on fish. The LFI-EN method is based on determining the weight shares (%) of bream (*Abramis Brama* L.), silver bream (*Abramis bjoerkna* L.), roach (*Rutilus rutilus* L.), bleak (*Alburnus alburnus* L.), and ruffe (*Gymnocephalus cernua* L.) (which show an increase with increasing environmental pressure indicators) and the weight share of tench (*Tinca tinca* L.), rudd (*Scardinius erythrophthalmus* L.), and perch (*Perca fluviatilis* L.) (which show a decrease with increasing environmental pressure indicators). Therefore, information about the *TS* structure of fish obtained using acoustic methods is completely unsuitable for carrying out assessment using the LFI-EN method. However, acoustic tests allow us to indicate changes in the size structure of fish. It can, therefore, be concluded that carrying out acoustic tests (as preliminary tests) may indicate the need to carry out an assessment using the LFI-EN fish index, or the lack of such a need.

Funding: This research was funded by the National Inland Fisheries Research Institute in Olsztyn as part of statutory research activity (topic no. Z-017).

Data Availability Statement: Data are contained within the article.

Acknowledgments: The author would like to thank the people involved in hydroacoustic collections, including: Bronisław Długoszewski (†), Lech Doroszczyk, and Jacek Tunowski.

Conflicts of Interest: The author declares no conflicts of interest.

References

1. Colby, P.J.; Spangler, G.R.; Hurley, D.A.; McCombie, A.M. Effects of eutrophication on salmonid communities in oligotrophic lakes. *J. Fish. Res. Board Can.* **1972**, *29*, 975–983. [CrossRef]
2. Hartmann, J. Fischereiliche Veranderungen in kulturbedingt eutrophierenden Seen. *Schweitz. Z. Hydrol.* **1977**, *39*, 243–254. [CrossRef]
3. Chybowski, Ł.; Białokoz, W.; Wołos, A.; Draszkiewicz-Mioduszewska, H.; Szlakowski, J. *Przewodnik Metodyczny do Monitoringu Ichtiofauny w Jeziorach [Methodological Guidelines for Ichthyofauna Monitoring in Lakes]*; Biblioteka Monitoringu Środowiska: Warszawa, Poland, 2016; p. 52.
4. Adamczyk, M.; Prus, P. Ryby w jeziorach [Fish in lakes]. In *Podręcznik do Monitoringu Elementów Biologicznych i Klasyfikacji Stanu Ekologicznego wód Powierzchniowych—Aktualizacja Metod [Manual for the Monitoring of Biological Elements and the Classification of Ecological Status of Surface Waters—Method Update]*; Kolada, A., Ed.; Biblioteka Monitoringu Środowiska: Warszawa, Poland, 2020; pp. 297–326.
5. Willemsen, J. Fishery-aspects of eutrophication. *Aquat. Ecol.* **1980**, *14*, 12–21. [CrossRef]
6. Karr, J.R. Assessment of biotic integrity using fish communities. *Fisheries* **1981**, *6*, 21–27. [CrossRef]
7. Argillier, C.; Causse, S.; Gevrey, M.; Pèdron, S.; De Bortoli, J.; Brucet, S.; Emmrich, M.; Jeppesen, E.; Lauridsen, T.; Mehner, T.; et al. Development of a fish-based index to assess the eutrophication status of European lakes. *Hydrobiologia* **2013**, *704*, 193–211. [CrossRef]
8. Ushakova, E.; Drobinina, E.; Puzik, A.; Mikheev, P. Impact of Wastewater on Water Quality and Fish Community in the Tolych River, Perm Krai, Russia a. *Pollution* **2023**, *9*, 459–476. [CrossRef]
9. Gassner, H.; Tischler, G.; Wanzenböck, J. Ecological integrity assessment of lakes using fish communities—Suggestions of new metrics developed in two Austrian Prealpine Lakes. *Int. Rev. Hydrobiol.* **2003**, *88*, 635–652. [CrossRef]
10. Gassner, H.; Wanzenböck, J.; Zick, D.; Tischler, G.; Pamminger-Lahnsteiner, B. Development of a fish based lake typology for natural Austrian lakes >50 ha based on the reconstructed historical fish communities. *Int. Rev. Hydrobiol.* **2005**, *90*, 422–432. [CrossRef]
11. Launois, L.; Veslot, J.; Irz, P.; Argillier, C. Development of a fish-based index (FBI) of biotic integrity for French lakes using the hindcasting approach. *Ecol. Indic.* **2011**, *11*, 1572–1583. [CrossRef]
12. Launois, L.; Veslot, J.; Irz, P.; Argillier, C. Selecting fish-based metrics responding to human pressures in French natural lakes and reservoirs: Towards the development of a fish-based index (FBI) for French lakes. *Ecol. Freshw. Fish* **2011**, *20*, 120–132. [CrossRef]
13. Kelly, F.L.; Harrison, A.J.; Allen, M.; Connor, L.; Rosell, R. Development and application of an ecological classification tool for fish in lakes in Ireland. *Ecol. Indic.* **2012**, *18*, 608–619. [CrossRef]
14. Shephard, S.; Reid, D.G.; Greenstreet, S.P.R. Interpreting the large fish indicator for the Celtic Sea. *ICES J. Mar. Sci.* **2011**, *68*, 1963–1972. [CrossRef]
15. *EN 14757*; Water Quality—Sampling of Fish with Multi-Mesh Gillnets. CEN (European Committee for Standardization): Brussels, Belgium, 2005.
16. Love, R.H. Dorsal-Aspect target strength of an individual fish. *J. Acoust. Soc. Am.* **1971**, *49*, 816–823. [CrossRef]
17. Love, R.H. Target strength of an individual fish at any aspect. *J. Acoust. Soc. Am.* **1977**, *62*, 1397–1403. [CrossRef]
18. Mehner, T. Prediction of hydroacoustic target strength of vendace (*Coregonus albula*) from concurrent trawl catches. *Fish. Res.* **2006**, *79*, 162–169. [CrossRef]
19. Foote, K.G. Fish target strengths for use in echo integrator surveys. *J. Acoust. Soc. Am.* **1987**, *82*, 981–987. [CrossRef]
20. Frouzova, J.; Kubecka, J.; Balk, H.; Frouz, J. Target strength of some European fish species and its dependence on fish body parameters. *Fish. Res.* **2005**, *75*, 86–96. [CrossRef]
21. Kubečka, J.; Frouzová, J.; Balk, H.; Čech, M.; Draštík, V.; Prchalová, M. Regressions for conversion between target strength and fish length in horizontal acoustic surveys. In *Underwater Acoustic Measurements, Technologies & Results*; Papadakis, J.S., Bjorno, L., Eds.; Foundation for Research and Technology: Heraklion, Greece, 2009; pp. 1039–1044.
22. Tušer, M.; Guillard, J.; Rustadbakken, A.; Mehner, T. Comparison of fish size spectra obtained from hydroacoustics and gillnets across seven European natural lakes. *Can. J. Fish. Aquat. Sci.* **2022**, *79*, 2179–2190. [CrossRef]

23. Emmrich, M.; Winfield, I.J.; Guillard, J.; Rustadbakken, A.; Vergès, C.; Volta, P.; Jeppesen, E.; Lauridsen, T.L.; Brucet, S.; Holmgren, K.; et al. Strong correspondence between gillnet catch per unit effort and hydroacoustically derived fish biomass in stratified lakes. *Freshw. Biol.* **2012**, *57*, 2436–2448. [CrossRef]
24. Dunalska, J.; Brzozowska, R.; Zdanowski, B.; Stawecki, K.; Pyka, J. Variability of organic carbon, nitrogen and phosphorus in the context of Lake Dejguny eutrophication (Mazurskie Lakes District). *Limnol. Rev.* **2006**, *6*, 79–86.
25. *Raport o Stanie Środowiska Województwa Warmińsko-Mazurskiego w 2008 Roku [Report on the State of the Environment in the Warmia and Masuria Voivodship in 2008]*; Biblioteka Monitoringu Środowiska: Olsztyn, Poland, 2009; p. 151.
26. 4.8.4_Zbiorcza Ocena wód_dane zestawione.xlsx [4.8.4_Collective Water Assessment_summarized data.xlsx]. In *Ocena Stanu Jezior w Latach 2010–2015—Tabele [Assessment of the Condition of Lakes in 2010–2015—Tables]*. Available online: https://wody.gios.gov.pl/pjwp/publication/LAKES/87 (accessed on 9 December 2023).
27. Klasyfikacja Wskaźników i Grup Wskaźników w Jednolitych Częściach wód Powierzchniowych Jezior za rok 2021 [Classification of Indicators and Indicator Groups in Lake Surface Water Bodies for 2021. Available online: https://wody.gios.gov.pl/pjwp/publication/LAKES/87 (accessed on 9 December 2023).
28. Aglen, A. Random errors of acoustic fish abundance estimates in relation to the survey grid density applied. *FAO Fish. Rep.* **1983**, *300*, 293–298.
29. Godlewska, M.; Długoszewski, B.; Doroszczyk, L.; Jóźwik, A. The relationship between sampling intensity and sampling error—Empirical results from acoustic surveys in Polish vendace lakes. *Fish. Res. Spec. Issue* **2009**, *96*, 17–22. [CrossRef]
30. Prchalová, M.; Kubečka, J.; Říha, M.; Mrkvička, T.; Vašeka, M.; Jůza, T.; Kratochvíl, M.; Peterka, J.; Draštíka, V.; Křížekd, J. Size selectivity of standardized multimesh gillnets in sampling coarse European species. *Fish. Res.* **2009**, *96*, 51–57. [CrossRef]
31. Hutorowicz, A.; Ulikowski, D.; Tunowski, J. Comparison of Size Distribution of Fish Obtained from Gill Netting and the Distributions of Echoes from Hydroacoustics in Lake Dejguny (Poland). *Water* **2023**, *15*, 1117. [CrossRef]
32. Walesiak, M. Zagadnienie oceny podobieństwa zbioru obiektów w czasie w syntetycznych badaniach porównawczych [On the problem of similarity of sets of the same objects in time in synthetic comparative studies]. *Przegląd Stat.* **1993**, *40*, 95–102.
33. Guiet, J.; Poggialec, J.-C.; Maurya, O. Modelling the community size-spectrum: Recent developments and new directions. *Ecol. Model.* **2016**, *337*, 4–14. [CrossRef]
34. Sprules, W.G.; Barth, L.E. Surfing the biomass size spectrum: Some remarks on history, theory, and application. *Can. J. Fish. Aquat. Sci.* **2016**, *73*, 477–495. [CrossRef]
35. Barange, M.; Hampton, I.; Pillar, S.C.; Soule, M.A. Determination of Composition and Vertical Structure of Fish Communities Using in situ Measurements of Acoustic Target Strength. *Can. J. Fish. Aquat. Sci.* **1994**, *51*, 99–109. [CrossRef]
36. Emmrich, M.; Brucet, S.; Ritterbusch, D.; Mehner, T. Size spectra of lake fish assemblages: Responses along gradients of general environmental factors and intensity of lake-use. *Freshw. Biol.* **2011**, *56*, 2316–2333. [CrossRef]
37. Murry, B.A.; Farrell, A.M. Resistance of the size structure of the fish community to ecological perturbations in a large river ecosystem. *Freshw. Biol.* **2014**, *59*, 155–167. [CrossRef]

water

MDPI

Article

Hydrology as a Determinant of Riparian Habitat Structure in Lowland River Floodplains

Paweł Burandt [1], Mirosław Grzybowski [2], Katarzyna Glińska-Lewczuk [1], Wojciech Gotkiewicz [3], Monika Szymańska-Walkiewicz [4] and Krystian Obolewski [5,*]

1 Department of Water Management and Climatology, University of Warmia and Mazury in Olsztyn, 10-719 Olsztyn, Poland; pawel.burandt@uwm.edu.pl (P.B.); kaga@uwm.edu.pl (K.G.-L.)
2 Department of Tourism, Recreation and Ecology, University of Warmia and Mazury in Olsztyn, 10-719 Olsztyn, Poland; grzybomi@uwm.edu.pl
3 Department of Agronomy and Agribusiness, University of Warmia and Mazury in Olsztyn, 10-719 Olsztyn, Poland; wgot@uwm.edu.pl
4 Department of Revitalization of Inland Waterways, University of Kazimierz Wielki in Bydgoszcz, 85-064 Bydgoszcz, Poland; szymanska.monika@ukw.edu.pl
5 Department of Hydrobiology, University of Kazimierz Wielki in Bydgoszcz, 85-064 Bydgoszcz, Poland
* Correspondence: obolewsk@ukw.edu.pl; Tel.: +48-52341-9184

Abstract: The objective of the study was to determine the relationship between the structure of phytocenoses in riparian wetland ecosystems and the hydrologic regime in a lowland river floodplain. The hydrobotanical study was conducted over three years—2017, 2018, and 2019—which differed in hydrological conditions (wet, average, and dry) in a middle section of the Supraśl floodplain (NE Poland) as a case study. The results showed that the structure and pattern of phytocenoses in the floodplain are primarily controlled by the hydrological regime of the river and the geomorphological features of the area. The reach and duration of the flood contributed to a specific pattern of riparian vegetation. Based on the plant community structure and riparian habitat indicators such as soil moisture, fertility, reaction pH, soil granulometry, and organic matter content, four habitat types were identified and supported by linear discriminant analysis (LDA): wet, semi-wet, semi-dry, and dry zones. The indicator species analysis (ISA) revealed species characteristic of the zones with the dominance of reed rush, reed canary grass, anthropogenic or partially natural herbaceous communities along watercourses or riparian meadows, respectively. Natural inundation of the river water is an important driver of site-specific vegetation elements and habitat types and determines habitat availability, biodiversity, and ecosystem functions of wetlands. This knowledge can serve as the basis for conservation efforts, sustainable management practices, and decision-making processes aimed at maintaining the biodiversity and ecological integrity of riparian ecosystems in similar regions.

Keywords: lowland river; hydrology; riparian vegetation structure; floodplain; natural flow regime; riparian zonation

1. Introduction

Natural floodplains are among the most productive and diverse ecosystems in the world. Due to increasing anthropogenic pressures and the influence of factors resulting from ongoing climate change, they are among the most threatened ecosystems [1].

Natural wet grasslands in the floodplains of lowland rivers are highly valuable for conservation [2,3]. They form important components of wetland habitats [4]. These grasslands provide multiple ecosystem services, including water regulation, carbon sequestration, landscaping, wildlife (native plants, birds, and invertebrates), recreation, and leisure [5]. In opinion of Grzybowski and Glińska-Lewczuk [6,7], traditional low-intensity management practices such as mowing or grazing are necessary to maintain their characteristic flora and promote species diversity. However, these semi-natural habitats have experienced rapid decline due to insensitive land use changes such as drainage, agricultural intensification, flood control, and neglect [5,8,9].

The structure of these dynamic environments, characterized by high diversity and habitat rotation [10,11], is primarily regulated by river flow regimes [12,13] and geomorphology [14]. Numerous studies have identified flood events as an important factor influencing plant distribution along elevation gradients in floodplains [15–17]. Therefore, floods and associated fluvial and geomorphic processes shape and maintain the spatial and temporal heterogeneity of floodplain areas and their plant communities [14,18,19].

The response of vegetation to hydrological changes depends on its ecophysiological characteristics [20]. Changes in habitat structure are influenced by species' environmental tolerance and the ability to compete with other species [21,22]. Plant species respond differently to direct flood or drought impacts, and these differences are reflected in species' zonation along elevation gradients in floodplains [17]. Species that are highly competitive under dry conditions [23] may not tolerate prolonged flooding and are easily damaged [24], whereas species characterized by greater resilience benefit from flooding [25]. Species tolerance to flooding determines their distribution in river valleys [26,27]. Studies by Van Eck et al. [28] have shown that species with more tolerant traits dominate at lower, more frequently flooded sites, while a greater number of less tolerant species are found at higher elevations that are flooded less frequently. Ultimately, survival to flooding depends on the timing, frequency, depth, and duration of the flood events [29,30]. Casanova and Brock [31] indicate that the duration of floods plays a critical role and is the most important factor in wetland plant community composition. Flooding events contribute to connectivity between patches of plant communities. Water can transport seeds from habitats in upper parts of the river network [32,33], altering local species composition [24,34]. Longitudinal fragmentation prevents the normal downstream transport of nutrients, fauna, flora, and organic matter, leading to the development of different environmental conditions along the river channel [35].

Regular droughts are seasonal environmental changes that are predictable and, when combined with regular flood events, increase ecosystem productivity (*flood pulse concept*) [36,37]. However, as climate changes, characterized by the increasing frequency and magnitude of extreme weather events [38], disruptions to natural processes are occurring, leading to a decrease in river water levels and posing a significant threat to the wetland ecosystems. Declining water levels in rivers disrupt lateral connections, causing water to retreat from floodplains [39]. This disruption of hydrological connectivity between riparian habitats and the increasing heterogeneity of the river channel result in changes in the species composition of the habitats.

The objective of our study was to evaluate the relationship between the species composition of vegetation in riparian meadow habitats and to determine the importance of environmental variables, particularly hydrologic variables, that determine it. Through changes in the ecological indicator values of vascular plants: soil moisture, fertility, pH, soil granulometry, and organic matter content, we investigated whether specific moisture zones determine the vegetation structure. We hypothesized that the reach and elevation of flooding influences the abundance and distribution of plant species, which is reflected in the structure and pattern of riparian habitats.

2. Materials and Methods

2.1. Study Site

The study was conducted on the lower course of the Supraśl River in NE Poland (Figure 1). To determine the different indicators of wetland biophysical habitats in relation to water level, we selected a fragment of the floodplain (2 km^2) adjacent to the main river.

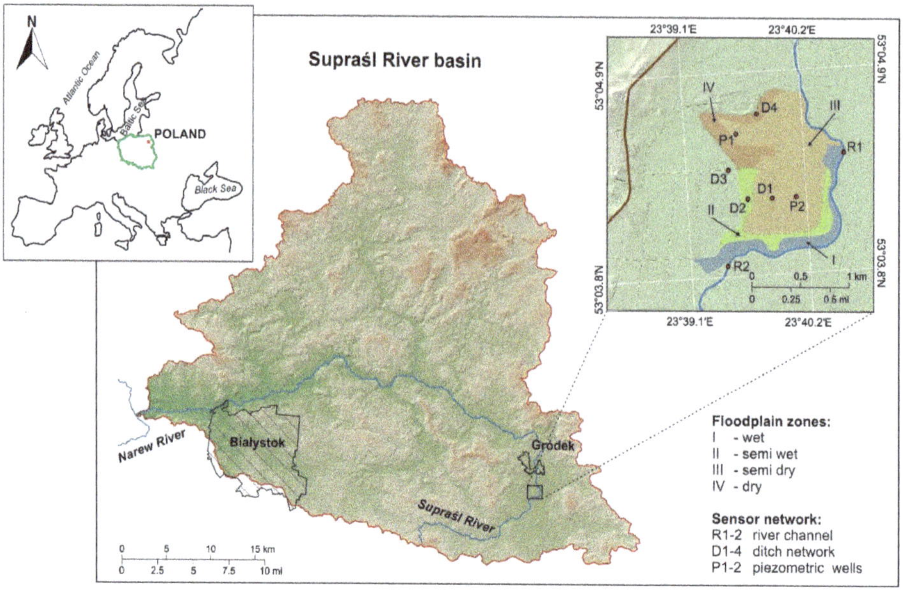

Figure 1. Location of the Supraśl River in northeastern Poland and study area.

The Supraśl River is one of the semi-natural ecosystems and therefore plays an important role in maintaining biodiversity in the region. The area hosts many bird species that are strongly linked to specific hydrological conditions and habitat types [40]. For these reasons, the upper part of the Supraśl floodplain was included in the Natura 2000 network as the "Puszcza Knyszyńska" Special Protection Area for birds (PLB 200003) and the "Ostoja Knyszyńska" (PLB 200006). In the wetlands of the study area, wading bird species such as black-tailed godwit (*Limosa limosa*), redshank (*Tringa totanus*), lapwing (*Vanellus vanellus*), snipe (*Gallinago gallinago*) and jack snipe (*Gallinago media*) breed.

The floodplain of the river is between 0.5 and 1.5 km wide and is characterized by a flat bottom and gentle slopes as it cuts through the frontal moraine zone. The geology of the valley is related to the Pleistocene glaciation, which shaped the area of the Białystok Upland and the Sokółka Foothills. As a result of terrain evolution, sandy depressions and extensive peat accumulations were formed in basin-like meltwater areas. The valley is characterized by organic soils such as peat, peat–marsh, and marsh–mineral soils. In the study area, the floodplain is well developed. The presence of peat–fibrous soils formed by the presence of near-surface groundwater plays an important role in shaping the hydrological regime in the study area, as they greatly increase groundwater retention. The hydrological regime plays an important role in structuring the habitat types in the floodplain. There is a certain spatial hydro-pedological pattern typical of natural central European floodplains that determines the biogeochemical relationships between the river hydrological regime, soil formation, and the corresponding plant communities [41]. Disturbance of this system, particularly by limiting flooding, can adversely affect the quality of wetland habitats.

The Supraśl is a third-order river and the largest right-bank tributary of the Narew River, belonging to the Vistula River basin. It has a length of 93.8 km and a catchment area of 1844.4 km^2. The Supraśl is a lowland river with a low channel gradient of 0.76‰, which is fed by numerous springs and groundwater [42]. This river exhibits a moderate hydrological regime characterized by a seasonal pattern of water flow. The highest flows occur in spring due to snowmelt and rainfall, while the lowest flows occur in summer due to evaporation and transpiration. Flow rates at Nowosiółki gauge (2009–2020) varied throughout the year, averaging between 0.23 and 2.18 m^3 s^{-1}. The high water retention in the river basin is reflected in the absence of sudden floods and in the relatively low ratio between the highest and lowest average monthly flows, which range from 10 to 12.

The qualitative and quantitative seasonality of water flow is influenced not only by morphological parameters or the size of the catchment area but also by the climatic conditions of the region. According to the climate classification of Koeppen and Geiger [43], the study area is located in the western part of the Dfb (Dfb: D = Continental; f = Fully humid; b = Warm summer) zone. The study area is characterized by a humid continental climate with certain subboreal features, resulting in long frosty winters, short early springs, a relatively short growing season, and warm summers. The average annual air temperature is 6.5 °C, and the growing season lasts 192 days. Winds from westerly (20.4%) and southwesterly (17.5%) directions prevail. The average annual precipitation in the region from 1971 to 2022 is 589 mm with a minimum of 456 mm and a maximum of 748.9 mm. Precipitation is not evenly distributed throughout the year. The highest rainfall occurs from May to August and peaks in August, while the lowest rainfall occurs between January and March. Snowfall accounts for about 21% of the annual precipitation. The average number of days with rainfall above 0.1 mm is 169, including about 63 days with snowfall. The study period (2017–2020) was characterized by variable moisture conditions. The year 2017 was exceptionally wet with a total precipitation of 935 mm. The years 2018 and 2019 were classified as dry with precipitation totals of 536 mm and 517 mm, respectively.

The Supraśl River valley is characterized by low population density and low levels of agricultural fertilization [44]. Although the riverbed follows a natural course, its hydrographic network was significantly modified by the introduction of a dense system of drainage ditches. These measures were aimed at adapting the area for agricultural use as pastures and meadows. The drainage system is currently used to irrigate the meadows through an extensive groundwater infiltration system. Currently, approximately 65% of the floodplain area in this section is occupied by wet meadows. These habitats have high ecological value and serve as important habitats for bird fauna. The dominant plant communities here belong to the *Molinio-Arrhenatheretea* class, which includes semi-natural or anthropogenic meadow and pasture communities of meso- and eutrophic habitats occurring on mineralizing and drying peatlands of low-lying peat [45].

2.2. Hydrological and GIS Data Sources

Hydrological conditions and spatial analysis of floods in the study area were performed with the help of a geographic information system (GIS) using QGIS 3.22.14 software (Białowieża) This allowed the visualization of data and simulation of water-level changes. The extent of river flooding in the valley delineating the river recharge zone was determined based on a digital elevation model (DEM) with a spatial resolution of 1 m, which was obtained from the resources of the Main Office of Geodesy and Cartography in Poland. A data set of surface and groundwater levels from hydrological years 2017–2019 was used for the hydrological analyses. The data were obtained from automatic pressure sensors programmed to record water levels with an accuracy of 1 cm per hour. The sensor network consisted of two devices installed in the river channel monitoring sites (R1-R2), four in the ditch network (D1-D4), and two in the piezometric observation wells (P1-P2), (Figure 1). Hydrological data were compared with water levels and flow rates of the Supraśl River measured at the nearest gauge in Nowosiółki by the Institute of Meteorology and Water Management (IMGW-PIB). Due to the long distance between the gauge and the study area, a data correction was applied. The determination of the extent of inundation was combined with the measurement of the height of the water table using GPS RTK (Real-Time Kinematic), which determines the coordinates of points in real time directly in the field.

2.3. Field Observations and Riparian Habitat Classification

The fieldwork included three survey cycles in 2017, 2018, and 2019 with surveys conducted from June to August. Initially, 4 types of riparian wetland habitats were distinguished based on an analysis of orthophoto maps (Geodesy and Mapping Center), field surveys, and hydrological data: I wet, II semi-wet, III semi-dry, and IV dry in the studied section of the floodplain (Figure 1).

To assess whether the four identified riparian habitat zones, corresponding to the influence of the duration and frequency of river floods, affect the distribution and composition of vegetation communities, we applied ecological indicator values as the most common and sufficiently effective method [46]. Thus, indicator values for soil moisture (SM_E), fertility (F_E), reaction pH (R_E), soil granulometry (SG_E), and organic matter content (OM_E) were adopted from Zarzycki et al. [47]. Indicator values were calculated based on in situ observations of plant species distribution/occurrence and assigned to a 5-point scale (weighted average). For species for which Zarzycki et al. [47] provided ranges of values, the average values from these ranges were taken. Only average indicator values from plots where at least three species with known indicator values were present were considered in the analysis. At each point where zone-specific vegetation was present, survey plots (10 × 10 m) were located to within 1 m using a GPS receiver. A total of 25 phytosociological relevés (vegetation plots) were conducted in 2017 and 25 phytosociological relevés (vegetation plots) were conducted in 2019. In the assemblages of *Agropyro-Rumicion crispi*, 16 phytosociological relevés (vegetation plots) were created, while *Phragmitetum australis*, *Phalaridetum arundinaceae*, and *Lysimachio vulgari—Filipenduletum* all had 3 plots each. A list of plant species was compiled for each plot, including a rating of the plants using a modified Braun–Blanquet 9-point scale. Plant community diversity was analyzed in four riparian habitat zones of the Supraśl floodplain using the species richness index (S), Shannon–Wiener diversity index (H') [48,49], Margalef index (R) [49], Simpson dominance index (D), and Pielou evenness index (J') [50]. The names of plant taxa were adopted after Szwed et al. [51]. The syntaxonomic nomenclature was adopted according to Ratyńska et al. [52].

It should be noted that the designated riparian habitat zones were also used as pastures for the Konik Polski, also known as the Polish horse (Konik means short horse), a semi-wild breed used mainly for conservation grazing. The studied section of the river floodplain is adjacent to extensively used cropland (cereals).

2.4. Statistical Analyses

Statistical analyses were performed to evaluate the effects of hydrologic factors on plant species distribution. The normality and homogeneity of variance assumptions were evaluated using Shapiro–Wilk and Levene's tests. For comparison between the means of biodiversity parameters in wet and dry years as well as riparian moisture zones, we performed an analysis of variance (one-way ANOVA) with Student's *t*-test and Tukey's HSD test as a post hoc procedure, respectively. Correlation analysis was used to evaluate the relationship between water level and vegetation occurrence.

To evaluate whether plant species could quantitatively represent different riparian habitat types (moisture zones) along a lateral moisture gradient, we ordinated riparian herbaceous vegetation data using linear discriminant analysis (LDA) in statistic package PAST 4.03 (Natural History Museum—University of Oslo, Norway) [53]. Major vegetation types were applied to establish the discriminant functions, and then the samples were classified into predicted vegetation zones. Based on the comparisons between a priori groups and predicted groups, riparian moisture zones were introduced to quantitatively interpret the results.

To identify vegetation communities with an affinity for a particular habitat type, indicator species analysis (ISA) [54] was used. The ISA value is calculated as the product of relative species abundance and frequency of occurrence to obtain a maximum indicator value (IV) for each species. Each species is then assigned to the group for which it has the highest indicator value. Indicator values range from 0 to 100, with a value of 100 representing a perfect indicator species, i.e., a species that occurs exclusively in one group, is found in all samples in that group, and has a high relative abundance within that group. A value of zero represents a species that has no indicator value for any group, is generally either rare in the data set, or occurs with a nearly uniform distribution in all or most groups [55]. Statistical significance of ISAs is assessed by a randomization procedure with 999 random permutations. Taxa that were significant at the 0.05 level were considered indicator species. The ISA was performed in PC–ORD version 6.0 developed and distributed by MjM Software (Gleneden Beach, OR, USA) [56].

3. Results

3.1. Hydrological Characteristics

Analysis of long-term hydrological data (2006–2020) for the Supraśl River showed a typical flood pulsing system (Figure 2). Due to intensive precipitation and snow melt, the year 2017 was characterized by extremely wet conditions, while 2018 was average and 2019 was classified as a dry year.

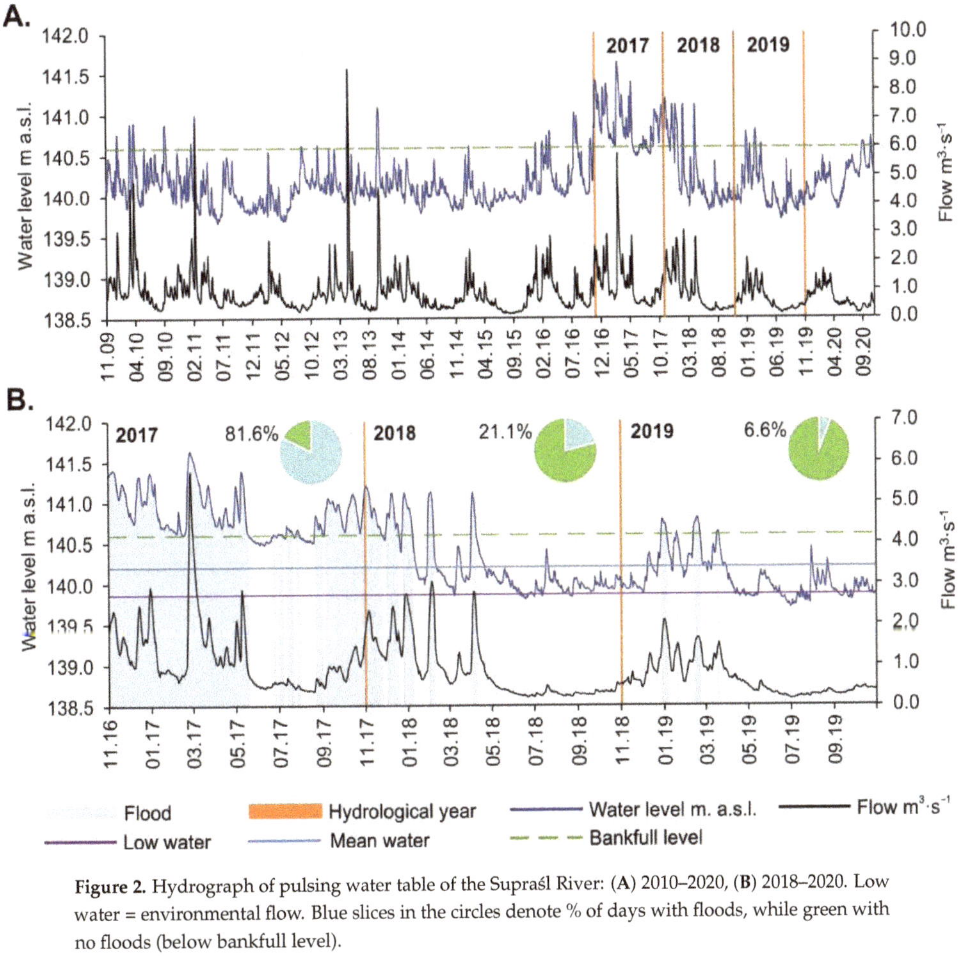

Figure 2. Hydrograph of pulsing water table of the Supraśl River: (**A**) 2010–2020, (**B**) 2018–2020. Low water = environmental flow. Blue slices in the circles denote % of days with floods, while green with no floods (below bankfull level).

Water levels of the Supraśl River exceeded the bankfull level in the wet year 2017 for almost 81.6% of days, while in 2018, it was 21.1%, and in the dry year 2019, they were only exceeded 6.6% of days. During 2017, the flood covered the entire study area and reached a maximum stage of 109 cm above the edge of the riverbed. In 2018, the maximal water stage was 64 cm above the riverbanks, and in 2019, it was only 26 cm.

3.2. Riparian Habitats

Within the study area, the dominant species were floodplain grasslands represented by species from the *Agropyro–Rumicion crispi* alliance. On the fringes of a large complex of floodplain grasslands, riverside herbaceous plants with *Filipendula ulmaria*—anthropogenic or partially natural herbal communities along watercourses developed. These are ecotone areas between the watercourse, which are moderately used for agriculture. Their species structure was unstable and most often corresponded to the characteristics of the *Lysimachio vulgaris–Filipenduletum* association. In the southern part of the area, the *Phalaridetum arundinaceae* rush was developing on marshy soils, touching the extensive *Phragmitetum* reed rush. Depending on the water conditions, they accompany the main species, mostly mud species from the *Calthion* alliance or herbaceous species from the *Filipendulion* alliance.

Various reaches and durations of the water table during flood pulsing across the naturally flat bottom of the Supraśl River valley contributes to specific riparian diversity patterns. The lateral zonation of vegetation formed by the reach of floods is presented in Figure 3.

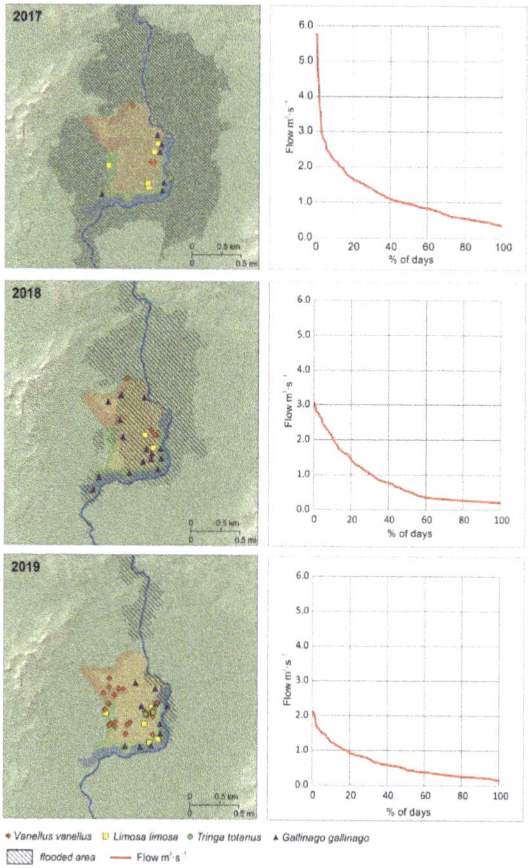

Figure 3. Left panel shows floodplain area flooded in 2017, 2018, and 2019 with the reach of riparian vegetation types and distribution of waterfowl species detected. Right panel shows diagrams with flow duration curves for the years of study, respectively. Data from 2017 to 2019 on the occurrence of valuable avifauna species are taken from the materials of the Polish Society for the Protection of Birds (PTOP, not published) and are included as a background characterizing the natural values of the studied section of the Supraśl River.

In consecutive years, flow duration curves for the wet 2017, average 2018 and dry 2019 years showed that the vegetation zone adjacent to the river bed is the most vulnerable to floods with flows as low as 2 $m^3 s^{-1}$, as noted in 2019. During the years of average conditions, when flow rates are as high as 3 $m^3 s^{-1}$ (2018), water covers the flat bottom. Flows as high as 3–6 $m^3 s^{-1}$ cause inundations of not only the flat floodplain area but also local convex forms as local sand bars. Under such conditions, certain phytocoenoses have adapted to wet and dry periods.

Based on the bird data for the study area provided by the Polish Society for the Protection of Birds (PTOP), we were able to determine a specific distribution of avifauna in relation to the reach of flood and vegetation zones (Figure 3). For example, during floods in 2017, when water depth exceeded 1.0 m, tall vegetation in zones I and II (shrubs and trees) was preferred by snipe (*Gallinago gallinago*), while black-tailed godwit (*Limosa limosa*) or lapwing (*Vanellus vanellus*) tended to prefer shallow waters in zone III. Waterfowl species were more widespread in the dry years of 2018 and 2019.

Based on hydrological and botanical data, four moisture habitat zones in the Supraśl river floodplain have been determined based on linear discriminant analysis (Figure 4a,b). LDA ordination yielded significant results when fitted with the percentage of flooded area at maximum inundation depth (h_{max}). The LDA best described the separation between moisture habitat zones. It indicated that the phytocenoses were distributed across the lateral floodplain gradient and the reach of floods, whereby zone I is more associated with riverbeds, zone II dominated by reed canary grass is associated with flat lands with shallow depressions that are flooded, zone III is principally overgrown by anthropogenic or partially natural herbaceous communities along watercourses and is associated with ditch networks, and zone IV creates floodplain riparian meadows on uplands flooded on extremely wet occasions.

A.

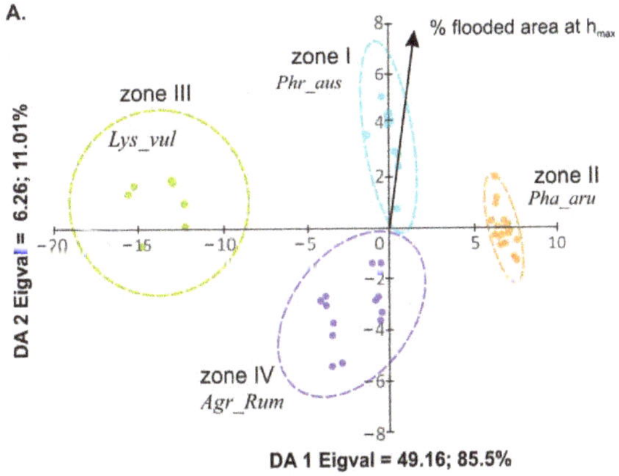

B.

Confusion matrix

Given groups:	Predicted groups: 4	3	2	1	Total
4	13	0	0	0	13
3	0	19	0	0	19
2	0	0	7	0	7
1	1	0	0	10	11
Total	14	19	7	10	50

100% correctly classified
98% jackknifed

Figure 4. Linear discriminant analysis biplot (**A**) and confusion matrix (**B**) for plant species distribution in moisture habitat zones in the Supraśl River floodplain. Denotations of dominant communities: *Agr_Rum: Agropyro-Rumicion crispi; Phr_aus: Phragmitetum australis; Pha_aru: Phalaridetum arundinaceae; Lys_vul: Lysimachio vulgari—Filipenduletum.* Zone: I moist (reed rush), II semi-moist (reed canary grass), III semi-dry (anthropogenic or partially natural herbaceous communities along watercourses), IV dry (floodplain riparian meadows) in the studied section of the Supraśl floodplain.

Biodiversity indicators showed statistical differences between zones (Figure 5) apart from the number of taxa and evenness. The highest differences in biodiversity were stated among the most remotely located zones I (H′ = 0.6) and IV (H′ = 1.45). In zone I, a clear dominance of the rush vegetation was observed and supported with dominance index D = 0.7. Differences between zones were also shown for indicator values of SM_E,

F_E, R_E, SG_E, as well as OM_E, and they are attached in the Supplementary Materials (Tables S1 and S2).

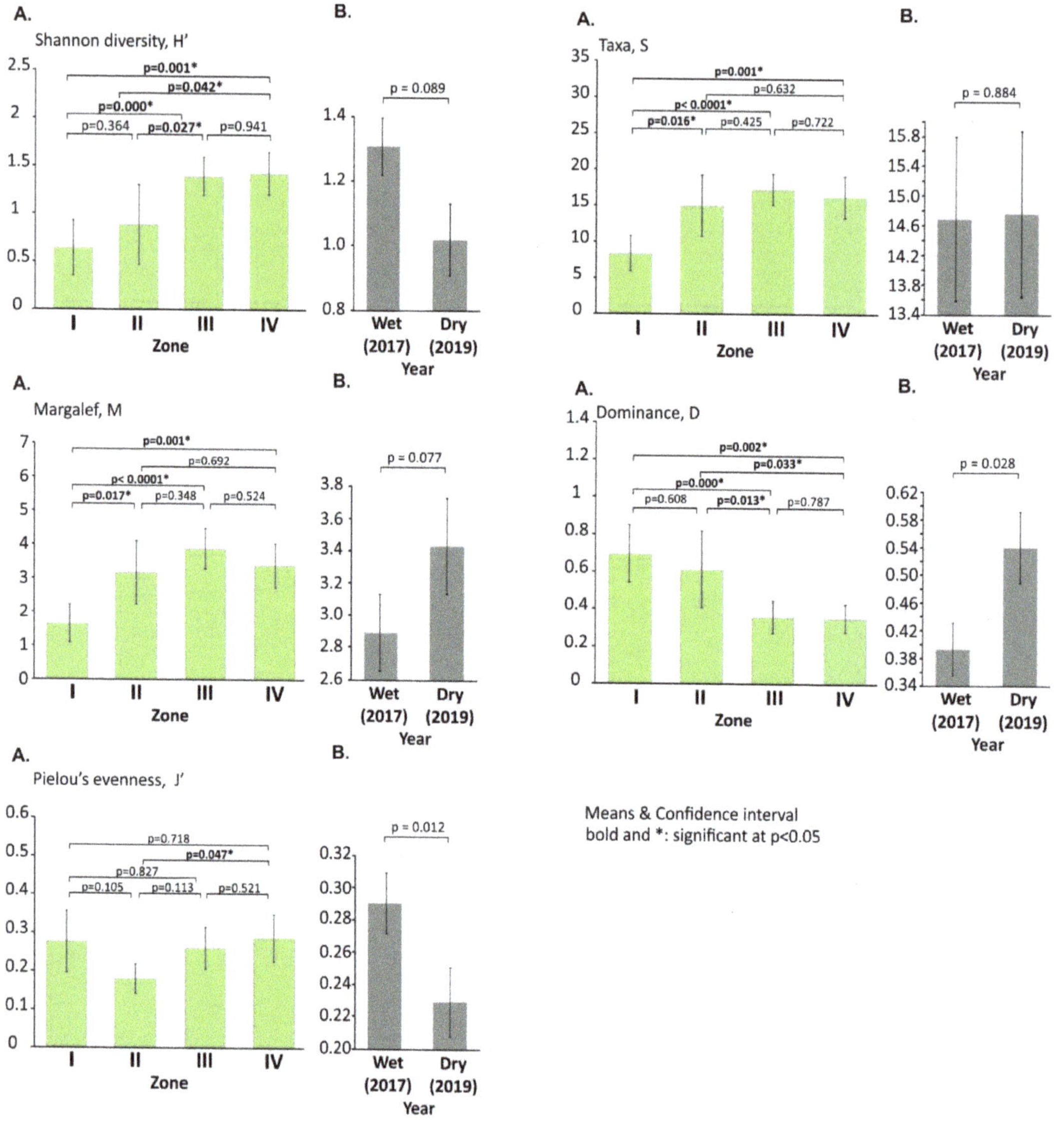

Figure 5. Biodiversity indices (mean ± SEM) of vegetation communities for (**A**) moisture zones and (**B**) wet and dry years in the Supraśl River floodplain (one-way ANOVA): (**A**) the Tukey's test as a post hoc procedure; (for (**B**) Student's *t*-test); *p*-value—Monte Carlo method after Bonferroni correction.

3.3. Indicator Species

Considering the relative abundance of 87 species in the four zones, we found species that indicate changes in moisture conditions (Table 1). The highest number of indicator species (9) was found in the dry zone (IV), representing floodplain herbaceous meadows and in the ecotone zone formed by semi-wet (II), which is represented by reed canary grass rushes. The number of indicator species was lower in the other two zones where the water conditions were more stable.

Table 1. The indicator species analysis (ISA). List of plant species characteristic for the different moisture conditions in the Supraśl river floodplain. Moisture riparian zones were defined by hydrology and plant communities criteria [54].

Moisture Riparian Zone	Species *	Observed Indicator Value (IV)	IV from Randomized Zones		p-Value **
			Mean	±SD	
I Wet	*Phalaris arundinacea*	80.5	32	8.98	0.0002
	Phragmites australis	55.3	16.7	7.3	0.0016
	Symphytum officinale	45.5	12.2	6.63	0.0008
	Urtica dioica	34.9	13.3	6.89	0.0132
	Alisma plantago-aquatica	27.3	10	5.48	0.0304
	Galium palustre	27.3	9.6	5.6	0.0112
	Carex rostrata	18.2	8.9	4.8	0.0720
II Semi-wet	*Filipendula ulmaria*	88.6	21.7	7.33	0.0002
	Galium palustre	71.4	11.9	6.49	0.0004
	Cirsium rivulare	70.4	16.4	6.88	0.0002
	Potentilla erecta	53.6	15.3	7.46	0.0010
	Veronica scutellata	52.7	13.7	6.83	0.0006
	Agrostis capillaris	42.9	9.6	5.55	0.0018
	Galium molugo	34.7	19.8	7.4	0.0444
	Achillea monticola	28.6	8.5	4.6	0.0204
	Galium aparine	28.1	9.9	5.5	0.0106
III Semi-dry	*Deschampsia caespitosa*	70.3	39.6	11.53	0.0134
	Festuca arundinacea	61.5	30.3	9.82	0.0092
	Lythrum salicaria	51.8	29.8	10.13	0.0296
	Epilobium palustre	44.5	21.1	7.25	0.0132
IV Dry	*Anthoxanthum odoratum*	76.1	22.7	9.03	0.0002
	Achillea millefolium	72.3	26.7	9.71	0.0010
	Lychnis flos-cuculi	56.5	23.9	6.76	0.0012
	Juncus compressus	52.1	23.2	7.4	0.0052
	Festuca pratensis	50.9	29.4	7.06	0.0114
	Potentilla anserina	49.2	23.9	7.28	0.0066
	Vicia cracca	41.6	16.1	7.55	0.0096
	Taraxacum officinale	30.8	10.1	5.68	0.0132
	Leontodon autumnalis	68	27.9	8.04	0.0002

Note(s): * only statistically significant species are presented; Monte Carlo test of significance of observed maximum indicator value for variable after 4999 permutations. ** proportion of randomized trials with indicator value equal to or exceeding the observed indicator value p-values = (1 + number of runs $\geq$ observed)/(1 + number of randomized runs). Zone = group identifier for group with maximum observed II and IV.

ISA analysis (Table 1) showed that among indicator species with high water requirements, *Phalaris arundinaceae* (IV = 80.5, p = 0.0002) and *Phragmites australis* (IV = 55.3, p = 0.0016) dominate. Both species in wet zone I form well-developed and low-impacted riparian galleries. At the other extreme (IV dry zone), remotely located from the riverbed were phytocoenoses composed of simpler communities characterized by species adapted to dry conditions and with occasional flood disturbances such as *Anthoxanthum odoratum* (IV = 76.1, p = 0.0002), *Achillea millefolium* (IV = 72.3, p = 0.001), *Leontodon autumnalis* (IV = 68.0, p = 0.0002), and *Lychnis flos-cuculi* (IV = 56.5, p = 0.0012). For semi-wet zone II, ISA produced the highest values for *Filipendula ulmaria* (IV = 88.6, p = 0.0002), *Cirsium rivulare* (IV = 70.4, p = 0.0002) or *Galium palustre* (IV = 71.4, p = 0.0004), while semi-dry zone III showed the highest values for *Deschampsia caespitosa* (IV = 70.3, p = 0.013), *Festuca arundinaceae* (IV = 61.5, p = 0.009), or *Lythrum salicaria* (IV = 51.8, p = 0.030).

4. Discussion

4.1. Patterns of Vegetation Diversity in Riparian Zones of Rivers

River floodplains are among the areas with the highest biodiversity, as these areas represent habitats with high levels of structural and functional dynamics, which are mainly induced by a hydrological regime [57]. Natural riparian landscapes are covered by mosaics of habitats differing in age, moisture, sediment characteristics, productivity, biota diversity, abundance, composition and successional status [58].

Considering the dynamic nature of river floodplains, we defined a tool that allows for understanding the relationship between the structure and functioning of a wetland ecosystem and hydrological changes in the river. The obtained results allowed us to simplify the complex ecosystem of river wetlands into a concept, combining the structure and pattern of wetland plants controlled mainly by both hydrological regime of the river and a pattern of hydrogeomorphic features of riparian area, as shown in Figure 6.

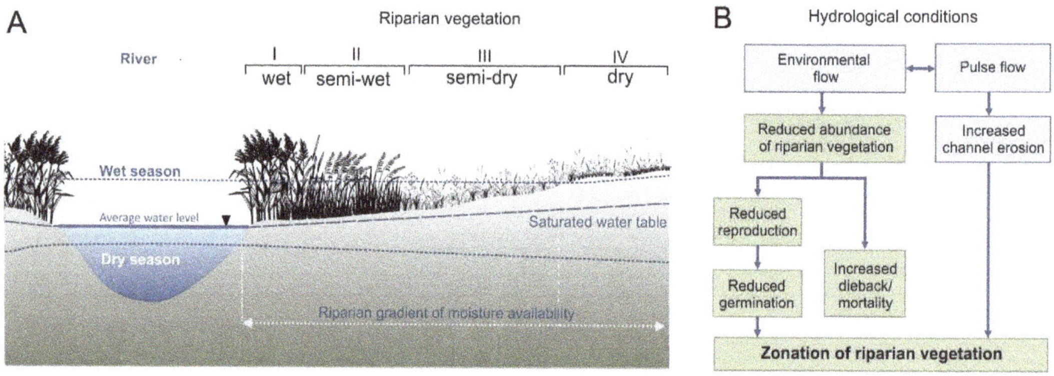

Figure 6. A sketch of riparian zonation in relation to river stage in a lowland river floodplain (**A**). Effects of hydrological conditions on vegetation in riparian area (**B**).

This concept takes into account both physical and biological elements that influence the dynamics and patterns of inundation along the river. Additionally, we considered various biological aspects of the wetland ecosystem, such as vegetation types, habitat types, and their interrelationships. Based on these data, we are able to forecast how changes in river water levels affect habitat availability, biological diversity, and the functions of the wetland ecosystem.

The natural decrease in the degree of hydration, determined by the relief of the terrain and the efficiency of the drainage network, resulted in variations in the floristic composition of floodplain grasslands. Within the river floodplain, flood disturbance and water availability vary along lateral gradients; over the length of the river, the intensity and frequency of flood disturbance decreases with increasing distance from (and above) the active channel, paralleling the increase in floodplain elevation resulting from sediment aggradation and riverbed incision processes [59]. Thus, changes in the hydrological regime set up patterns of plant species diversity in river riparian zones, which is in line with the works of other authors [60–62]. Our hydroecological analysis performed on the example of the Supraśl River floodplain (NE Poland) presents significant results by coupling the wetland landscape ecology into hydrological units.

Based on them, the composition of wetland communities in riparian biophysical habitats varies along a lateral hydrological gradient. The distribution, extent, and abundance of wetland communities increase progressively from herbaceous communities through marshes within the riverine zone, reflecting a moisture gradient. Simultaneously, the predominant patch area and quantity shift as transitional meadow communities replace them beyond the riverine zone. Floodwater plays a pivotal role in regulating the distribution of wetland communities within the studied floodplain. The typical marsh communities in the saturated zone give way to a broader distribution of wet meadows in the seasonally saturated zone, which is eventually supplanted by the predominant presence of wet meadows in the yearly unsaturated zone. Additionally, a cluster analysis of quantitative spatial characteristics reveals numerical distinctions between marshes and non-marshes in the core of the floodplain, further emphasizing the influence of hydrological factors on wetland community composition.

Vegetation locally reduces stream gradient stress across the floodplain. Disturbance regimes influence plant species diversity at local scales [63] through increasing groundwater depth and decreasing the replenishment of shallow soil moisture by overbank floods [64]. A zonal vegetation in floodplains is shaped by broad hydrological gradients, periodic flooding [65] and, to a lesser extent, soil composition [66]. Species that prefer wet microhabitats grow and survive longer compared to those in dry microhabitats [66].

4.2. Changes in Vegetation Cover Due to Hydrological Variations

Our results (Figure 5) confirmed that riparian vegetation is a suitable environmental change indicator that responds directly to the flow regime in an inter-annual timeframe [67,68]. We have shown that vegetation cover changes as a result of flooding and subsequent drying in the higher parts of the floodplain, which is consistent with the results of Kleinhans et al. [59].

The intermediate intensity or frequency of disturbance (or length of time since disturbance) should result in the highest levels of diversity by allowing a mixture of species adapted to stable environments and species adapted to frequent disturbance to coexist, although the magnitude of the effect may be small [69,70]. Plant indicator systems in the riverbed are well known [71–73]; however, they do not cover the vegetation of the flood zones of the river valley. This is due to the limitations of river monitoring systems, which do not include assessments of river valley biodiversity. As a result of our research, we have also identified a set of indicator species that represent the variability of the river's impact on floodplain vegetation. Low flows were mostly regarded as negative both for aquatic organisms [74] and for riparian vegetation [75]. It could be proven that the species is able to outlast unsuitable conditions as long as seeds are remnant in the seed bank and that the negative effects of the new floodplain stream on the target species can be compensated sufficiently. Literature data [66,76] indicate that the duration of the withdrawal period must be long enough (here, germination is observed after 3 weeks, and there is at least another 3 weeks for rooting) to ensure successful establishment of the target species. If this period is too short, or if the water level rises significantly again, the seedlings may be fatally damaged [76] and the seed bank may be depleted without new seeds being produced. On the other hand, if the period of drought is too long, more competitive reed species may become established and outcompete the target species. Vegetation is in a fluctuating equilibrium that changes over time [77]. In a dry year, there may be large areas of typical terrestrial vegetation, whereas in wet years, reed species or aquatic plants may dominate [78]. Floodplain vegetation is able to recover from disturbances (floods or dry years), and a resilient system needs even these extreme events from time to time to reduce strong competitors [79,80]. Thus, if humans want to restore floodplains that suffer from static water conditions, dynamically changing water conditions need to be restabilized at irregular intervals, just as nature would do. If the release of water into the floodplain is controlled by humans through sluices (rather than automatically through spillways in the embankment; [81]), the hydrological objectives and target habitats and species must be clearly defined. Plant species characteristic of the various moisture conditions in the floodplain (Table 1) should be monitored to evaluate the effectiveness of potential restoration efforts.

4.3. Impact of Unstable River Valley Environment on Other Natural Values

The unstable environment of the river floodplain, combining the high dynamics of plant succession with frequent habitat restoration due to flooding, is inhabited by unique bird communities adapted to such conditions (Figure 6). Changes in the river environment, which limit the extent and frequency of flooding, are usually associated with climate change or anthropogenic transformations [6,7]. We observed that changes in the moisture content of the Supraśl Valley and the introduction of the Konik polski population in 2019 caused a reconstruction of meadow-breeding waders. Horses on a pasture influence the grass not only by grazing but also by trampling, rolling and leaving droppings. The year-round grazing of Konik polski horses, kept at a low stocking rate in the Supraśl Valley, has little effect on the grassland communities. The impact of the presence of horses was mainly manifested by changes in the structure of the sward and a reduction in the sward cover [82]. The droughts in the Supraśl Valley favored the growth of the populations of lapwing and black-tailed godwit, but they did not cause changes in the use of the Supraśl Valley by redshank, and they reduced the abundance of snipe. The black-tailed godwit and the redshank seem to have special needs, i.e., large wet meadows with large areas flooded in spring, where extensive cattle grazing takes place and where no additional fertilizer is used [83]. Our studies show that the lapwing has the highest tolerance to changes in meadow moisture, while the snipe has the lowest, which is also confirmed by field observations of other authors [84–86].

The results of this study provide valuable information regarding the relationships between wetland vegetation structure and hydrological changes in rivers. Riparian wetland ecosystems play a crucial role in maintaining biological diversity. By examining the impact of water level changes in rivers on habitat availability and the biological diversity of wetland ecosystems, we aim to better understand the complex interactions between riparian ecosystems and river hydrology. Our research, as well as that of Osmundson et al. [87] and Obolewski et al. [88], suggests that river valley restoration should begin with the identification of physical and biological pathways linking flow changes with ecological functions and species responses.

This knowledge can serve as a basis for conservation efforts, sustainable management practices, and decision-making processes aimed at preserving the biological diversity and ecological integrity of riparian ecosystems in similar regions. Monitoring and detailed solutions for the use of the floodplain should be included in the management of nature conservation forms (in case of the Supraśl floodplain, the task plan for the protection of the Natura 2000 site) after public consultation and then implemented in the water management plan (a mandatory document for water management authorities). By understanding the dynamics of riparian wetland ecosystems and their response to hydrological changes, we can work toward the conservation and sustainable management of these important ecological systems.

5. Conclusions

The results of our study on the impact of hydrological conditions on the functioning of wetland ecosystems, using the example of the Supraśl floodplain located in central Europe, show that the natural flow regime, defined as flood pulse, is responsible for the development of site-specific phytocenoses adapted to local hydrological conditions. Maintaining the river's natural flow regime with seasonal and annual water changes is crucial to ensure optimal preservation of biodiversity in wetland ecosystems. The spatial heterogeneity of flooding along rivers and streams affects local species diversity. The composition of riparian vegetation depends on the extent and timing of flooding and, as such, can indicate the development of strategies for riparian area management, biodiversity conservation, and planning actions.

The example of the Supraśl River shows that dynamically fluctuating water conditions, floods and droughts that occur at irregular time intervals require a long-term observation period for reliable ecological monitoring of riparian wetlands. Therefore, a thorough understanding of the functioning of such ecosystems and the relationships within them is essential. Therefore, it is important to select and clearly define target groups from among the many competing floodplain habitat types. All these factors make the Supraśl River floodplain an important natural area that requires special protection and sustainable management. Preserving biodiversity, maintaining appropriate hydrological conditions, and using natural resources wisely are crucial for effective conservation and meeting the needs of current and future generations.

Then, the intended measures have to be conducted, and the resilient system will react. This approach should further minimize detrimental effects on other habitats which are not of first priority (in case of groundwater drawdown the aquatic habitats) but whose conservation is also compulsory, e.g., due to the EU Water Framework Directive. All stakeholders should be involved in this process in order to avoid discussions or even the suspension of measures during appropriate periods (e.g., natural floods or low water conditions).

Supplementary Materials: The following supporting information can be downloaded at https://www.mdpi.com/article/10.3390/w16010164/s1, Table S1: List of ecological indicators and plant species; Table S2: Analysis of variance (two-way ANOVA: year and zone) for ecological indicator values.

Author Contributions: Conceptualization, P.B. and M.G.; methodology M.G.; software, P.B.; validation, K.G.-L., K.O. and M.S.-W.; formal analysis, P.B.; investigation, P.B., W.G. and M.G.; data curation, M.S.-W., W.G.; writing—original draft preparation, P.B.; writing—review and editing, K.G.-L., K.O. and P.B.; visualization, P.B.; supervision, K.O.; funding acquisition, K.O. All authors have read and agreed to the published version of the manuscript.

Funding: The study was financially supported by Minister of Education and Science in the range of the program entitled "Regional Initiative of Excellence" for the years 2019–2023, Project No. 008/RID/2018/19, amount of funding 12,000,000 PLN. The results presented in this paper were obtained as part of a comprehensive study partially financed by the University of Warmia and Mazury in Olsztyn, grant No. 30.610.008–110. The data used in the study were obtained with the kind permission of the Polish Society for the Protection of Birds from the project "Active protection of the Common Snipe in peat bogs and river valleys in the Natura 2000 special bird protection area PLB200003" (grant no. POIS.02.04.00-00-0031/16), co-financed by the EU from the Cohesion Fund under Measure 2.4, Priority Axis II of the Operational Program Infrastructure and Environment 2014–2020 and the Voivodeship Fund for Environmental Protection and Water Management in Białystok.

Data Availability Statement: The data presented in this study are available upon request from the corresponding author.

Conflicts of Interest: The authors declare no conflicts of interest.

References

1. Tockner, K.; Pusch, M.; Borchardt, D.; Lorang, M.S. Multiple Stressors in Coupled River-Floodplain Ecosystems. *Freshw. Biol.* **2010**, *55*, 135–151. [CrossRef]
2. Critchley, C.N.R.; Chambers, B.J.; Fowbert, J.A.; Bhogal, A.; Rose, S.C.; Sanderson, R.A. Plant Species Richness, Functional Type and Soil Properties of Grasslands and Allied Vegetation in English Environmentally Sensitive Areas: Species Richness and Type in Relation to Soil Properties of Grassland. *Grass Forage Sci.* **2002**, *57*, 82–92. [CrossRef]
3. Rodríguez-Rojo, M.P.; Font, X.; García-Mijangos, I.; Crespo, G.; Fernández-González, F. An Expert System as an Applied Tool for the Conservation of Semi-Natural Grasslands on the Iberian Peninsula. *Biodivers. Conserv.* **2020**, *29*, 1977–1992. [CrossRef]
4. Petermann, J.S.; Buzhdygan, O.Y. Grassland Biodiversity. *Cur. Biol.* **2021**, *31*, R1195–R1201. [CrossRef] [PubMed]
5. Bengtsson, J.; Bullock, J.M.; Egoh, B.; Everson, C.; Everson, T.; O'Connor, T.; O'Farrell, P.J.; Smith, H.G.; Lindborg, R. Grasslands—More important for ecosystem services than you might think. *Ecosphere* **2019**, *10*, 02582. [CrossRef]
6. Grzybowski, M.; Glińska-Lewczuk, K. Principal Threats to the Conservation of Freshwater Habitats in the Continental Biogeographical Region of Central Europe. *Biodivers. Conserv.* **2019**, *28*, 4065–4097. [CrossRef]
7. Grzybowski, M.; Glińska-Lewczuk, K. The Principal Threats to the Peatlands Habitats, in the Continental Bioregion of Central Europe—A Case Study of Peatland Conservation in Poland. *J. Nat. Conserv.* **2020**, *53*, 125778. [CrossRef]
8. Wehn, S.; Burton, R.; Riley, M.; Johansen, L.; Hovstad, K.A.; Rønningen, K. Adaptive Biodiversity Management of Semi-Natural Hay Meadows: The Case of West-Norway. *Land Use Policy* **2018**, *72*, 259–269. [CrossRef]

9. Johansen, L.; Westin, A.; Wehn, S.; Iuga, A.; Ivascu, C.M.; Kallioniemi, E.; Lennartsson, T. Traditional Semi-Natural Grassland Management with Heterogeneous Mowing Times Enhances Flower Resources for Pollinators in Agricultural Landscapes. *Glob. Ecol. Conserv.* **2019**, *18*, e00619. [CrossRef]
10. Bates, A.J.; Sadler, J.P.; Henshall, S.; Hannah, D.M. Ecology and Conservation of Arthropods of Exposed Riverine Sedi-Ments (ERS). *Terr. Arthropod Rev.* **2009**, *2*, 77–98.
11. Kimberley, A.; Hooftman, D.; Bullock, J.M.; Honnay, O.; Krickl, P.; Lindgren, J.; Plue, J.; Poschlod, P.; Traveset, A.; Cousins, S.A.O. Functional Rather than Structural Connectivity Explains Grassland Plant Diversity Patterns Following Landscape Scale Habitat Loss. *Landsc. Ecol.* **2021**, *36*, 265–280. [CrossRef]
12. Gurnell, A.; Surian, N.; Zanoni, L. Multi-Thread River Channels: A Perspective on Changing European Alpine River Systems. *Aquat. Sci.* **2009**, *71*, 253–265. [CrossRef]
13. Zeiringer, B.; Seliger, C.; Greimel, F.; Schmutz, S. River Hydrology, Flow Alteration, and Environmental Flow. *Riverine Ecosyst. Manag.* **2018**, *8*, 67–89. [CrossRef]
14. Ward, J.V.; Tockner, K.; Arscott, D.B.; Claret, C. Riverine Landscape Diversity: Riverine Landscape Diversity. *Freshw. Biol.* **2002**, *47*, 517–539. [CrossRef]
15. Lenssen, J.P.M.; De Kroon, H. Abiotic Constraints at the Upper Boundaries of Two Rumex Species on a Freshwater Flooding Gradient. *J. Ecol.* **2005**, *93*, 138–147. [CrossRef]
16. van Eck, W.H.J.M.; Lenssen, J.P.M.; van de Steeg, H.M.; Blom, C.W.P.M.; de Kroon, H. Seasonal Dependent Effects of Flooding on Plant Species Survival and Zonation: A Comparative Study of 10 Terrestrial Grassland Species. *Hydrobiologia* **2006**, *565*, 59–69. [CrossRef]
17. Arias, M.E.; Wittmann, F.; Parolin, P.; Murray-Hudson, M.; Cochrane, T.A. Interactions between Flooding and Upland Disturbance Drives Species Diversity in Large River Floodplains. *Hydrobiologia* **2018**, *814*, 5–17. [CrossRef]
18. Stanford, J.A.; Lorang, M.S.; Hauer, F.R. The Shifting Habitat Mosaic of River Ecosystems. *SIL Proc. 1922–2010* **2005**, *29*, 123–136. [CrossRef]
19. Mouw, J.E.B.; Stanford, J.A.; Alaback, P.B. Influences of Flooding and Hyporheic Exchange on Floodplain Plant Richness and Productivity. *River Res. Applic.* **2009**, *25*, 929–945. [CrossRef]
20. Chen, Z.; Wang, W.; Woods, R.A.; Shao, Q. Hydrological Effects of Change in Vegetation Components across Global Catchments. *J. Hydrol.* **2021**, *595*, 125775. [CrossRef]
21. Härdtle, W.; Redecker, B.; Assmann, T.; Meyer, H. Vegetation Responses to Environmental Conditions in Floodplain Grasslands: Prerequisites for Preserving Plant Species Diversity. *Basic Appl. Ecol.* **2006**, *7*, 280–288. [CrossRef]
22. Hodapp, D.; Roca, I.T.; Fiorentino, D.; Garilao, C.; Kaschner, K.; Kesner-Reyes, K.; Schneider, B.; Segschneider, J.; Kocsis, Á.T.; Kiessling, W.; et al. Climate Change Disrupts Core Habitats of Marine Species. *Glob. Chang. Biol.* **2023**, *29*, 3304–3317. [CrossRef] [PubMed]
23. Grant, K.; Kreyling, J.; Heilmeier, H.; Beierkuhnlein, C.; Jentsch, A. Extreme Weather Events and Plant–Plant Interactions: Shifts between Competition and Facilitation among Grassland Species in the Face of Drought and Heavy Rainfall. *Ecol. Res.* **2014**, *29*, 991–1001. [CrossRef]
24. Richards, D.R.; Moggridge, H.L.; Warren, P.H.; Maltby, L. Impacts of Hydrological Restoration on Lowland River Floodplain Plant Communities. *Wetlands Ecol. Manage.* **2020**, *28*, 403–417. [CrossRef]
25. Lenssen, J.P.M.; Van De Steeg, H.M.; De Kroon, H. Does Disturbance Favour Weak Competitors? Mechanisms of Chang-Ing Plant Abundance after Flooding. *J. Veg. Sci.* **2004**, *15*, 305–314. [CrossRef]
26. Gattringer, J.P.; Ludewig, K.; Harvolk-Schöning, S.; Donath, T.W.; Otte, A. Interaction between Depth and Duration Matters: Flooding Tolerance of 12 Floodplain Meadow Species. *Plant Ecol.* **2018**, *219*, 973–984. [CrossRef]
27. Lan, Z.; Huang, H.; Chen, Y.; Liu, J.; Chen, J.; Li, L.; Li, L.; Jin, B.; Chen, J. Testing Mechanisms Underlying Responses of Plant Functional Traits to Flooding Duration Gradient in a Lakeshore Meadow. *J. Freshwat. Ecol.* **2019**, *34*, 481–495. [CrossRef]
28. van Eck, W.H.J.M.; van de Steeg, H.M.; Blom, C.W.P.M.; de Kroon, H. Is Tolerance to Summer Flooding Correlated with Distribution Patterns in River Floodplains? A Comparative Study of 20 Terrestrial Grassland Species. *Oikos* **2004**, *107*, 393–405. [CrossRef]
29. Obolewski, K.; Skorbiłowicz, E.; Skorbiłowicz, M.; Glińska-Lewczuk, K.; Astel, A.; Strzelczak, A. The Effect of Metals Accumulated in Reed (Phragmites Australis) on the Structure of Periphyton. *Ecotoxicol. Environ. Saf.* **2011**, *74*, 558–568. [CrossRef]
30. Garssen, A.G.; Baattrup-Pedersen, A.; Voesenek, L.A.C.J.; Verhoeven, J.T.A.; Soons, M.B. Riparian Plant Community Responses to Increased Flooding: A Meta-analysis. *Glob. Chang. Biol.* **2015**, *21*, 2881–2890. [CrossRef]
31. Casanova, M.T.; Brock, M.A. How Do Depth, Duration and Frequency of Flooding Influence the Establishment of Wetland Plant Communities? *Plant Ecol.* **2000**, *147*, 237–250. [CrossRef]
32. Gerard, M.; Kahloun, M.E.; Mertens, W.; Verhagen, B.; Meire, P. Impact of Flooding on Potential and Realised Grassland Species Rich-Ness. *Plant Ecol.* **2008**, *194*, 85–98. [CrossRef]
33. Soomers, H.; Karssenberg, D.; Soons, M.B.; Verweij, P.A.; Verhoeven, J.T.A.; Wassen, M.J. Wind and Water Dispersal of Wetland Plants Across Fragmented Landscapes. *Ecosystems* **2013**, *16*, 434–451. [CrossRef]
34. Moggridge, H.L.; Gurnell, A. Hydrological Controls on the Transport and Deposition of Plant Propagules within Riparian Zones. *Riv. Res. Appl.* **2010**, *26*, 512–527. [CrossRef]

35. Lake, P.S. *Drought and Aquatic Ecosystems: Effects and Responses: Lake/Drought and Aquatic Ecosystems: Effects and Responses*; John Wiley & Sons, Ltd.: Chichester, UK, 2011.
36. Junk, W.J.; Bayley, P.B.; Sparks, R.E. The Flood Pulse Concept in River-Floodplain Systems. *Can. J. Fish Aquat. Sci.* **1989**, *106*, 110–127.
37. Davidson, T.A.; Mackay, A.W.; Wolski, P.; Mazebedi, R.; Murray-Hudson, M.; Todd, M. Seasonal and Spatial Hydrological Variability Drives Aquatic Biodiversity in a Flood-Pulsed, Sub-Tropical Wetland: Aquatic Biodiversity in Tropical Flood-Pulsed Wetlands. *Freshw. Biol.* **2012**, *57*, 1253–1265. [CrossRef]
38. Seneviratne, S.I.; Nicholls, N.; Easterling, D.; Goodess, C.M.; Kanae, S.; Kossin, J.; Luo, Y.; Marengo, J.; McInnes, K.; Rahi-Mi, M.; et al. *Managing the Risks of Extreme Events and Disasters to Advance Climate Change Adaptation: Special Report of the Intergovernmental Panel on Climate Change*; Field, C.B., Barros, V., Stocker, T.F., Dahe, Q., Eds.; Cambridge University Press: Cambridge, NY, USA, 2012; pp. 109–230.
39. Blom, C.W.P.M.; Bögemann, G.M.; Laan, P.; van der Sman, A.J.M.; van de Steeg, H.M.; Voesenek, L.A.C.J. Adaptations to Flooding in Plants from River Areas. *Aquat. Bot.* **1990**, *38*, 29–47. [CrossRef]
40. Żmihorski, M.; Krupiński, D.; Kotowska, D.; Knape, J.; Pärt, T.; Obłoza, P.; Berg, Å. Habitat Characteristics Associated with Occupancy of Declining Waders in Polish Wet Grasslands. *Agric. Ecosyst. Environ.* **2018**, *251*, 236–243. [CrossRef]
41. Wassen, M.J.; Joosten, J.H.J. In Search of a Hydrological Explanation for Vegetation Changes along a Fen Gradient in the Biebrza Upper Basin (Poland). *Vegetatio* **1996**, *124*, 191–209. [CrossRef]
42. Górniak, A. Ecohydrological Determinants of Seasonality and Export of Total Organic Carbon in Narew River with High Peatland Contribution (North-Eastern Poland). *Ecohydrol. Hydrobiol.* **2019**, *19*, 1–13. [CrossRef]
43. Kottek, M.; Grieser, J.; Beck, C.; Rudolf, B.; Rubel, F. World Map of the Köppen-Geiger Climate Classification Updated. *Metz* **2006**, *15*, 259–263. [CrossRef] [PubMed]
44. Giełczewski, M. *The Narew River Basin: A Model for the Sustainable Management of Agriculture, Nature and Water Supply*; Netherlands Geographical Studies; Utrecht University Repository: Utrecht, The Netherlands, 2003.
45. Kiryluk, A. *Zmiany Siedlisk Pobagiennych i Fitocenoz w Dolinie Supraśli*; IMUZ Falenty: Falenty, Poland, 2007; pp. 12–14.
46. Zolotova, E.; Ivanova, N.; Ivanova, S. Global Overview of Modern Research Based on Ellenberg Indicator Values. *Diversity* **2022**, *15*, 14. [CrossRef]
47. Zarzycki, K.; Trzcińska-Tacik, H.; Różański, W.; Szeląg, Z.; Wołek, J.; Korzeniak, U. *Ecological Indicator Values of Vascular Plants of Poland. Ekologiczne Liczby Wskaźnikowe Roślin Naczyniowych Polski*; Szafer Institute of Botany, Polish Academy of Sciences: Kraków, Poland, 2002.
48. Simpson, E.H. Measurement of Diversity. *Nature* **1949**, *163*, 668. [CrossRef]
49. Nkoa, R.; Owen, M.D.K.; Swanton, C.J. Weed Abundance, Distribution, Diversity, and Community Analyses. *Weed Sci.* **2015**, *63*, 64–90. [CrossRef]
50. Głowacka, A.; Flis-Olszewska, E. The Biodiversity of Weed Communities of Dent Maize, Narrow-Leaved Lupin and Oat in Relation to Cropping System and Weed Control. *Agron. Sci.* **2022**, *77*, 123–137. [CrossRef]
51. Szwed, W.; Hennekens, S.M.; Pelsma, T.A.H.M.; Ratyńska, H.; Rusińska, A. A Numerical Data Base and Checklist of Taxa of Polish Flora Applicable in Phytosociology, Particularly for the TURBOVEG. *Zesz. Nauk. Wyższej Szkoły Pedagog. W Bydg.* **1999**, *14*, 5–18.
52. Ratyńskia, H.; Wojterska, M., Brzeg, A.; Kołacz, M. *Multimedialna Encyklopedia Zbiorowisk Roślinnych Polski*; Computer File, edition ver. 1.1, OCLC Number 82377940; Instytut Edukacyjnych Technologii Informatycznych: Bydgoszcz, Polish, 2010.
53. Hammer, O.; Harper, D.A.T.; Ryan, P.D. PAST: Paleontological Statistics Software Package for Education and Data Analysis. *Palaeont. Electr.* **2001**, *4*, 9.
54. Dufrêne, M.; Legendre, P. Species Assemblages and Indicator Species:The Need for a Flexible Asymmetrical Approach. *Ecol. Monogr.* **1997**, *67*, 345–366. [CrossRef]
55. Severns, P.M.; Sykes, E.M. Indicator Species Analysis: A Useful Tool for Plant Disease Studies. *Phytopathology* **2020**, *110*, 1860–1862. [CrossRef]
56. Mccune, B.; Mefford, M.J. *PC-ORD. Multivariate Analysis of Ecological Data. Version 5.32*; MjM Software: Gleneden Beach, OR, USA, 2006.
57. Mccabe, D.J. Rivers and Streams: Life in Flowing Water. *Nat. Educ. Knowl.* **2011**, *1*, 4.
58. Schindler, S.; O'Neill, F.H.; Biró, M.; Damm, C.; Gasso, V.; Kanka, R.; van der Sluis, T.; Krug, A.; Lauwaars, S.G.; Sebesvari, Z.; et al. Multifunctional Floodplain Management and Biodiversity Effects: A Knowledge Synthesis for Six European Countries. *Biodivers. Conserv.* **2016**, *25*, 1349–1382. [CrossRef]
59. Kleinhans, M.G.; de Vries, B.; Braat, L.; van Oorschot, M. Living Landscapes: Muddy and Vegetated Floodplain Effects on Fluvial Pattern in an Incised River: Effects of Mud and Vegetation On River Pattern. *Earth Surf. Process. Landforms* **2018**, *43*, 2948–2963. [CrossRef] [PubMed]
60. Tabacchi, E.; Tabacchi, A.-M.P. Functional Significance of Species Composition in Riparian Plant Communities. *J. Am Water Resources Assoc.* **2001**, *37*, 1629–1637. [CrossRef]
61. Brown, R.L.; Peet, R.K. Diversity and Invasibility of Southern Appalachian Plant Communities. *Ecology* **2003**, *84*, 32–39. [CrossRef]
62. Mligo, C. Diversity and Distribution Pattern of Riparian Plant Species in the Wami River System, Tanzania. *J. Plant Ecol.* **2016**, *10*, 259–270. [CrossRef]

63. Huston, M. A General Hypothesis of Species Diversity. *Am. Nat.* **1979**, *113*, 81–101. [CrossRef]
64. Bendix, J.; Hupp, C.R. Hydrological and Geomorphological Impacts on Riparian Plant Communities. *Hydrol. Process.* **2000**, *14*, 2977–2990. [CrossRef]
65. Zerbe, S. Rivers and Floodplains. In *Restoration of Ecosystems—Bridging Nature and Humans*; Springer: Berlin/Heidelberg, Germany, 2023; pp. 209–233. [CrossRef]
66. Gattringer, J.P.; Donath, T.W.; Eckstein, R.L.; Ludewig, K.; Otte, A.; Harvolk-Schöning, S. Flooding Tolerance of Four Floodplain Meadow Species Depends on Age. *PLoS ONE* **2017**, *12*, e0176869. [CrossRef]
67. Benjankar, R.; Egger, G.; Jorde, K.; Goodwin, P.; Glenn, N.F. Dynamic Floodplain Vegetation Model Development for the Kootenai River, USA. *J. Environ. Manage.* **2011**, *92*, 3058–3070. [CrossRef]
68. Rivaes, R.; Boavida, I.; Santos, J.M.; Pinheiro, A.N.; Ferreira, T. Importance of Considering Riparian Vegetation Requirements for the Long-Term Efficiency of Environmental Flows in Aquatic Microhabitats. *Hydrol. Earth Syst. Sci.* **2017**, *21*, 5763–5780. [CrossRef]
69. Amoros, C.; Bornette, G. Connectivity and Biocomplexity in Waterbodies of Riverine Floodplains: Connectivity and Biocomplexity in Riverine Floodplains. *Freshwat. Biol.* **2002**, *47*, 761–776. [CrossRef]
70. Mackey, R.L.; Currie, D.J. A Re-Examination of the Expected Effects of Disturbance on Diversity. *Oikos* **2000**, *88*, 483–493. [CrossRef]
71. Szoszkiewicz, K.; Jusik, S.; Pietruczuk, K.; Gebler, D. The Macrophyte Index for Rivers (MIR) as an Advantageous Approach to Running Water Assessment in Local Geographical Conditions. *Water* **2019**, *12*, 108. [CrossRef]
72. Kaijser, W.; Birk, S.; Hering, D. Environmental Ranges Discriminating between Macrophytes Groups in European Rivers. *PLoS ONE* **2022**, *17*, e0269744. [CrossRef] [PubMed]
73. Połeć, K.; Grzywna, A.; Tarkowska-Kukuryk, M.; Bronowicka-Mielniczuk, U. Changes in the Ecological Status of Rivers Caused by the Functioning of Natural Barriers. *Water* **2022**, *14*, 1522. [CrossRef]
74. Walters, A.W.; Post, D.M. How Low Can You Go? Impacts of a Low-Flow Disturbance on Aquatic Insect Communities. *Ecol. Appl.* **2011**, *21*, 163–174. [CrossRef] [PubMed]
75. Stromberg, J.C.; Beauchamp, V.B.; Dixon, M.D.; Lite, S.J.; Paradzick, C. Importance of Low-Flow and High-Flow Characteristics to Restoration of Riparian Vegetation along Rivers in Arid South-Western United States. *Freshw. Biol.* **2007**, *52*, 651–679. [CrossRef]
76. Jensch, D.; Poschlod, P. Germination Ecology of Two Closely Related Taxa in the Genus Oenanthe: Fine Tuning for the Habitat? *Aquat. Bot.* **2008**, *89*, 345–351. [CrossRef]
77. Wang, D.; Liu, Y.; Zheng, L.; Li, D. Growing Impacts of Low-Flow Events on Vegetation Dynamics in Hydrologically Connected Wetlands Downstream Yangtze River Basin after the Operation of the Three Gorges Dam. *J. Geogr. Sci.* **2023**, *33*, 885–904. [CrossRef]
78. Finger-Higgens, R.; Bishop, T.B.B.; Belnap, J.; Geiger, E.L.; Grote, E.E.; Hoover, D.L.; Reed, S.C.; Duniway, M.C. Droughting a Megadrought: Ecological Consequences of a Decade of Experimental Drought atop Aridification on the Colorado Plateau. *Glob. Chang. Biol.* **2023**, *29*, 3364–3377. [CrossRef]
79. Ilg, C.; Dziock, F.; Foeckler, F.; Follner, K.; Gerisch, M.; Glaeser, J.; Rink, A.; Schanowski, A.; Scholz, M.; Deichner, O.; et al. Long-Term Reactions of Plants and Macroinvertebrates to Extreme Floods in Floodplain Grassland. *Ecology* **2008**, *89*, 2392–2398. [CrossRef] [PubMed]
80. Larson, D.M.; Carhart, A.M.; Lund, E.M. Aquatic Vegetation Types Identified during Early and Late Phases of Vegetation Recovery in the Upper Mississippi River. *Ecosphere* **2023**, *14*, e4468. [CrossRef]
81. Tockner, K.; Schiemer, F.; Baumgartner, C.; Kum, G.; Weigand, E.; Zweimüller, I.; Ward, J.V. The Danube Restoration Project: Species Diversity Patterns across Connectivity Gradients in the Floodplain System. *Regul. Rivers Res. Mgmt.* **1999**, *15*, 245–258. [CrossRef]
82. Chodkiewicz, A.; Stypiński, P.; Studnicki, M.; Borawska-Jarmułowicz, B. The Influence of Konik Horses Grazing and Meteorological Conditions on Wetland Communities. *Agriculture* **2023**, *13*, 325. [CrossRef]
83. Joyeux, E.; Haie, S.; Le Rest, K.; Quaintenne, G.; Francesiaz, C. Meadow-Breeding Waders in France: Population Sizes, Distribution and Conservation Challenges. *Wader Study* **2023**, *129*, 166–176. [CrossRef]
84. Christen, W. Population Trend and Migration of the Northern Lapwing Vanellus Vanellus in the Aare Plain near Solo-Thurn. *Ornithol. Beob.* **2007**, *104*, 173–188.
85. Meltofte, H.; Amstrup, O.; Petersen, T.L.; Rigét, F.; Tøttrup, A.P. Trends in Breeding Phenology across Ten Decades Show Varying Adjustments to Environmental Changes in Four Wader Species. *Bird Study* **2018**, *65*, 44–51. [CrossRef]
86. Jellesmark, S.; Ausden, M.; Blackburn, T.M.; Hoffmann, M.; McRae, L.; Visconti, P.; Gregory, R.D. The Effect of Conservation Interventions on the Abundance of Breeding Waders within Nature Reserves in the United Kingdom. *Ibis* **2023**, *165*, 69–81. [CrossRef]

87. Osmundson, D.B.; Ryel, R.J.; Lamarra, V.L.; Pitlick, J. Flow–Sediment–Biota Relations: Implications for River Regulation Effects on Native Fish Abundance. *Ecol. Appl.* **2002**, *12*, 1719–1739. [CrossRef]
88. Obolewski, K.; Glińska-Lewczuk, K.; Bąkowska, M. From isolation to connectivity: The effect of floodplain lake restoration on sediments as habitats for macroinvertebrate communities. *Aquat. Sci.* **2018**, *80*, 4. [CrossRef]

MDPI AG
Grosspeteranlage 5
4052 Basel
Switzerland
Tel.: +41 61 683 77 34

Water Editorial Office
E-mail: water@mdpi.com
www.mdpi.com/journal/water

www.ingramcontent.com/pod-product-compliance
Lightning Source LLC
Chambersburg PA
CBHW040825250926
51850CB00007BA/1259
* 9 7 8 3 7 2 5 8 5 2 6 1 1 *